ADVANCES IN IRRIGATION

Volume 1

CONTRIBUTORS

M. G. Bos
D. A. Bucks
Gaylon S. Campbell
Melvin D. Campbell
A. J. Clemmens
A. R. Dedrick
L. J. Erie
Ray D. Jackson
William A. Jury
J. T. Musick
F. S. Nakayama
J. A. Replogle
B. A. Stewart
A. W. Warrick

ADVANCES IN IRRIGATION

Edited by

DANIEL HILLEL

Department of Plant and Soil Sciences
University of Massachusetts
Amherst, Massachusetts

VOLUME 1

ACADEMIC PRESS 1982
A Subsidiary of Harcourt Brace Jovanovich, Publishers

New York London
Paris San Diego San Francisco São Paulo Sydney Tokyo Toronto

ACADEMIC PRESS, INC.
111 Fifth Avenue, New York, New York 10003

United Kingdom Edition published by
ACADEMIC PRESS, INC. (LONDON) LTD.
24/28 Oval Road, London NW1 7DX

ISBN 0-12-024301-6

ISSN 0275-7915
This publication is not a periodical and is not subject to copying under CONTU guidelines.

PRINTED IN THE UNITED STATES OF AMERICA

82 83 84 85   9 8 7 6 5 4 3 2 1

CONTENTS

Canopy Temperature and Crop Water Stress

RAY D. JACKSON

Use of Solute Transport Models to Estimate Salt Balance below Irrigated Cropland

WILLIAM A. JURY

Level-Basin Irrigation

A. R. DEDRICK, L. J. ERIE, and A. J. CLEMMENS

Flow Measurement Flumes: Applications to Irrigation Water Management

J. A. REPLOGLE and M. G. BOS

Principles, Practices, and Potentialities of Trickle (Drip) Irrigation

D. A. BUCKS, F. S. NAKAYAMA, and A. W. WARRICK

LIST OF CONTRIBUTORS

Numbers in parentheses indicate the pages on which the authors' contributions begin.

M. G. Bos, International Institute for Land Reclamation and Improvement, 6700 AA Wageningen, The Netherlands (147)

D. A. Bucks, U.S. Department of Agriculture, Agricultural Research Service, U.S. Water Conservation Laboratory, Phoenix, Arizona 85040 (219)

Gaylon S. Campbell, Department of Agronomy and Soils, Washington State University, Pullman, Washington 99164 (25)

Melvin D. Campbell, Department of Agronomy, U and I, Inc., Kennewick, Washington 99336 (25)

A. J. Clemmens, U.S. Department of Agriculture, Agricultural Research Service, U.S. Water Conservation Laboratory, Phoenix, Arizona 85040 (105)

A. R. Dedrick, U.S. Department of Agriculture, Agricultural Research Service, U.S. Water Conservation Laboratory, Phoenix, Arizona 85040 (105)

L. J. Erie, U.S. Department of Agriculture, Agricultural Research Service, U.S. Water Conservation Laboratory, Phoenix, Arizona 85040 (105)

Ray D. Jackson, U.S. Department of Agriculture, Agricultural Research Service, U.S. Water Conservation Laboratory, Phoenix, Arizona 85040 (43)

William A. Jury, Department of Soil and Environmental Sciences, University of California, Riverside, California 92521 (87)

J. T. MUSICK, U.S. Department of Agriculture, Agricultural Research Service, Conservation and Production Research Laboratory, Bushland, Texas 79012 (1)

F. S. NAKAYAMA, U.S. Department of Agriculture, Agricultural Research Service, U.S. Water Conservation Laboratory, Phoenix, Arizona 85040 (219)

J. A. REPLOGLE, U.S. Department of Agriculture, Agricultural Research Service, U.S. Water Conservation Laboratory, Phoenix, Arizona 85040 (147)

B. A. STEWART, U.S. Department of Agriculture, Agricultural Research Service, Conservation and Production Research Laboratory, Bushland, Texas 79012 (1)

A. W. WARRICK, Department of Soils, Water and Engineering, University of Arizona, Tucson, Arizona 85721 (219)

PREFACE

Water constitutes one of the principal constraints to increasing food production in our hungry world. So tenuous and delicate is the balance between the demand for water by crops and its supply by precipitation that even short-term dry spells often reduce production significantly, and prolonged droughts can cause total crop failure and mass starvation. Irrigation is the artificial application of water to agricultural crops, designed to permit farming in arid regions and to offset drought in semiarid or semihumid regions. As such, it plays a key role in feeding an expanding population and seems destined to play an even greater role in the foreseeable future. Even in areas where total rainfall is ample, it is often unevenly distributed during the year so that only with irrigation is multiple cropping possible. In fact, the potential productivity of irrigated land can exceed that of unirrigated land severalfold, particularly in dry regions, due both to increased yields per crop and to the possibility of multiple cropping.

The process of irrigation generally consists of introducing water into the part of the soil profile that serves as the root zone, for the subsequent use of the crop. A well-managed irrigation system is one that optimizes the spatial and temporal distribution of water, not necessarily to obtain the highest yields or to use the lowest amount of water possible, nor always to attain the highest yield per unit amount of water, but to maximize the benefit-to-cost ratio. Although the problem and its solution are site specific in each case, the principles involved are indeed universal. From its early and primitive antecedents in the great river valleys of the Middle East some seven millennia ago, irrigation has envolved into a highly sophisticated operation, involving the simultaneous monitoring and at least partial control of weather, soil, and crop variables. Yet progress continues. What with the steepening costs of energy and water, the search for means to increase the efficiency of irrigation and of water utilization is becoming all the more urgent.

This volume is the first of a new serial publication entitled *Advances in Irrigation*. Designed to follow the format of such existing serials as *Advances in Agronomy* and *Advances in Hydroscience*, also published by Academic Press, the new serial is aimed at fulfilling a perceived global need for periodically updated comprehensive elucidations of the various topics of contemporary interest and importance related to the rapidly advancing science and engineering practice of irrigation. Scientific and professional journals report highly specific developments and are necessarily limited to short, entirely technical, communications. At the opposite end of the spectrum of publications are the textbooks on irrigation, which tend to be very general and avoid in-depth analyses of specific topics. Between the journals and the textbooks, there is room, and need, for a regular ongoing forum wherein to present the developments arising out of irrigation research in the form of critical reviews of selected topics that from time to time appear to be ripe for such presentation. Such reviews do not merely summarize and encapsulate what is precisely known at present, but assess its significance in relation to alternative approaches, analyze the trends, and point to future prospects. The aim is thus both to inform of progress to date, and to spur continued progress. The contributions envisaged will include, but not be limited to, state-of-the-art reports, critiques of current practices, economic analyses, case studies of irrigation in different locations, changing concepts of irrigation, and water-use efficiency. Considerations will be given to such topics as energy and irrigation, controlled-environment crop production, conjuctive use of rainfall and irrigation, and the precise role of irrigation in regional and global food production.

Toward the aim described, we have invited the participation of several of the leading researchers to help launch this new serial. As is well known, excellent scientists are, by definition, very busy people, notoriously (and understandably) reluctant to be harnessed to any fixed schedule. Some of the people invited to contribute to this volume will therefore appear only in a subsequent volume. However, we are fortunate to have been granted the timely participation of a sufficient number of outstanding contributors to make this an even better volume than we had dared hope for initially. As Editor, I wish to express my gratitude for their exemplary cooperation.

DANIEL HILLEL

And the Lord God planted a garden in the east, in Eden.
. . . And out of the ground the Lord God raised
Every tree that is pleasing to the sight and good for food;
And the tree of life was also within the garden,
And the tree of the knowledge of good and evil.
And a river emerged from Eden to water the garden,
And from thence it parted and became four heads.
. . . And the Lord God placed man in the garden of Eden
To work it and to preserve it.

GENESIS 2 : 8–10, 15

INTRODUCTION

The Book of Genesis tells the story most succinctly: the first tree plantation[1] was irrigated by a river, which was divided into four channels. And man, so lacking—then as now—in the knowledge of good and evil, was appointed custodian and entrusted with the protection of God's garden. Alas, greedy man soon abused the trust, succumbing to momentary temptation and becoming more a consumer (beyond his needs) than a preserver. He, and his descendants, must thenceforth and forever suffer the consequences of this folly. Man's entire relationship to the environment is illustrated therein.

In the specific context of irrigation, the lesson is even more startlingly clear. Injudicious management of soil and water—out of ignorance or shortsighted greed—can destroy the bountiful garden on which we all depend for our sustenance. Since its inception seven or more millennia ago in the river valleys of eastern, southern, and southwestern Asia, and northeastern Africa, irrigation development has in some places become ultimately self-destructive. All too often, the short-term gain in production, leading to intensive settlement, was followed inexorably by long-term loss, in the form of water resource depletion and pollution, as well as of soil erosion and degradation. The problem is not inherent in the principle of irrigation per se, but in its frequently careless practice. What is at fault is the unmeasured application of water to land with little control over

[1]The word "paradise," incidentally, is derived from the Old Persian word *pairi-daēza*, meaning an enclosed park, or orchard.

quantity or quality, without knowledge or regard for the potentially insidious processes thereby set in motion.

It is the universal fallacy of man to assume that if a little of something is good, then the more the better. In irrigation (and indeed in so many other walks of life, as we tend to discover only too late), just enough is best, and by that we mean a measured quantity of water applied at a rate calibrated to meet the continuous requirements of the crop, no less and certainly no more. The application of too little water is an obvious waste, as it might fail to produce the sought-after benefit. Excessive flooding of the land, is, however, even more destructive, as it saturates the soil, impedes aeration, leaches nutrients, induces greater evaporation and salinization, and ultimately raises the water table to a level that can only be drained at an enormous (and at times prohibitive) expense. Even where groundwater drainage of irrigated land is feasible, there remains the problem of its disposal. Dumping it back into the stream merely serves to salinize the water supply—diminished in any case—upon which depend other irrigators further downstream.[2]

Advances in the science and art of irrigation necessarily imply the gradual acquisition and utilization of knowledge leading to more precise control over the quantity and quality of water applied. The aim of such control must be to optimize the practice of irrigation in conjunction with all the other interacting variables which together govern crop production.

Control of irrigation should properly begin at the source—the river, reservoir, or aquifer. Withdrawal of water from a source in excess of the rate of replenishment eventually depletes the source[3] and might deprive the crops of their water supply at the very time of greatest need. The measured withdrawal of water should be calibrated to answer seasonal crop needs on a continuous, time-variable basis in accordance with crop growth stages, changing evaporative demand, and space-and-time-variable soil conditions. Too many irrigation projects have been (and indeed still are being) designed by hydraulic engineers alone, without sufficient regard for the utilization of water in the field. Irrigation systems

[2]The word "rival," now taken to mean "adversary" was originally used in Roman law as a term for those who shared the water of a *rivus*, or irrigation channel.

[3]Especially serious is the depletion of groundwater from an aquifer. It might cost the user little to drill a well and begin pumping water out of the ground. If pumping exceeds recharge, however, the water table is gradually lowered to where it can no longer be pumped profitably. Having already realized his profit, the groundwater exploiter might then abandon the land or sell it to others. The cost of restoring the water table afterward is likely to be enormously greater than the cost of lowering it in the first place, but that problem is generally left for future generations.

which deliver water to the field on a fixed, imposed schedule, leave the irrigator with little choice or discretion, and no incentive, to economize his water use. Ideally, water should be available *on demand,* and be priced in proportion, or even in progressive disproportion, to the quantity used per unit area of land.

Proper control of irrigation continues beyond the source of the water, and extends through its conveyance and delivery to the field. The prevalent use of open and unlined canals too often entails large losses of water by seepage, which causes waterlogging of adjacent lands and exacerbates the drainage problem. Unlined canals are also subject to frequent breaching of their banks, because of scouring by flowing water or uncontrolled crossings by animals.[4] Still another problem in unlined canals is the flow-hindering effect of riparian vegetation. Ideally, the conveyance of water should be carried out in closed conduits, and under sufficient pressure to permit delivery to fields of various elevations. However, the capital costs of closed conveyance systems, and the energy costs of pressurizing water, are considered prohibitive in many situations.

Ultimately, the major control of irrigation rests in the hands of the irrigator in the field. Unfortunately, it is here that we can still witness the greatest inefficiency. Even the terms used to characterize efficiency are obscure and confusing. In principle, any measure of utilization or application efficiency should refer to the quantity of a resource actually utilized beneficially as a fraction of the total quantity applied. "Irrigation efficiency" has variously been defined as the quantity of water added to the root zone, or the quantity of evapotranspiration, or just the quantity of transpiration, each as a fraction of the total quantity of water delivered to the field.

One of the problems here is that the root zone of a growing crop is practically impossible to define with any degree of precision (as it varies in time, with the crop, and indeed with the method of irrigation). Two processes which strongly affect the profile moisture storage and the disposition of applied water are in effect invisible to the onlooker, and measurable only with much difficulty and imprecision by the scientist: evaporation and deep percolation. In the absence of universal, simple and reliable ways for the practical irrigator to measure these time-variable and space-variable processes, and given the uncertainty in the very definition of irrigation efficiency, it is not surprising that in so many (perhaps most)

[4]Water buffaloes are a particular problem in South Asia, where they habitually wade into the canals to wallow in the water, incidentally churning up the bottom and damaging the banks.

cases, irrigation water is applied very inefficiently indeed. The problem is compounded by the difficulty of achieving uniform distribution and supply to the root zone in surface-irrigation systems (still the most prevalent of irrigation systems) where the irrigator depends on the hydraulic properties of the soil itself, which tends to be spatially heterogeneous, to distribute, absorb, and retain the applied water.

Even more baffling than "irrigation efficiency" is the term "water use efficiency," commonly defined as the ratio of crop yield (total above-ground dry matter in some cases, or grain yield in other cases) to the total quantity of water "consumed" by evapotranspiration. An alternative definition proposed lately is to relate the yield increment resulting from irrigation (i.e., the yield of the irrigated crop minus the yield without irrigation) to the amount of irrigation, or to the increment of evapotranspiration resulting from irrigation. By either of these definitions, the qualities related to each other (weight of crop per volume of water, in effect "apples divided by oranges") do not provide us with any universal criterion of efficiency ranging, say, between 0 and 100%.

The ultimate test of efficiency, in realistic terms, is that of economic profitability. And that, evidently, is the least objective and universal of all measures of efficiency. The economics of each country or region differs from that of all others, and much depends on local policies and priorities; indeed on the whole complex system of incentives and rewards. The cost of water relative to the costs of other inputs—energy, labor, capital—is an extremely important factor. Water is kept artificially cheap in too many places, through policy manipulation or maintenance of anachronistic vested "water rights". Individual farmers may find it profitable, and hence "smart," to waste water even knowingly, as the means to save water may be more expensive than the water to be saved. All too often, society as a whole bears the costs or, worse yet, the deleterious consequences of today's "smart" practices (e.g., depletion of water resources) as they are passed on to future generations.

The widespread habit of lumping evaporation from the soil with transpiration by the crop in the single term "evapotranspiration," however convenient, actually adds to confusion. Transpiration is a necessary, one might even say a productive, expenditure of water in crop production. Evaporation, on the other hand, is often a partially avoidable loss. The two processes interact, of course, and may be mutually compensatory, but only to a limited degree. An irrigation system which minimizes evaporation from the soil (e.g., drip irrigation) can in fact produce a significant savings of water.

Various names are being developed and improved in the continuing effort to achieve a higher level of control in irrigation. Among these are the use of laser guidance systems for precision shaping of the land surface; precision tillage and mulching to promote infiltration and reduce evaporation while minimizing the expenditure of energy; metering valves and pressure control devices to regulate water delivery; flumes and weirs to measure the amount of water used (it is amazing, and disconcerting, to discover in how many cases water is still being applied without any measure!). The conjunctive use of precipitation (rainfall) and irrigation can result in a large savings of water and an increase of the land area which can be effectively irrigated. Above all, there is steady progress in developing methods to monitor and control the various components of the dynamic balance of water in the field. As we develop the ability to measure the pertinent variables more reliably, the degree of uncertainty associated with water management diminishes and greater control becomes possible.

A crucial problem in irrigation management is to determine just when, where, and in what amount to apply the water for greatest beneficial effect. To determine the optimal timing and quantity of irrigation, there are at least three alternative or mutually complementary approaches. The answer can be sought in the soil, in the plant, and in the microclimate. The soil "reservoir" can be monitored for its moisture content or potential with a view to determining the root zone's deficit to, say "field capacity." Additionally, measurements of the salt concentraton of the soil solution and of the groundwater (if present not far below the root zone) can indicate the adequacy, or inadequacy, of irrigation. Or, the plant can be observed in an attempt to detect early signs of incipient stress, so as to be able to irrigate in time to forestall any substantial yield reduction due to stress. Finally, the weather can be monitored to determine the evaporative demand. As shown in the chapters of this book, all these approaches are being pursued concurrently.

For many years, the predominant approach to irrigation has been (and still is in many places) to apply water as infrequently as practicable, and only when the soil moisture reservoir in the root zone is almost completely depleted of so-called "available" (or, to use a more recently coined term, "extractable") water. Soil moisture is then replenished by a massive application of water sufficient to saturate the soil to some depth. Such periodic and abrupt saturation of the soil's surface zone tends to impeded aeration, and the subsequent gradual desiccation of that zone could be further impediment to root growth and function there. In the last decade, however, evidence has been mounting that a better approach might be to attempt to establish and continuously maintain a wet (but unsaturated) soil moisture regime in order to prevent the plants from experiencing stress and thus promote their maximal, uninterrupted growth. There are indications that in many (though not all) cases the increased yield may be more than proportionate to the increased water use. In optimal combination with improved fertilization, soil amendments, tillage, and pest control, high frequency irrigation, if calibrated precisely to answer the transpirational needs of the crop while preventing any possible accumulation of salts in the root zone, may hold the best promise at present to help us achieve a higher level of beneficial control over irrigation. However, no technical system, however sophisticated, is a panacea. The efficacy of a system depends on how it is used. Improperly used, even the most advanced system can fail. Progress demands that we continue to work painstakingly to achieve better, more comprehensive, and more precise knowledge of the complex processes we are attempting to control, and then to determine how to achieve such control economically under various circumstances.

In irrigation, as in all science, the so-called "last word" is only the "latest word," and must never be allowed to become the "final word."

DANIEL HILLEL

ADVANCES IN IRRIGATION

Volume 1

CONJUNCTIVE USE OF RAINFALL AND IRRIGATION IN SEMIARID REGIONS*

B. A. Stewart and J. T. Musick

U.S. Department of Agriculture
Agricultural Research Service
Conservation and Production Research Laboratory
Bushland, Texas

I. Introduction

Irrigated agriculture is very important in meeting the food and fiber needs of an increasing global population. In the United States, irrigation has increased over 300% since 1939 (Jensen, 1980). In 1974, 18% of the cropland in the

*Contribution from Agricultural Research Service, U.S. Department of Agriculture, in co-operation with The Texas Agricultural Experiment Station, Texas A & M University, College Station, Texas 77843.

ISBN 0-12-24301-6

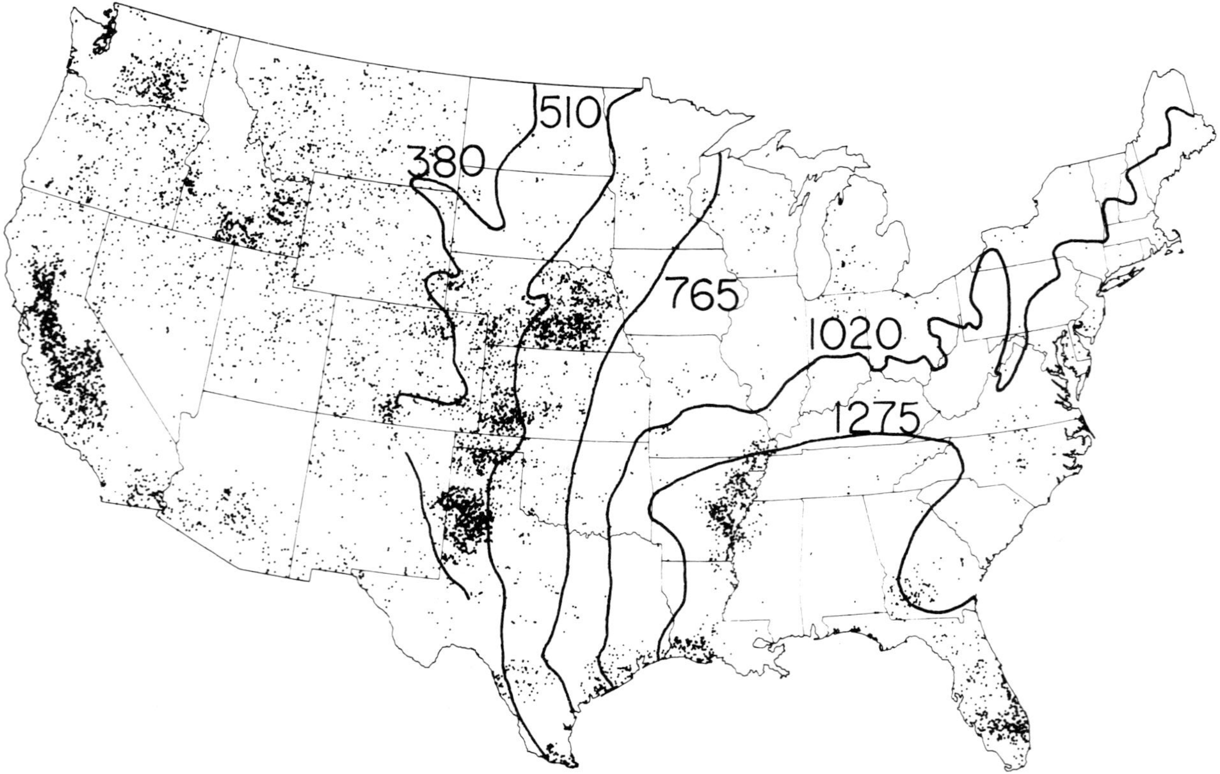

FIG. 1. Irrigated cropland, 1977. One dot equals 3240 ha. Total for the United States was 25.5 million hectares. Average annual precipitation contour lines are shown for areas east of the Rocky Mountains.

United States was irrigated, but because of higher yields and higher value crops grown under irrigation, 27% of agricultural products sold were produced on irrigated cropland. On a worldwide scale, about 12% of the cropland is irrigated (van Schilfgaarde and Rawlins, 1981). Of the 170×10^6 ha of irrigated land, 25×10^6 ha are in the United States, and over 40% of the latter area is in the Great Plains.

Irrigation will continue to increase, but supplies and costs of energy will become major constraints in future irrigation development. Irrigated agriculture is highly energy intensive, particularly in areas using groundwater that requires high pumping lifts. Also, increasing population and industrial development will compete for available water resources.

Growing demands for water increase the need for irrigated agriculture to become more efficient. On-farm conveyance and application efficiencies in the United States were estimated to be about 50%, and ranged from 40% in the Northwest and Rocky Mountain regions to about 60% in the Southern Great Plains region (Interagency Task Force Report, 1979). Future irrigation management systems will have to utilize water and energy resources more efficiently. One way to increase efficiency is to improve the conjunctive use of rainfall and irrigation water.

The distribution of irrigated land in the United States is illustrated in Fig. 1, and the regional changes which took place between 1974 and 1979 are shown in Table I. The largest increase in irrigated cropland occurred in the semiarid Central and Southern Great Plains, but the largest percentage increases occurred in the subhumid Corn Belt, and the subhumid and humid South and Southeast. Average precipitation isohyetal lines are also shown in Fig. 1. The precipitation contour lines for the western parts of the United States are not shown, but the irrigated

TABLE I

IRRIGATED LAND IN THE UNITED STATES[a]

Region	State	Hectares (1000s) 1974	Hectares (1000s) 1979	Change (%)
Arid Southwest	AZ, CA	4010	4470	+11
Arid Pacific Northwest	ID, OR, WA	2963	3158	+7
Semiarid Central Mountains	CO, MT, NV, UT, WY	4587	4280	−7
Semiarid Central and Southern Great Plains	KS, NE, NM, OK, TX	7343	8987	+22
Subhumid Corn Belt	IL, IN, MN, MO, WI	261	602	+131
Subhumid and humid South and Southeast	AR, FL, GA, LA, MS, NC, SC	1993	2511	+26

[a] Jensen (1980).

land in the Southwest and Pacific Northwest regions generally receive less than 250-mm annual precipitation, and the Central Mountains region averages less than 500 mm.

The conjunctive use of rainfall and irrigation offers considerable potential for increasing water-use efficiency and reducing energy needs for pumping. The traditional method of irrigating in arid regions is to apply fixed amounts of irrigation water at fixed time intervals. In essence, this method tends to ignore precipitation. More recently, irrigation scheduling, based on computer models to estimate daily evapotranspiration and soil water depletion, has become common. This allows predicting the optimal timing and amount of irrigation while taking into account previous and expected precipitation, available soil water, the evapotranspiration rate for the particular crop, and the capacity of the irrigation system.

The most common irrigation management objective is to eliminate water as the production-limiting variable while minimizing excessive application. In most arid regions where irrigation is practiced, sufficient water is made available for the land area irrigated, so rainfall during the irrigation season is not generally considered an essential portion of the water requirements. Only when significant amounts of precipitation occur, are irrigations delayed. This will, of course, conserve both water and energy, but it usually has no major effect on yields or on the amount of land irrigated.

Successful irrigation in humid regions will require systems that conjunctively utilize precipitation and irrigation. Soils in humid regions are often shallow and have relatively low water-storage capacities. Where irrigation is practical, water resources are generally adequate. The objective is to add sufficient irrigation water to eliminate water as a limiting factor in crop production. This requires frequent irrigation and greatly increases the chances of rainfall occurring soon after irrigation, which may result in excessive erosion, leaching of plant nutrients, and excessively wet soils. Therefore, surface or subsurface drainage systems are often required.

Semiarid regions, receiving between 225- and 510-mm average annual precipitation, offer the greatest potential for the conjunctive use of irrigation and rainfall. This is because land resources are generally much more abundant than water resources. Rainfall in semiarid regions is sufficient to allow some crop production without irrigation, but yields are normally low, and crop failures often occur when less than average precipitation occurs. Irrigation in the semiarid regions accounts for about 55% of all irrigated areas in the United States. Unlike the arid regions, where cropland was usually not developed until water resources were assured, millions of hectares of prime farmland were cropped in semiarid regions before irrigation became common. Irrigation was developed in these areas to increase yields and to stabilize production.

Rainfall is the most important part of the total water resources base in the semiarid regions of the United States. Much of the irrigated land in the Central and Southern Great Plains closely parallels the 510-mm precipitation contour line (Fig. 1). Because of rainfall and relatively high system efficiencies, irrigation water requirements are lower than in arid regions.

This article deals primarily with semiarid regions, and much of the data presented is from the Great Plains of the United States. The Great Plains can be considered a case study. The principles presented are applicable throughout the world where similar conditions exist. Also, some of the principles discussed are deemed to be applicable to arid and humid regions as well.

II. Rainfall Distributions in Semiarid Regions

Management systems for the conjunctive use of rainfall and irrigation water cannot be developed without an understanding of the general character of rainfall in a given area. We shall look primarily at the Central and Southern Great Plains regions, because these areas contain a large concentration of irrigated land and receive significant rainfall. Although much of the irrigated land in these regions parallels the 510-mm average annual precipitation line (Fig. 1), the variations in rainfall are indeed large. Some insight into the rainfall pattern in these areas can be gained by using a few "rules of thumb." One is that average annual precipitation increases about 25 mm for every 40 km of distance east of the 510-mm isohyetal line. Conversely, annual precipitation decreases at a similar rate west of the isohyetal line. A second "rule of thumb" is that the maximum annual precipitation for a specific area in these regions approaches 200% of the average, and the minimum is about 50% of the average. The distribution of annual rainfall totals is therefore skewed. Since an unusually wet year raises the average more than an unusually dry year lowers it, more years have below-average than above-average rainfall. For the semiarid regions in the Central and Southern Great Plains, rainfall is less than average in about 55% of the years. Skewness of rainfall distribution declines as rainfall amounts increase. The frequency distribution of annual rainfall at several locations in the Central Great Plains, shown in Fig. 2, illustrates the skewed distribution discussed above.

Perhaps even more important than average annual rainfall is the amount that occurs during the irrigation season. For the Central and Southern Great Plains, the most prevalent period of irrigation is June through September. During this period, precipitation averages about 50–55% of the average annual precipitation. However, yearly variations can be very great. The accumulated rainfall at Bushland, Texas, for 42 years is illustrated in Fig. 3. Also shown is the seasonal water use for grain sorghum grown during the same period and irrigated for high yields. For wet and dry seasons having an average occurrence of one in three, the dry seasons contribute less than 25% of the seasonal water use needed for high

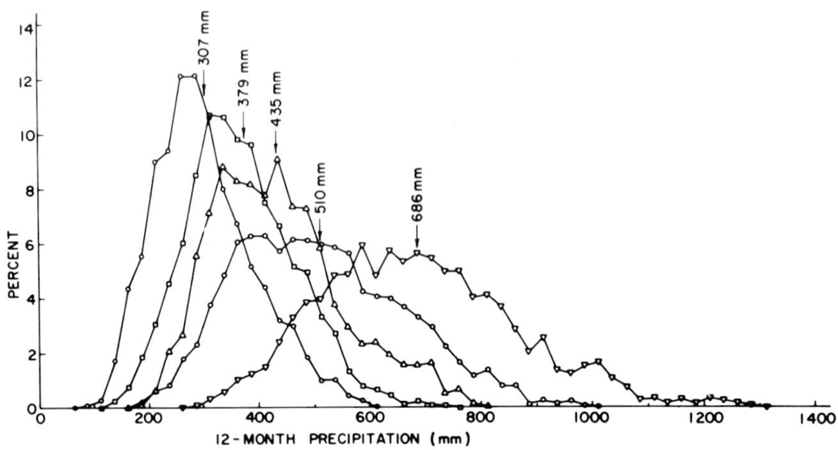

FIG. 2. Percent occurrence distribution of continuous 12-month precipitation amounts by 25-mm class increments across the Central Great Plains. The curves are average data for three to six selected sites in Kansas, Nebraska, Colorado, or Wyoming having nearly identical long-term means, shown on the graph, for length of record ranging from 56 to 75 years. Data are plotted for the full distribution range of the driest to the wettest consecutive 12-month period on records.

yields, whereas the wet seasons contribute over 60% of the needs, with minor losses to surface runoff. When a wet season with reasonably good rainfall distribution is combined with a soil profile that is initially wet at planting time, irrigation may not be needed for high yields. However, since rainfall cannot be accurately predicted, seasonal irrigations are commonly applied even during wet seasons. In drier seasons, rainfall may be largely ignored since it contributes little to crop production. Rainfall may also contribute very little to yields from dryland (unirrigated) fields during the much drier seasons, which may cause crop failure.

The amount of rainfall that occurs during the irrigation season in semiarid regions is indeed important because of the large amount of irrigable land in relation to amount of water available for irrigation. Since land resources are usually much more abundant than water resources, the tendency has been to irrigate the maximum amount of land with the available irrigation water. In designing such systems, it is generally assumed that near-average rainfall will be received. This approach, however, can be unsatisfactory in both drought years and wet years because the irrigation water supply (when the source is groundwater) is determined by the well capacity, and it is no greater during a drought year than during an average or a wet year. Also, in addition to providing less rain than average seasons, drought seasons have higher evaporative demand because of higher air temperatures, more wind advection, and larger vapor pressure deficits. Daytime temperatures are 3–6°C warmer than in average seasons, and advective energy becomes more significant. Consequently, irrigation systems designed to

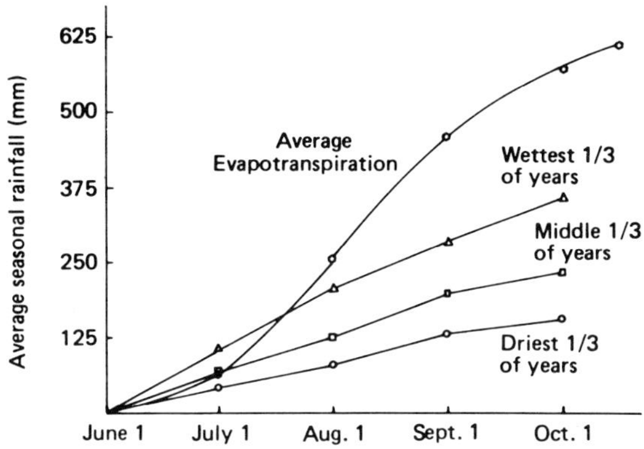

FIG. 3. Accumulated rainfall at Bushland, Texas, for the wettest 14 years, the middle 14 years, and the driest 14 years from 1939 to 1980.

maximize the amount of irrigated land can become inadequate during very dry seasons. On the other hand, systems designed to supply adequate irrigation water for high yields during dry seasons are not efficiently used during average-to-wet seasons. Maximizing the use of rainfall is becoming ever more important in pump-irrigated regions because of the rapid depletion of the groundwater supply, the Ogallala aquifer. The rapidly rising cost of energy for lifting the groundwater is also making it much more important to use rainfall efficiently so as to lower the demand for irrigation.

III. Limited Irrigation

The objective of irrigation, particularly in arid regions, has been to prevent water from becoming a limiting factor in crop production. However, where the water supply is inadequate in relation to the land area that can be irrigated, limited irrigation is often practiced with a view to obtaining less than maximal yields. The management strategy of limited irrigation is to optimize production per unit of applied water, rather than to maximize yield per unit of land. A major advantage of limited irrigation is the increase of total farm yield and water-use efficiency by reducing the area allocated to dryland cropping. Water-use efficiency of dryland crops such as wheat and sorghum are traditionally low because of normally low yields associated with soil water depletion during the vegetative growth stage, and because of water stresses developing later during even more critical stages. For example, long-term average values at Bushland, Texas, of seasonal water-use efficiencies for dryland grain sorghum and wheat were 0.45 and 0.24 kg/m^3, respectively (Unger, 1972). Well-timed seasonal irrigations at the same location generally result in irrigation water-use efficiencies of about 1.40 kg/m^3 for grain sorghum and about 0.55 kg/m^3 for wheat (Musick, 1973). Consequently, limited irrigation warrants serious consideration as a means of increasing efficient use of water resources for drought-tolerant crops such as sorghum, wheat, cotton, sunflower, and sugar beet.

Limited irrigation was a practice originated in the Texas High Plains. Dryland sorghum and cotton farmers started drilling wells and irrigating during the major drought of the 1930s. Jones and Gaines (1941) showed that a preplant irrigation and one or two growing-season irrigations resulted in acceptable yield levels, greatly exceeding yields from dryland. However, increased yields associated with inexpensive energy for pumping, low-cost nitrogen fertilizers, development of high-yielding hybrid sorghums, and a tremendous increase in the number of irrigation wells, caused a change in irrigation practices from limited irrigation to full irrigation, maximizing per-hectare yields. A major drought lasting six years during the 1950s accelerated and completed the shift to full irrigation, which then continued as a common practice, even following a major climatic shift to increased rainfall. Lately, however, the declining water supplies and increasing cost of energy already mentioned are once again focusing attention on limited irrigation. The primary question is now: "Is a fixed amount of irrigation water utilized more efficiently by the full irrigation of a small area or by the limited irrigation of a larger area that would otherwise be in dryland production? Also: Should limited irrigation be practiced on some crops to permit full irrigation on other crops?"

IV. Yield, Transpiration, Evapotranspiration, and Seasonal Water Application

In considering the conjunctive use of rainfall and irrigation to improve water-use efficiency, we need to distinguish among various definitions of this concept. When water-use efficiency is defined as yield per unit of transpiration, the conjunctive use of rainfall and irrigation water will probably not have any major effect. Furthermore, the use of limited irrigation as opposed to full irrigation will often not greatly increase the efficiency, except for critically timed irrigations on such crops as grain sorghum, cotton, and sunflower.

Water-use efficiency is more frequently defined as yield per unit amount of evapotranspiration, because evapotranspiration is more easily measured than transpiration. Since transpiration and evapotranspiration are usually closely correlated, their relationships to yield are generally similar. However, the yield per unit of evapotranspiration efficiency can be improved by partitioning more of the evapotranspiration into transpiration and avoiding losses by evaporation.

A third way that water-use efficiency can be defined is by calculating the yield increase from irrigation (irrigated crop yield minus dryland crop yield) per unit amount of irrigation water applied. When this meaning is used, both "limited" irrigation and "conjunctive use of rainfall and irrigation" can contribute to major improvements of water-use efficiency.

The relationships between crop yields and water use have been studied by many workers. Literature can be selected that shows relationships ranging from linear to curvilinear (both concave and convex). Reviews by De Wit (1958), Arkley (1963), and Stegman *et al.* (1980) state that when yields are transpiration limited, strong linear correlations usually occur between cumulative seasonal dry-matter production and cumulative seasonal transpiration. Furthermore, since transpiration and evapotranspiration are closely associated, dry-matter yields versus cumulative evapotranspiration usually also plot as a straight-line relationship. In the case of grain crops, a considerable quantity of transpiration or evapotranspiration must take place during vegetative growth before the first increment of grain yield is produced. Again, the relationship between grain yield and seasonal evapotranspiration is often linear, but the slope of the line is different than the one obtained for total dry matter versus evapotranspiration. Stegman *et al.* (1980) reviewed the literature and observed that when the relationship was curvilinear, the nonlinearity was most pronounced at the higher evapotranspiration levels. They suggested that profile drainage losses probably contributed to the nonlinearity. In addition, aeration deficiencies, plant lodging, and higher disease incidence may have limited yields.

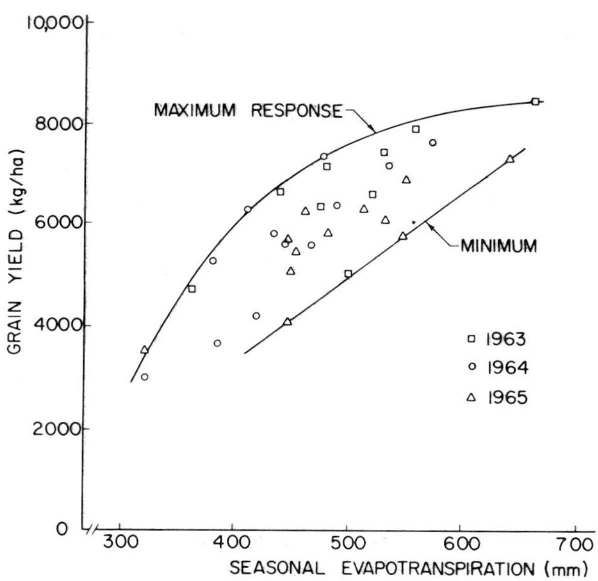

FIG. 4. Grain sorghum yields as a function of seasonal water use (Musick and Dusek, 1971).

Grain sorghum yield as a function of seasonal water use (Musick and Dusek, 1971) is shown in Fig. 4. The data are somewhat variable, with points falling between a maximum and minimum response curve. Although yield levels varied somewhat among growing seasons, the major cause of data variability was associated with number, timing, and size of seasonal irrigations, and the amount and distribution of seasonal rainfall.

A straight-line relationship can logically fit the data plotted in Fig. 4. However, the higher yielding treatments indicate that under conditions of good water management, the seasonal water-use–yield curve is a curvilinear diminishing-return relationship within the range of 300–650 mm of water use. These data indicate that the water-use–yield function is not an explicit relationship and that it can vary considerably, depending on various factors which affect both yield and water use. Therefore, the use of limited irrigation, particularly when the water can be added during the more critical growth periods, shows considerable promise for increasing water-use efficiency.

V. Increased Efficiency of Soil Water Storage

The preceding discussion suggested that water-use efficiency often decreases at higher water-use levels. Data by Malm and Hsi (1968), shown in Table II, indicated that grain sorghum yields may increase even when irrigation water is added late in the growing season. The increase they obtained for the last irrigation, however, was much smaller than for irrigations applied earlier in the sea-

TABLE II

GRAIN YIELDS OF GRAIN SORGHUM AT CLOVIS, NEW MEXICO, WHEN IRRIGATED AT PLANTING AND 0–4 TIMES DURING THE SEASON FOR 1966–1967[a]

	Stage of development when irrigated					
		Height (cm)				
Treatment	Planting	15–20	60–75	Bloom	Soft dough	Grain yield (kg/ha)
T-1	×	—	—	—	—	5610
T-2	×	×	—	—	—	6020
T-3	×	×	×	—	—	7060
T-4	×	×	×	×	—	10,010
T-5	×	×	×	×	×	10,530

[a] Malm and Hsi (1968).

son. This suggests that the last increment of water was not used with maximum efficiency. Irrigation water added late in the season may contribute to water losses by runoff or deep percolation, or result in significant quantities of water remaining unused in the soil profile when the crop is harvested.

Water remaining in the soil profile is later subject to losses by evaporation (especially following major tillage), slow profile drainage, and use by weeds or crop volunteer plants. Also, subsequent precipitation during the fallow period is stored less efficiently if the soil profile is left relatively wet at the time of harvest. Data in Table III, adapted from Jensen and Sletten (1965), further illustrate that grain yields are increased by irrigations late in the growing season. However, the amount of water left in the soil profile at harvest is greatly increased by late irrigations. When the crop was irrigated at the dough stage of grain, there were 145 mm of available water left in the soil profile to a depth of 120 cm. Consequently, there was very little opportunity for storing any precipitation received prior to the time of seeding the next crop. Musick (1970) showed a significant curvilinear relationship between antecedent soil water after harvest and preseason storage efficiency. As antecedent soil water increased, storage efficiency decreased (Fig. 5). Highest storage-efficiency values were in the 40–50% range when the soil profile was near the wilting point at harvest time. These findings emphasize the importance of limiting irrigation, particularly late in the season, to prevent ending the growing season with a relatively wet profile, which minimizes the potential for efficient storage of the next crop's preseason rainfall. Such storage is essential for maximizing the conjunctive use of rainfall and irrigation water.

TABLE III

GRAIN YIELDS AND SOIL WATER AT HARVEST FOR GRAIN SORGHUM AS RELATED TO IRRIGATION CUTOFF DATE FOR 1959[a]

Timing of last seasonal irrigation	Grain yield (kg/ha)	Soil water at harvest (mm in 120-cm soil profile)	
		Available	Deficit[b]
Late boot (Aug. 4)	5100	37	125
Late flower (Aug. 19)	6680	68	94
Milk stage (Sept. 2)	8320	87	75
Dough stage (Sept. 15)	8400	145	17

[a] Jensen and Sletten (1965).

[b] Difference between available soil water and approximate field capacity of 162 mm for 120-cm soil profile.

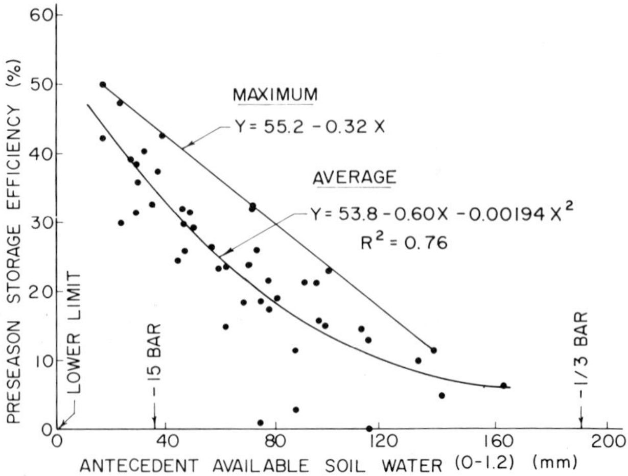

FIG. 5. Effect of antecedent available soil water on the storage efficiency of precipitation (Musick, 1970).

VI. Conservation Tillage

The conjunctive use of irrigation water and rainfall can be improved indirectly by conservation tillage. Conservation tillage decreases runoff, increases infiltration, and reduces evaporation by leaving more crop residues on the soil surface. Consequently, storage of rainfall in the soil is increased during periods between crops or during summer fallow. In field studies in Colorado, Montana, and Nebraska, the amount of precipitation stored as soil water during fallow ranged from 16% of the total with no residues to 34% with 11 tons/ha of wheat straw on the surface (Greb *et al.*, 1967). When Unger (1978) placed various amounts of wheat straw on Pullman clay loam at Bushland, Texas, he obtained improved precipitation storage during fallow, ranging from 23% of rainfall with no straw mulch to 46% with 12 tons mulch/ha. Dryland grain sorghum after the fallow yielded 1780 and 3990 kg/ha with the 0- and 12-ton/ha mulch treatments, respectively. Although residues from dryland crops are often quite limited in semiarid regions, substantial amounts remain from irrigated crops. Unger and Wiese (1979) in a four-year study at Bushland, Texas, used no-tillage, sweep tillage, and disk tillage for residue management and weed control from the time of irrigated wheat harvest until the time of dryland sorghum planting 11 months later. The precipitation storage values during the fallow period averaged 35, 23, and 15% for the no-tillage, sweep-tillage, and disk-tillage plots, respectively. Subsequent dryland grain sorghum yields averaged 3140, 2500, and 1930 kg/ha, respectively. It is evident from the results that residues maintained on the surface as a mulch can greatly increase precipitation storage and crop yields.

Musick *et al.* (1977) tried no-tillage fallow combined with chemical weed control following irrigated wheat. The no-tillage treatment increased irrigation water-use efficiency in all four years of the study when compared to conventional disk tillage. In one season, a no-tillage treatment with only two in-season irrigations yielded almost as much as a disk-tillage treatment receiving four irrigations.

A generalized approach to conservation tillage systems on bed-furrow irrigated land under limited irrigation is suggested as follows:

(1) Perform the major loosening operation soon after harvest by undercutting existing bed-furrows with sweep tillage;

(2) restrict subsequent tillage operations to shallow cultivations which firm the moist soil below tillage depth and "mulch" the loose surface soil;

(3) maintain bed-furrows in place and thus minimize the need to re-form beds from dry surface soil just prior to seeding time.

Such tillage systems, involving a reduction in the number of operations and the intensity of soil disturbance, in combination with the use of herbicides, are projected to spread rapidly in future years. This should result in more efficient use of the limited resources of rainfall and irrigation water for crop production.

VII. Specific Systems for the Conjunctive Use of Rainfall and Irrigation

The difficulty in optimizing the conjunctive use of irrigation and rainfall in semiarid regions lies in combining the use of limited irrigation water resources with the inherently unpredictable rainfall. Since rainfall is so highly variable, there is no way to determine in advance the optimum amount of land to be irrigated in a given year. This is because land is abundant in relation to water, and the amount of land that can be irrigated depends on the water resources. In semiarid regions the total water supply (irrigation water plus rainfall) varies greatly from year to year, as already discussed. When irrigation water is fairly abundant, the tendency is to ignore rainfall. However, during good rainfall years the wells can be shut off for longer periods. With groundwater resources being depleted, as in the southern portion of the Great Plains, the desire is to utilize the wells continuously during the irrigation season to maintain as much irrigated land as feasible. Maximizing the irrigated land area requires the efficient management of rainfall on irrigated land.

Furrow irrigation practices generally involve running excess water off the end of fields for a considerable period, depending on the infiltration rate of the soil, to ensure adequate wetting of the lower portion of the field. If the time allowed is too short, the yields on the lower portion of the field may be drastically reduced, but if the time allowed is too great, excessive runoff occurs and the probability of percolation losses on the upper end of the field is greatly enhanced. Although irrigation runoff water can be caught in reuse systems, some losses do occur. The U.S. Department of Agriculture Soil Conservation Service assumes an irrigation efficiency of 65% in designing reuse systems for the Texas High Plains. When substantial rainfall occurs within a few days following irrigation, stormwater runoff results. Reuse systems are not ordinarily adequate to capture this runoff, so in-season rainfall is often of limited benefit to the irrigated crop. In addition, furrow irrigation of cropland that has had the surface wetted with rainfall often results in much faster advance of the irrigation water, which increases the probability of incurring excessive irrigation water runoff. To improve the management of irrigation water, sprinkler systems have been installed on some land that was previously furrow irrigated. Use of sprinkler systems allows the application of smaller amounts of irrigation water and eliminates the need for reuse systems. Their use does not, however, substantially reduce stormwater runoff whenever rainfall closely follows an irrigation.

The amount of stormwater runoff that occurs from irrigated land is not known, but there is no question that it is substantial, particularly during wetter than average years. We pointed out earlier that 50–55% of the average annual rainfall in the Central and Southern Great Plains occurs during the summer irrigation season. Long-term studies at Bushland, Texas (Jones, 1975), showed that stormwater runoff from nonirrigated land managed in a wheat–grain sorghum–fallow system (two crops in three years with 11-month fallow between crops) averaged 38 mm annually and ranged from 0–150 mm. Runoff amounts from irrigated land are certainly greater.

In the past, the emphasis in irrigation research in the semiarid regions was mainly on managing the irrigation water. Although this is essential, maximum efficiency of irrigation water use in these regions will require a much greater emphasis on managing rainfall on irrigated land. Only in recent years has serious attention been given to this matter. Examples of such efforts are discussed in the following sections.

A. Limited-Irrigation–Dryland Farming System

Stewart *et al.* (1981) designed and tested a limited-irrigation–dryland (LID) farming system. The LID farming system utilizes seasonal rainfall and limited amounts of irrigation water. The unique feature of their system is the flexible adjustment during the crop-growing season of the amount of land irrigated, allowing more land to be irrigated during wet years with a limited amount of irrigation water than during a dry year. Additional objectives are to use furrow irrigation without causing runoff from the field and to prevent runoff from in-season rainfall, even when it occurs simultaneously with, or closely following, an irrigation.

In this system (Fig. 6), a 600-m-length field is divided into three water management sections. The upper half of the field is managed as "fully irrigated." The next one-fourth is managed as a "tailwater runoff" section that utilizes furrow runoff from the fully irrigated section. Finally, the lowermost one-fourth is managed as a "dryland" section capable of receiving and utilizing any runoff which might result from either irrigation or rainfall on the wetter, fully irrigated and tailwater runoff sections. Plant densities are reduced down the field by seeding 6 kg/ha on the fully irrigated section, 3 kg/ha on the tailwater section, and 1.5 kg/ha on the dryland section. Furrow dams (Clark and Hudspeth, 1976) are placed every 3 m or so throughout the length of the field. In the work reported by Stewart *et al.* (1981), alternate furrows were irrigated and the dams in these furrows were cupped to ensure that the irrigation water moved over the dams and down the furrow rather than across the bed (Fig. 7). The dams were partially washed out by the irrigation water to the distance that the water had advanced down the furrows. The remaining furrow dams on the lower part of the irrigated furrows, and the dams in the alternate furrows for the entire length of the field, stopped rainfall runoff and prevented any water from leaving the field. Also, as soil water was depleted in the lower, drier field sections, shrinkage cracks de-

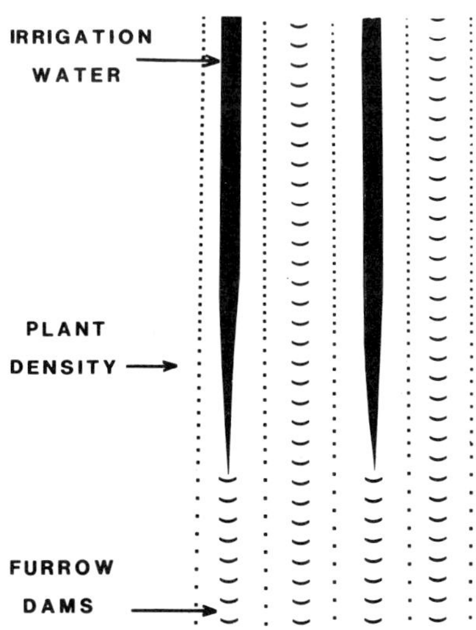

FIG. 6. Schematic of a limited-irrigation–dryland farming system that self-adjusts the amount of irrigated land, depending on seasonal precipitation (Stewart *et al.*, 1981).

FIG. 7. LID farming system trial: Dams were placed in all furrows, but every second furrow was irrigated and the dams in these furrows were cupped to facilitate irrigation-water advance within the furrow (Stewart *et al.*, 1981).

veloped in the swelling clay soil profile, and these cracks helped in absorbing rainfall and preventing runoff.

A predetermined amount of irrigation water (190 mm) was applied by irrigating five times at 14-day intervals. Equal amounts of irrigation water (38 mm) were added each time. With either a 12- or 24-hr irrigation set, the system required only a minimum amount of labor and management (and could have been easily automated) because the water application was changed by a predetermined convenient schedule rather than by monitoring when the entire field had been adequately wetted.

The 190 mm of seasonal irrigation water, in conjunction with average seasonal rainfall and some initially stored soil water, is generally sufficient to produce a fairly substantial yield of grain sorghum. In some years (Fig. 3), the addition of 190 mm of irrigation water would be adequate to irrigate practically the entire field. Rainfall for any given year, however, is quite variable. The theory behind this system is that the field will of itself "integrate" the rainfall and irrigation water supplies and the portion of the total field that is irrigated will vary from season to season. The wetter the season, the further the limited amount of applied water will move down the furrow. Therefore, more of the field will be irrigated. Conversely, during drier than average years, less of the field will be irrigated.

The reduction of plant densities with added length of field is meant to moderate water stress effects where water is more limited. However, even the lowest plant density used is sufficient to produce a reasonable yield level in wet years when some irrigation water may reach the lower part of the field. Some crops, and grain sorghum in particular, can adjust yield by tillers that produce grain and by increased head size to provide relatively high grain yields even under otherwise favorable conditions when plant density is low. Such crops may not produce grain when plant density is too high while water is severely limited.

Results obtained during 1979 for the LID system are compared to fully irrigated and dryland systems in Fig. 8 and Table IV. The amount of seasonal rainfall that occurred during 1979 was near average and only a small amount of irrigation water moved into the dryland section of the field (Fig. 8). Extremely wet years will no doubt result in some loss of water from the field, but indications are that for most years the system can successfully retain and utilize all seasonal rainfall and the limited amounts of irrigation water without releasing runoff. In the long run, this could well result in a substantial improvement of water-use efficiency (Table IV).

The primary disadvantage of the system is that plant density and soil fertility level need to be decreased with length of field. We are confident, however, that seeding equipment can be modified to easily handle seeding-rate changes. Fertilizer additions can be varied by fertilizing the field at right angles to the row direction prior to bedding the land. An alternative that may be more efficient is to

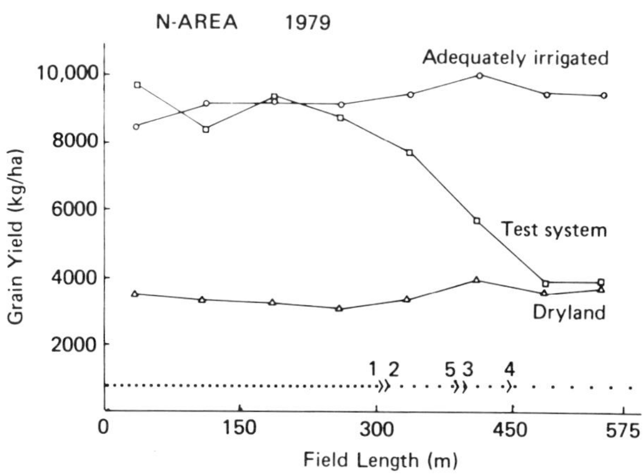

FIG. 8. LID farming system trial: yield levels at various locations down the field. The dotted line at the bottom illustrates the seeding-rate changes that correspond to the "adequately irrigated," "tailwater runoff," and "dryland" sections of the test system. The numbers show the distance of advance down the field of each of the irrigations (Nos. 1–5) (Stewart *et al.*, 1981).

TABLE IV

YIELD OF GRAIN SORGHUM AND WATER-USE EFFICIENCY OF A SYSTEM DESIGNED FOR THE CONJUNCTIVE USE OF IRRIGATION AND RAINFALL WITHOUT SURFACE RUNOFF[a]

Treatment	Irrigation water (mm)		Seasonal rainfall (mm)	Rainfall runoff (mm)	Grain yield (kg/ha)	Irrigation water-use efficiency (kg/m^3)	
	Applied	Runoff				Applied	Net intake
Adequately irrigated	602	191	224	22	9300	0.98	1.43
Test system	185	0	224	0	7300	2.09	2.09
Dryland	—	—	224	13	3430	—	—

[a] Stewart *et al.* (1981).

apply the fertilizer in the irrigation water. This would distribute the fertilizer in direct proportion to water distribution and could increase the use efficiencies of both.

The concept of adjusting the amount of land irrigated during a given year, depending on the amount of seasonal rainfall, may also be applicable to center-pivot systems. One approach might be to have "inner circle" and "outer circle" crop areas. The inner circle could have a high plant density, or perhaps a crop that required a relatively high soil water level such as corn. The outer circle could be a drought-tolerant crop such as grain sorghum planted at a low density. The center-pivot system would need to be modified so that either the inner circle, or the entire circle, could be irrigated. During extremely dry years, irrigation would be concentrated on the inner circle with the outer circle depending entirely on rainfall. During normal-to-wet years, some water would be available for application to the outer circle.

B. Low-Energy Precision-Application System

Lyle and Bordovsky (1980) developed a new concept in irrigation system design which has the potential for significant savings of both water and energy. This is a moving low-energy precision-application (LEPA) system which, rather than spraying water into the air at moderate-to-high pressures, distributes it directly to the furrow at very low pressures through drop tubes and orifice-controlled emitters. This occurs as the system continuously moves through the field in a rectilinear fashion (Fig. 9). The system is used in conjunction with furrow dams designed to retain all applied water and stormwater on the field, thus also optimizing the utilization of rainfall. Although the system has had only limited field evaluation, indications are that significant savings of both water and

FIG. 9. A low-energy precision-application system used in conjunction with microbasin land preparation to retain all applied water and stormwater (Lyle and Bordovsky, 1980).

energy can be achieved. Application efficiencies in the range of 95–99% have been measured. The system allows application of irrigation water without (1) percolation losses that often occur with furrow irrigation, particularly on medium- and coarse-textured soils; (2) irrigation water runoff; and (3) droplet evaporation and drift losses associated with sprinkler irrigation. The furrow dams also prevent stormwater runoff, even when rainfall occurs immediately following irrigation. Runoff could occur in the event of an unusually large rainstorm, particularly on steeper slopes. The disadvantages of the system are that the necessary equipment requires considerable capital investment and a higher level of management than do many other irrigation systems.

C. Alternating Strips of Grain Sorghum with Wheat

Musick and Dusek (1975) conducted a six-year study in which grain sorghum and winter wheat were grown in alternating 4.5-m-wide drill strips. Since the two crops have completely different growing seasons, outside crop rows could benefit from a border effect when no crop was growing on adjacent strips. Each strip contained six beds separated by five inside furrows. In one irrigation scheme, water was applied to two furrows in each six-bed strip. Each of the four inside beds had an irrigated furrow on one side and a nonirrigated furrow on the other. The two outside rows functioned as border rows and were not adjacent to an irrigated furrow. In other irrigation schemes, the number of furrows irrigated and the frequency of applications could be adjusted, depending on the amount of irrigation water available. The objective of the system was to utilize the rainfall during the nongrowing period of one crop more efficiently than it could be stored in the soil for the next crop. For example, rainfall during the nongrowing period of repeatedly grown wheat averaged 207 mm per season during the study. Storage efficiency measured at wheat-planting time in another study averaged only 21% (Johnson and Davis, 1972). In principle, alternating strips of grain sorghum with winter wheat should allow for more efficient use of much of the rainfall. The 4.5-m-strip width was selected because of its compatibility with available field equipment. During the study, applied irrigation water was reduced about 45% and yields were reduced about 25% in comparison with fully irrigated crops. Thus, the system offers considerable potential for increasing the use efficiency of both rainfall and limited irrigation water. The disadvantages of the system are: (1) It necessitates the management and movement of irrigation equipment over twice the area being irrigated during a growing season; (2) it requires greater precision of cultural practices, especially weed control; and (3) the probability for runoff from rainfall still exists.

D. Other Methods

There are many other methods of enhancing the conjunctive use of rainfall with irrigation, and it is beyond the scope of this article to discuss all of them. Some notable methods that have been studied include the use of tensiometers, runoff pits, snow management, crop breeding, antitranspirants, and irrigated–dryland crop rotations. The use of center-pivot sprinklers has become increasingly popular in areas previously watered by furrow systems. The reasons for their increased use include improved water-use efficiency and labor savings. Sprinkler systems, however, require greater capital investment and consume more energy. There has been a large shift in recent years to lower pressure systems for energy conservation. These systems use a different type of nozzling, and while they do indeed require less energy, they often create runoff problems on some soils because the irrigation water is applied to a smaller area at a higher intensity than when a high-pressure system is used. Residue-management practices and other cultural practices are being investigated as ways to alleviate these problems.

The most serious long-range problem with center-pivot systems in areas of declining groundwater is how to deal with the requirements of the constant-capacity sprinkler system in view of the reduction in well yields. As water supplies diminish, adjustments such as alternative cropping systems, decreased application rates, and lower crop water use will be required.

Hughes and Harman (1969) suggested that well yields in the southern Great Plains decreased in relation to declining water levels according to the following equation:

$$Q = \mathrm{IC} \times (\mathrm{PT}/\mathrm{IT})^2$$

where Q is the well capacity, IC is the initial well capacity, PT is the present thickness of the aquifer, and IT is the initial thickness of the aquifer.

The aquifer thickness of many areas in the Southern Great Plains has decreased by 50% in the past 20–30 years. Based on the above equation, the well yields would be only 25% their initial capacities.

VIII. Economic Considerations

Should producers who have more land than can be irrigated with the limited amount of water available irrigate part of the land at optimal water application rates, or irrigate more land at suboptimal rates? The data presented show that limited irrigation in a semiarid climate is potentially the more efficient alternative. This is largely due to more effective use of rainfall. The most efficient alternative, however, is not always the most profitable. The producer must carefully analyze economic parameters such as profits, return on investment, and returns on the water applied. When new land is being developed, the cost of development for full irrigation is less. Consequently, full irrigation will usually look better in an economic analysis. In areas where irrigation systems are already installed and irrigation supplies are declining, limited irrigation may be more feasible. This is the case in portions of the Southern Great Plains of the United States. Many irrigation-well yields in this area have decreased to the point that the capacity of the irrigation delivery system exceeds the water supply, and little additional expense is required to apply the water over a larger area. In many of these cases, it may indeed be more profitable to use limited irrigation. The management systems used for limited irrigation must be carefully planned, however, to ensure that increased gains in yields due to higher water-use efficiencies are not offset by increased costs of tillage, seeding, fertilization, pest control, and harvesting.

Limited irrigation could be a national goal in some areas or countries. This could allow the use of incentives or subsidies to use fixed sources of water over a much longer period than would be otherwise done. The limited-irrigation strategy might well result in increased production of food and fiber over the long run.

IX. Summary

Efficient use of water for crop production has been a major concern for centuries. Today this concern is greater than ever because of the rising needs of food and fiber, coupled with the decreasing supplies of water in several major irrigation regions. Also, the tightening supply and increasing cost of energy for irrigation makes efficient use of water doubly important because efficient use of water automatically improves the efficiency of energy use in irrigation.

Water-use efficiencies, as defined in this article, can be improved in all climatic regions by reducing losses due to surface runoff, evaporation (e.g., by mulching), and excessive percolation. Computerized irrigation scheduling (based on expected precipitation, available soil water, evapotranspiration of the crop as predicted by climatic data and crop coefficients, and capacity of the irrigation system), has proved effective in preventing overirrigation and in increasing water-use efficiency. Hence this practice will undoubtedly be expanded and improved in the future.

In most arid regions, cropland was not developed until water resources were developed. So, with the ratio of water availability to cropland, presumed to be optimal, limited irrigation is not usually practiced. Since rainfall supplies only a minute portion of the total crop water requirements in arid regions, the best strategy there will probably continue to be to irrigate sufficiently to prevent water from being a limiting factor in crop production while reducing excessive application. The conservation of water and energy by improved scheduling to promote better utilization of rainfall will be beneficial, though in many cases it may not lead to increased irrigated acreages.

In humid areas, improved scheduling will result in a more consistent yield response to the water applied, less energy use, and less soil erosion. The control of erosion in cropland irrigated in humid areas presents a major challenge that will require new approaches. Since water resources in these regions are generally more than adequate, the tendency will be to maintain the soil profile sufficiently wet to prevent water from being a limiting factor in crop growth. Consequently, rainfall on these wet soils can accelerate erosion. The development of both surface and subsurface drainage systems on irrigated lands and the development of erosion-control practices compatible with the irrigation system are essential in these regions.

Semiarid regions offer a greater potential for limited irrigation and the conjunctive use of rainfall with irrigation. This is because rainfall is a major part of the total water supply, even on irrigated land. Also, there is generally an abundance of cropland in semiarid regions and a shortage of irrigation water. Maximum water-use efficiency can only be achieved when both rainfall and irrigation water are adequately managed. In the past, the tendency in many cases was to manage the irrigation water and largely ignore the rainfall. Recently, however, an increased awareness of the depleting groundwater supplies and of increasing energy costs has caused a shift in emphasis to more efficient conjunctive water use.

Maximizing water-use efficiency in semiarid regions will generally involve limited irrigation. When full irrigation is practiced, the probability of runoff from rainfall is very high, and this will, of course, result in lowered water-use efficiency. The real challenge in these regions is to design innovative and flexible systems that prevent runoff, decrease evaporation, reduce percolation, and manage water for crops that can tolerate some stress and still utilize limited amounts of water productively. Such systems will allow more cropland to be irrigated with the limited water resources available.

References

Arkley, R. J. (1963). Relationship between plant growth and transpiration. *Hilgardia* **34,** 559–584.

Clark, R. N., and Hudspeth, E. B. (1976). Runoff control for summer crop production in the Southern Plains. *Pap.—Am. Soc. Agric. Eng.* **76-2008.**

De Wit, C. T. (1958). Transpiration and crop yields. *Versl. Landbouwkd. Onderz.* **646.**

Greb, B. W., Smika, D. E., and Black, A. L. (1967). Effect of straw-mulch rates on soil water storage during summer fallow in the Great Plains. *Soil Sci. Soc. Am. Proc.* **31,** 556–559.

Hughes, W. F., and Harman, W. L. (1969). Projected economic life of water resources, Subdivision Number 1 High Plains Underground Water Reservoir. *Tex., Agric. Exp. Stn.,* [Tech. Monogr.] **TM-6,** 1–82.

Interagency Task Force Report (1979). "Irrigation Water Use and Management." U.S. Dept. of Agriculture and Environmental Protection Agency, Washington, D.C.

Jensen, M. E. (1980). Introduction. *In* "Design and Operation of Farm Irrigation Systems" (M. E. Jensen, ed.), Monogr. No. 3, Chapter 1. *Am. Soc. Agric. Engl.,* St. Joseph, Michigan.

Jensen, M. E., and Sletten, W. H. (1965). Evapotranspiration and soil moisture-fertilizer interrelations with irrigated grain sorghum in the Southern High Plains. *Conserv. Res. Rep. (U. S. Dep. Agric.)* **5,** 1–27.

Johnson, W. C., and Davis, R. G. (1972). Research on stubble-mulch farming of winter wheat. *Conserv. Res. Rep. (U.S. Dep. Agric.)* **16.**

Jones, D. L., and Gaines, F. (1941). Pump irrigation on the South Plains. *Prog. Rep.—Tex., Agric. Exp. Stn.* **PR-728.**

Jones, O. R. (1975). Yields and water-use efficiencies of dryland winter wheat and grain sorghum production systems in the Southern High Plains. *Soil Sci. Soc. Am. J.* **39,** 98–103.

Lyle, W. M., and Bordovsky, J. P. (1980). New system for precision irrigation and energy conservation. *Pap.—Am. Soc. Agric. Eng.* **80-2069.**

Malm, N. R., and Hsi, D. C. H. (1968). Irrigation management. *Sorghum Newsl.* **11,** 89.

Musick, J. T. (1970). Effect of antecedent soil water on preseason rainfall storage in a slowly permeable irrigated soil. *J. Soil Water Conserv.* **25,** 99–101.

Musick, J. T. (1973). Supplementing rainfall with limited irrigation in minimum tillage systems. "Texas Tech University Conservation Workshop on Maximizing Rainfall in Semiarid Areas for Plant Production." Proc. Soil. Conserv. Serv., Lubbock, Texas.

Musick, J. T., and Dusek, D. A. (1971). Grain sorghum response to number, timing, and size of irrigations in the Southern High Plains. *Trans. ASAE* **14,**401-404, 410.

Musick, J. T., and Dusek, D. A. (1975). Limited irrigation of grain sorghum in alternating strips with wheat. *Trans. ASAE* **18,**544-548.

Musick, J. T., Wiese, A. F., and Allen, R. R. (1977). Management of bed-furrow irrigated soil with limited- and no-tillage systems. *Trans. ASAE* **20,**666-672.

Stegman, E. C., Musick, J. T., and Stewart, J. I. (1980). Irrigation water management. *In* "Design and Operation of Farm Irrigation Systems" (M. E. Jensen, ed.), Monogr. No. 3. Am. Soc. Agric. Eng., St. Joseph, Michigan.

Stewart, B. A., Dusek, D. A., and Musick, J. T. (1981). A management system for the conjunctive use of rainfall and limited irrigation of graded furrows. *Soil Sci. Soc. Am. J.* **45,**413-419.

Unger, P. W. (1972). Dryland winter wheat and grain sorghum cropping systems—Northern High Plains of Texas. *Tex., Agric. Exp. Stn.* [*Bull.*] **B-1126,**1-20.

Unger, P. W. (1978). Straw-mulch rate effect on soil water storage and sorghum yield. *Soil Sci. Soc. Am. J.* **42,**486-491.

Unger, P. W., and Wiese, A. F. (1979). Managing irrigated winter wheat residues for water storage and subsequent dryland grain sorghum production. *Soil Sci. Soc. Am. J.* **43,**582-588.

van Schilfgaarde, J., and Rawlins, S. L. (1981). Water resources in a growing society. *In* "Efficient Water Use in Crop Production" (H. M. Taylor, W. Jordan, and T. Sinclair, eds.). Am. Soc. Agron., Madison, Wisconsin (in press).

IRRIGATION SCHEDULING USING SOIL MOISTURE MEASUREMENTS: THEORY AND PRACTICE

Gaylon S. Campbell

Department of Agronomy and Soils
Washington State University
Pullman, Washington

Melvin D. Campbell

Department of Agronomy
U and I, Inc.
Kennewick, Washington

I. Theory of Irrigation Scheduling

A. Introduction

There are no records to show when an irrigator first "sampled" the soil to determine whether it was time to irrigate, but we may assume that this practice is

ISBN 0-12-24301-6

almost as old as the practice of irrigation. Irrigation scheduling by soil moisture "measurement" is probably the oldest method in existence. The fact that irrigators still trust the "feel" and color of a shovelful of soil in preference to many other irrigation-scheduling procedures indicates that scheduling by soil moisture is still the most popular of any of the methods. Newer methods for scheduling irrigation by monitoring crop or atmospheric factors have provided additional understanding of the system and guidance for irrigators lacking experience, but have not fully satisfied the need for an unambiguous criterion for scheduling irrigation.

Proper scheduling of irrigation requires the answers to two questions: When should the water be turned on, and when should it be turned off? Since the soil is the primary recipient of the irrigation water, it seems reasonable that the answers to these questions should come from monitoring the soil. Exceptions might occur when the monitoring of soil moisture is too expensive or of insufficient accuracy to satisfy irrigation-scheduling needs. In this article, we describe recent advances in both the theory and practice of irrigation scheduling using soil moisture measurement which allow accurate, quantitative scheduling using these methods.

B. Factors Affecting Irrigation Scheduling

Soil water has a direct effect on plant growth through its effect on water status. Soil water also affects plant growth indirectly through its effect on soil temperature, soil aeration, and nutrient leaching. Moisture conditions which minimize water stress may therefore result in suboptimal soil temperature or aeration, or may leach nutrients which are necessary for optimum plant growth. To further complicate the picture, maximum crop production may not be the only goal of irrigation scheduling. It is often necessary to minimize environmental impacts of irrigation or to maximize production from a scarce or suboptimal water source. These factors dictate on the one hand that the goals and principles for irrigation scheduling be clear in the mind of the scheduler, and on the other hand, that quantitative methods be employed for achieving these goals.

Recognition of the fact that the soil water system is dynamic is of fundamental importance. It is often the rate of water movement within the system, not the water content or potential that determines whether or not irrigation is necessary. Hillel (1972) stressed this point by comparing the plant's water supply with a bank account that is "robbed and embezzled almost daily." Any measure of soil water content or potential is useful mainly as an index of the rate at which water is taken up by plants or lost from the root zone. A given water content or potential is therefore most useful in conjunction with other information about the soil–plant–atmosphere system. The theory section of this article will attempt to outline the principles behind these dynamic concepts and to show what additional information is needed in order to use measurements of soil water content or potential to determine when irrigation should commence or cease.

C. Dynamics of Soil Water Redistribution

We first examine those factors which determine how much water should be applied in an irrigation. When water infiltrates soil, the wetting pattern generally consists of soil at almost uniform water content near saturation extending from the wetted surface to the wetting front. The wetting front is generally sharp, with the water content below being that which existed before irrigation. After irrigation ceases, the water in this nearly saturated zone moves deeper into the profile through redistribution. As the transmission zone drains, its hydraulic conductivity decreases until the rate of drainage becomes small. The water content at which this rate of drainage is considered to be negligibly small has been called the "field capacity" of the soil. Soluble salts, including mobile nutrients, move with the water during infiltration and redistribution. These processes must therefore be managed so that salinization of the root zone is avoided (adequate leaching fraction maintained) while nutrient leaching is minimized. Minimizing drainage below the root zone increases irrigation efficiency and reduces waterlogging and soil aeration problems.

A simple analytic expression for the redistribution process, which is sufficient for our purposes here, was given by Gardner (1970). The analysis begins with the Richards equation for transient, unsaturated flow:

$$\frac{\partial \theta}{\partial t} = \frac{\partial}{\partial z}\left(k\,\frac{\partial \psi_{\mathrm{m}}}{\partial z} - k\right) \tag{1}$$

where θ is the water content, ψ_{m} is the matrix potential, k is the unsaturated conductivity function, z is the vertical dimension, and t is time. During infiltration, the gradient in water content, and therefore the gradient in matrix potential are nearly zero in the transmission zone. In addition, during redistribution the change in water content at all depths in the transmission zone occurs at about the same rate, so $d\theta/dt$ is approximately constant with depth. If we assume that $d\psi_{\mathrm{m}}/dz = 0$ and $d\theta/dt$ is not a function of depth, then Eq. (1) can be simplified and integrated from the soil surface to the depth of initial wetting L to obtain

$$L(d\theta/dt) = -k \tag{2}$$

If the hydraulic conductivity is approximated as a power function of water content

$$k = k_s(\theta/\theta_s)^m \tag{3}$$

then the field capacity water content is

$$\theta_{fc} = \theta_s(\epsilon/k_s)^{1m} \tag{4}$$

where the subscript "s" indicates saturation values, m is an empirical constant, and ϵ is the drainage rate which one considers negligible for the purpose of the calculation. A rough estimate of the depth to which the wetting front will penetrate after redistribution (assuming no plant uptake during redistribution) is obtained from

$$L_f = L(\theta_S - \theta_i)/(\theta_{fc} - \theta_i) \tag{5}$$

where θ_i is the initial water content of the soil before infiltration.

An implicit assumption in this simple analysis is that the soil profile is uniform. Layering in the profile reduces the rate of water transport in the soil profile and can substantially increase the field capacity water content. Clothier *et al.* (1977) gave an analysis similar to the one presented here for soil underlain by coarse-textured materials.

In practice, field capacity water content is usually measured in the field, rather than calculated from these models. The value of the models however, is to establish the following points:

(1) Field capacity is determined by the hydraulic properties of the soil and the drainage rate that is chosen as negligible.

(2) Drainage does not cease, but only decreases over time to some small value.

(3) Water is retained by soil after irrigation simply because it cannot run out fast enough to be lost over the time span of interest to the irrigator. It is not retained because of the attraction of the soil matrix for water, and any water content or potential which is associated with field capacity is therefore only an index of the rate of water flow in the soil profile which results from the gravitational potential gradient.

The principles that have been established in this simple analysis can be applied to the problem of irrigation scheduling. We see that when the soil water content is maintained at values above field capacity, drainage from the profile is rapid, and both water and nutrients are lost. Figure 1 shows that this transition from negligible drainage to appreciable drainage occurs over a narrow range of water contents. The actual water content at which this transition occurs depends on the particular soil, but a rapid increase in drainage when water content exceeds a given value is a characteristic of all soils. One task of the irrigator is to control

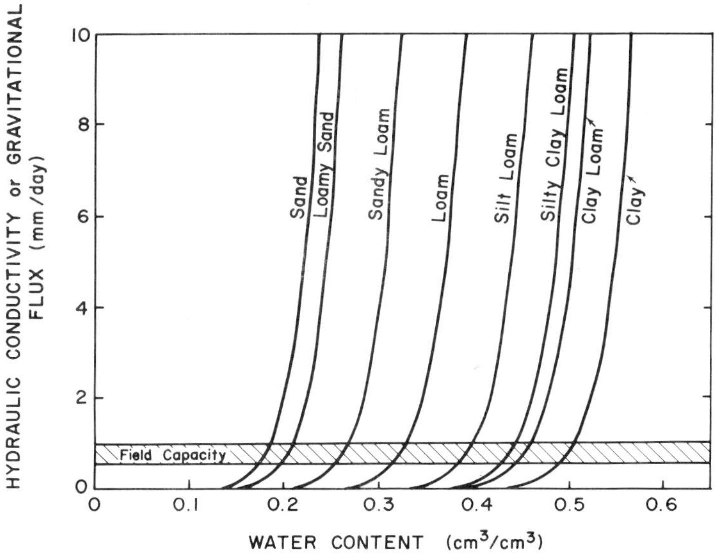

FIG. 1. Typical hydraulic conductivity functions for soils of different texture. These are also the steady gravitationally induced fluxes of water through a uniform profile of soil. A typical range for field capacity is shown.

the soil water content so that drainage remains at the value required to meet the objectives of the irrigation.

D. Dynamics of Soil Water Uptake by Plants

By use of a simple model we were able to find a "full point" for the soil: the water content above which drainage out of the root zone becomes appreciable. We can use another simple model to establish the soil's "refill point," and then consider how this value might be influenced by environmental factors.

Water flow through plants is often illustrated using a simple electrical analog, such as that shown in Fig. 2a. Transpiration is shown as a current generator which, except for stomatal closure, is largely uninfluenced by the plant. The water potentials in the roots and leaves of the plant are determined by the magnitude of the transpiration rate, the resistances to flow in the soil and plant, and by the soil water potential. The leaf water potential is given by

$$\bar{\psi}_L = \bar{\psi}_S - E(\bar{R}_{rs} + R_p) \tag{6}$$

where $\bar{\psi}_S$ is an appropriate mean soil water potential, $\bar{R}_{rs}$ is a mean resistance to water flow into roots, R_p is a resistance to water flow from roots to the evaporating surfaces in leaves, and E is the transpiration rate. The resistances are large

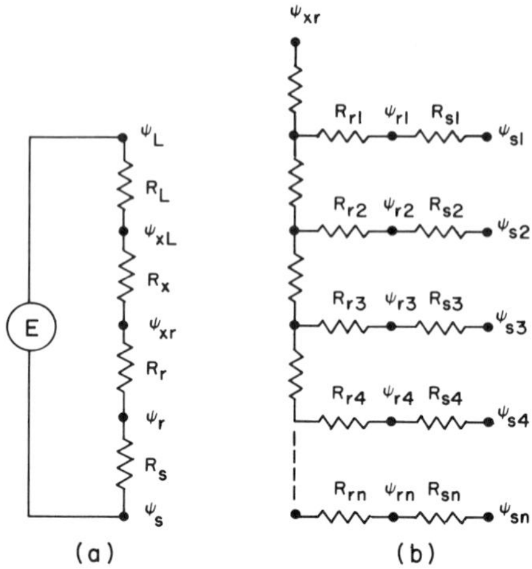

FIG. 2. Resistances and potentials in the soil-plant-atmosphere system for (a) the entire plant, and (b) the plant root system; E is transpiration rate.

enough so that the leaf water potential changes primarily in response to changes in evaporative demand.

The purpose of irrigation is to maximize leaf water potential. Irrigation can increase leaf water potential by increasing soil water potential or by decreasing the resistance to water flow from the soil to the plant. Irrigation can also adversely affect leaf water potential by cooling the soil or decreasing root aeration, thus decreasing root respiration and increasing root resistance. The refill point should be set at the lowest water content which allows $\bar{\psi}_s$ and $\bar{R}_{rs}$ to remain negligible in comparison with the potentials and resistances which already exist in the system. In order to do this, it is necessary to understand the nature of the resistances in greater detail.

Figure 2b shows an electrical analog of the rooting system of a plant. Here the soil is divided up into layers, each of which is represented by a given water potential, hydraulic conductivity, and rooting density. Two resistances are shown for flow from soil into the root, the resistance to water in the soil, and that for water movement into the root (primarily endodermis resistance). Resistance to flow through root xylem is assumed negligible so that the root xylem forms an equipoential surface through the soil. The root resistance for the entire plant can be determined by noting the xylem water potential and transpiration rate when the soil is wet so that $\bar{\psi}_s$ and $\bar{R}_s$ are negligible (Cowan, 1965). The total root resistance is then divided between layers using

$$R_{ri} = R_r/\phi_i \tag{7}$$

where ϕ_i is the fraction of the total root system in the ith soil layer.

A steady-state model similar to that of Gardner (1960), Cowan (1965), and Campbell *et al.* (1976) can be used to determine the soil resistance in each soil layer:

$$R_{si} = B_i(\psi_{si} - \psi_{ri})/[(\psi_e/\psi_{si})^{n-1} - (\psi_e/\psi_{ri})^{n-1}] \quad (8)$$

where the water potential subscripts s, r, and e refer to water potential in the bulk soil, at the root surface, and at air entry (a soil property); n is an empirical constant in the hydraulic conductivity function for the soil

$$k = k_s(\psi_e/\psi)^n \quad (9)$$

which generally has a value ranging from 2 to 4; and B_i is a constant for the soil layer given by

$$B_i = (1 - n)\ \ln(\pi r_1^2 L_i)/(4\pi L_i\, \Delta z_i k_s \psi_e) \quad (10)$$

Here r_1 is the root radius, L_i is the root density (length of roots per unit volume of soil) in the ith layer, and Δz is the depth of the layer. Since the water potential at the root surface is a function of transpiration rate and soil resistance, Eq. (8) cannot be solved explicitly for soil resistance, and an iterative technique must be employed.

Following the analysis of Hillel *et al.* (1976), we can sum the water uptake from each soil layer to obtain

$$E = \sum_i (\psi_{si} - \psi_x)/(R_{si} + R_{ri}) \quad (11)$$

This equation may be rearranged to give the xylem water potential:

$$\psi_x = \frac{\sum_i \psi_{si}/(R_{si} + R_{ri}) - E}{\sum_i 1/(R_{si} + R_{ri})} \quad (12)$$

Using the additional relationship for potential drop between the xylem and evaporating surfaces of the leaf

$$\psi_L = \psi_x - ER_p \quad (13)$$

with Eq. (12) we obtain

$$\psi_L = \frac{\sum_i \psi_{si}/(R_{si} + R_{ri})}{\sum_i 1/(R_{si} + R_{ri})} - E\left[R_L + \frac{1}{\sum_i 1/(R_{si} + R_{ri})}\right] \quad (14)$$

We note that Eq. (14) is similar in form to Eq. (6), indicating that an appropriate mean soil water potential for the root zone is given by

$$\bar{\psi}_s = \frac{\sum_i \psi_{si}/(R_{si} + R_{ri})}{\sum 1/(R_{si} + R_{ri})} \tag{15}$$

and an appropriate mean resistance for the root-soil system is given by

$$\bar{R}_{rs} = \left[\sum_i 1/(R_{si} + R_{ri})\right]^{-1} \tag{16}$$

This description of water uptake by plants may be used to gain some insight into considerations which determine the refill point for irrigation. We will consider first conditions similar to those of Cowan (1965) and Campbell *et al.* (1976), where the root density is uniform throughout a soil layer and there is no water flow into the soil layer. Figure 3 shows the ratio of soil to total resistance for two rooting densities and three transpiration rates. The potential at which irrigation must occur to keep soil resistance negligible (Fig. 3) is somewhat dependent on transpiration rate and strongly dependent on rooting density. The refill point for this system would also depend on the hydraulic properties of the soil. At low rooting densities irrigation apparently must occur at potentials above −1 bar to avoid excessive soil resistance. The increase in water availability to the plant which would result from increasing this potential to field capacity potential by irrigation is probably inconsequential. The refill point for the high rooting density, however, is around −3 bars. If transpiration rate is high, so that leaf water potentials are near values which close stomates, then an increase in soil water potential of 3 bars could markedly improve plant water status. Typical rooting densities are usually between the extremes shown in Fig. 3 (Lesczynski

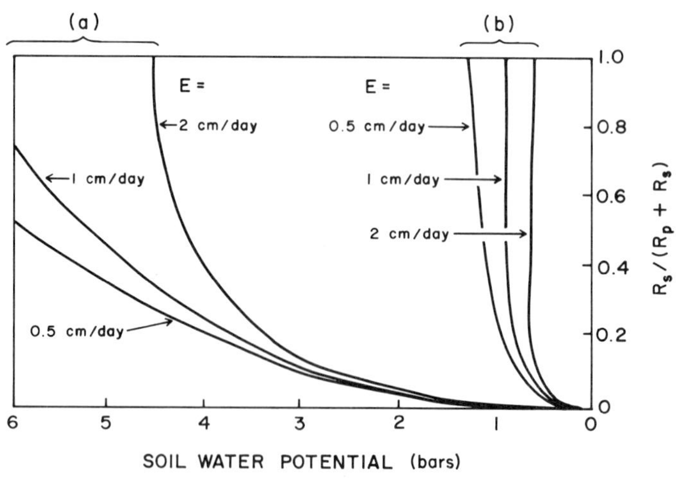

FIG. 3. Ratio of soil resistance to soil-plus-plant resistance as a function of soil water potential for (a) 15-cm-deep soil with root density of 10 cm^{-2} and (b) 150-cm rooting depth with root density of 0.1 cm^{-2}. Soil hydraulic properties are as in Campbell *et al.* (1976).

and Tanner, 1976; Robertson *et al.*, 1980), so we might expect some benefit to result from increasing water potential and some from decreasing resistance to water flow in soil.

A more realistic description of plant water uptake takes into account changes in root density with depth and water transport between layers (Hillel *et al.*, 1976). We ran a simulation similar to that of Hillel *et al.* (1976), except that soil resistance was calculated using Eq. (8), and stomatal resistance was allowed to control transpiration so that leaf water potentials were maintained above a preset limit. Rooting densities for the simulation were those given by Robertson *et al.* (1980) for corn. We simulated water loss from a crop starting at field capacity and ending at the point where transpiration was 90% of potential transpiration. A variety of soil textures were used, and transpiration rates of 5, 10, and 15 mm/day were simulated. Results of these simulations indicate that soil resistance remains negligible until transpiration is well below the value at which irrigation should occur. Plant water potential is influenced primarily by the soil water potential in the upper, high-density portion of the root zone, and the primary effect of irrigation is to increase soil water potential rather than reduce resistance. Transpiration rate does influence the choice of a refill point, however, because high transpiration rates will drop leaf water potential to the critical value at a higher soil water potential than will low transpiration rates [Eq. (6)]. It appears then that the refill point should be the maximal soil water potential which does not increase root resistance or leaching of nutrients.

II. Practical Aspects of Irrigation Scheduling

In order to schedule irrigation using soil moisture measurements, one must select a site for monitoring soil moisture, choose a device for determining the soil water content or potential, set full and refill points for the particular soil being monitored, and establish some kind of record-keeping scheme which will show when irrigation is necessary and whether overirrigation has occurred. The following description will cover these points with the objective of explaining how the principles in Section I have been applied in the field.

A. Measurement-Site Selection

Spatial variability in both soils and crops is high. If one were to try to obtain an accurate representative average water content or potential for a field, many samples would be required. Fortunately, irrigation scheduling does not require a knowledge of the field average. Since the field is irrigated as a unit, a single representative monitoring site can be used to indicate the water status for the entire field.

It is desirable to choose a site for convenient access that is more than 10–20 m from any edge of the field and that is among healthy, vigorous plants, in soil of above-average water capacity for the field, and in a location exposed to normal climatic and irrigation variations. The number of moisture measurements made or sensors installed at a given site is determined by the method chosen for making the measurement. For row crops, sampling or sensor placement should be in the row within the active root zone of the plants. Measurements for other crops should be from among actively growing plants where normal water consumption may be expected.

B. Moisture-Monitoring Procedures

Several devices and procedures have been used for obtaining soil moisture measurements necessary for irrigation scheduling (Haise and Hagan, 1967). Frequently soil samples are taken. These are weighed and dried for water content determination. Soil moisture had also been monitored using tensiometers (Taylor, 1965), gypsum resistance sensors (Fishbach, 1971), and neutron scattering (Gear *et al.*, 1977). Each method has advantages and disadvantages. The best method to use is one which gives measurements with the required precision at a cost that is feasible for field use.

Soil sampling for direct determination of water content is simple but time consuming, and it can therefore be prohibitively expensive. Spatial variability of soil water content is a problem with this method since one cannot return to exactly the same location each time for the measurement. An indication of the number of samples needed for a given precision can be calculated (Warrick and Nielsen, 1980). If we wish to know soil water content to within 0.01 cm^3/cm^3 at least 90% of the time, and if the standard deviation in water content is around 0.03 cm^3/cm^3 (a conservative value from Warrick and Nielsen, 1980), then we would need to take more than 20 samples. It is obviously impractical to take this number of samples from each field which is to be scheduled.

Tensiometers are frequently used to schedule irrigation. They have the advantage that they are relatively simple and inexpensive. Their disadvantage is that they are limited to potentials wetter than around 0.8 bar and require frequent servicing for proper function. Another disadvantage is that they measure potential only in the immediate vicinity of the unit, so that several tensiometers are needed to give a reliable spatial average.

Gypsum blocks are simple and inexpensive. The main difficulty in using them is the uncertainty in calibration. One reads an electrical resistance of the block and infers a water content or potential. The relationship between electrical resistance and water potential varies from block to block and may also vary from soil to soil and with time, due to changes in the electrical conductivity of the soil solution. The electrical resistance–water content relationship varies from soil to soil because moisture characteristics differ among soils. It is therefore difficult to interpret readings quantitatively in terms of water stored or water lost. They can, however, be useful as a relative measure of crop water use. Several gypsum blocks are needed to adequately represent the water status of a measurement site since each gives only a point measurement.

The neutron probe is widely used for soil moisture measurement, but its use for irrigation scheduling is relatively recent. Haise and Hagan (1967) stated that "... radiation hazards, high cost, and maintenance problems generally restrict its use to a research tool." Recent improvements in neutron probe design have changed this, and the neutron probe is now widely used to schedule irrigation. It operates by producing fast neutrons from a radioactive source in a probe. The emitted neutrons are scattered and moderated by the water in the soil. The probe is inserted into the soil through an access tube, and the water content is directly related to the number of neutrons scattered back to a detector. The measurement is therefore an average for the soil volume surrounding the access tube, so a single measurement gives the same information that several measurements from tensiometers or gypsum blocks would give. The calibration for the probe is relatively constant from soil to soil, and once it is known, it should not change with time. Readings are rapid, and the instruments are reasonably portable. The main disadvantage of the neutron probe is the expense of the equipment. Expenses associated with the time required to make measurements and the interpretation of measurements are generally less than with other methods. It is this reduction in measurement cost that has made large-scale application of soil moisture monitoring schemes for irrigation scheduling feasible. Our experience indicates that whereas any of the various monitoring methods may be used, the neutron probe offers the best combination of features for irrigation scheduling.

C. Setting the Full and Refill Points

In Section I we analyzed the considerations which should go into setting the full and refill points. The full point is the field capacity for the soil. A rough approximation of field capacity is obtained by determining a typical hydraulic conductivity for a uniform profile of soil having the texture one finds at the measurement site. The refill point is the potential below which crop production is measurably decreased. In the previous analysis, we assumed that production and transpiration were related, and that the refill point was the potential at which the transpiration rate was decreased by 10% from its potential value. Field measurements have been used to indicate soil water potentials at which irrigation should be applied for maximum production (Taylor, 1965; Haise and Hagan, 1967; Taylor and Ashcroft, 1972). If irrigation is scheduled using a tensiometer, and the refill point for the crop is within the range of the tensiometer, then this point may be read directly. If a water content method is used, then the potentials in these references must be converted to water contents to know the refill point for a given soil. Rough approximations of the water content for any given potential can be obtained from textural information. More exact estimates come from measurement of water content and corresponding water potential values in the field.

The field capacity water content can be calculated from Eq. (4) if k_s and m are known. An approximate value for k_s can be obtained if the soil texture is known. Data from Bloemen (1980) were used to correlate k_s with the silt plus clay fraction of soils. The following relationship was obtained:

$$k_s \text{ (cm/day)} = 180 \exp[-4.26(x_s + x_c)] \tag{17}$$

where x_s and x_c are silt and clay fractions. Structure obviously plays an important role in determining saturated conductivity. This is not taken into account in Eq. (17), but the equation does account for the gross differences which result from differences in texture.

The value for m can be related to the moisture release curve for a soil (Campbell, 1974). If the moisture release function can be approximated by

$$\psi = a\theta^{-b} \tag{18}$$

then $m = 2b + 3$.

Values for b were obtained using the correlations of Gupta and Larsen (1979). It was found that for the water potentials of interest in irrigation scheduling, the constant a in Eq. (18) can be taken as $a = -4 \times 10^{-4}$ bars when water content is measured in cm^3/cm^3. The 1-bar water content for midrange bulk density and low organic matter can be predicted from Gupta and Larsen (1979):

$$\theta_1 \text{ (cm}^3/\text{cm}^3) = 0.2x_s + 0.6x_c + 0.09 \tag{19}$$

The b value is calculated from

$$b = -7.82/\ln \theta_1 \tag{20}$$

Figure 1 shows the gravitational flux as a function of water content for uniform profiles of the soils shown. The flux equals k, where k is calculated from Eq. (3). A typical range for field capacity is also shown in the figure. A first estimate of the full point is obtained by entering Fig. 1 with a known texture and reading the water content from the abscissa. This measurement can be refined through measuring the water content in the field following a thorough irrigation at a time when plant uptake can be neglected.

In practice, water content of the soil usually is not monitored during irrigation to determine when irrigation should cease. Overirrigation can be monitored by measuring the water content in the soil below the root zone of the crop. Figure 1 may be used directly with this measurement to estimate the flux out of the bottom of the root zone. If this flux is higher than the leaching requirement, then the amount of water applied in an irrigation should be reduced.

The refill point for a soil could be set using the analysis in Section I, but available data on resistances, rooting densities, etc., usually are not sufficient to make reliable estimates of the refill point. More reliable estimates are obtained by referring to tables in Haise and Hagan (1967) or in Taylor and Ashcroft (1972). These tables give water potentials at which various crops are to be irrigated. If tensiometers are to be used, then these tensions can be used directly, as long as they are within the range of the tensiometer. If some water content method is used to schedule irrigation, then one needs to convert the potentials in the tables to equivalent water contents. Equations (18)–(20) can be used to obtain rough estimates of these water contents. Water content–water potential relations are shown in Fig. 4 for representative soils using these functions. Again, factors such as bulk density, hysteresis, and uncertainties in the empirical relationships from which these equations were derived enter in, but the equations will give first-order estimates of the refill points. Improved data may be obtained by measuring the water content and water potential of the soil simultaneously. For neutron probe measurements, a probe may be installed in the field and water content measurement made. Water potential measurements are made at several locations around the access tube with tensiometers after the soil has been wetted and is desorbing. The water content measurement and the average of the potential readings can be used to determine the b value for the soil using

$$b = \ln(a/\psi)/\ln \theta \tag{21}$$

where $a = -4 \times 10^{-4}$ bar, ψ is the average of the tensiometer readings (bars), and θ is the water content (cm^3/cm^3) from the probe count. Equation (18) can

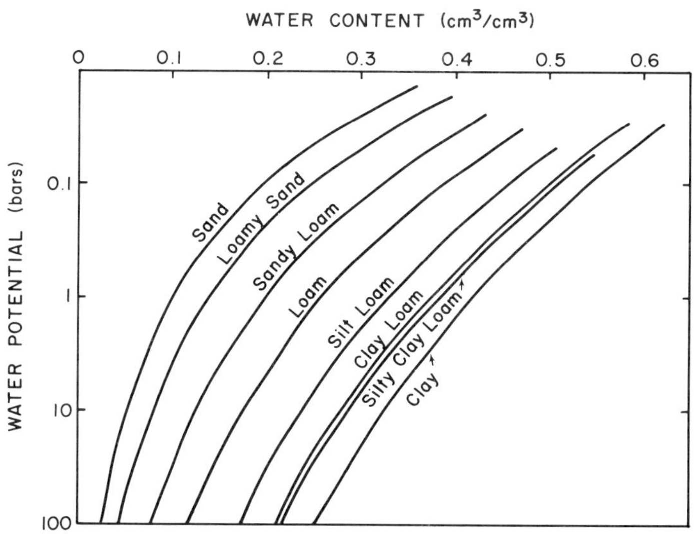

FIG. 4. Typical moisture release curves for soils of various textures.

then be used with this value of b to obtain improved estimates of water content for any other water potential around the value at which the measurement was made.

The tables in Taylor and Ashcroft and those in Haise and Hagen give a range of water potentials which reflect the effect of evaporative demand on the refill point. If evaporative demand is low, irrigation can be delayed to a lower potential than if evapotranspiration (ET) is high. Refill points for both high and low evaporative demand should be determined.

D. Measurements and Graphical Interpretation for Irrigation Scheduling

Once the monitoring site has been selected, the full and refill points determined for the field, and the monitoring method selected, one is ready to start monitoring soil water and scheduling irrigation. Measurements of soil water should be made as frequently as possible. For high-value crops which are irrigated frequently, daily measurements are advisable. For other crops and with lower frequency irrigation, measurements could be made less frequently. Measurements should be made at the depth of maximum root activity (Haise and Hagan, 1967). This is generally between 15 and 30 cm for irrigated crops. A measurement from below the root zone is also necessary to determine whether too much water is being applied and leaching is occurring. These measurements are plotted as functions of time on a graph similar to Fig. 5. The full, low-ET refill, and high-ET refill points should also be indicated on the diagram. Irriga-

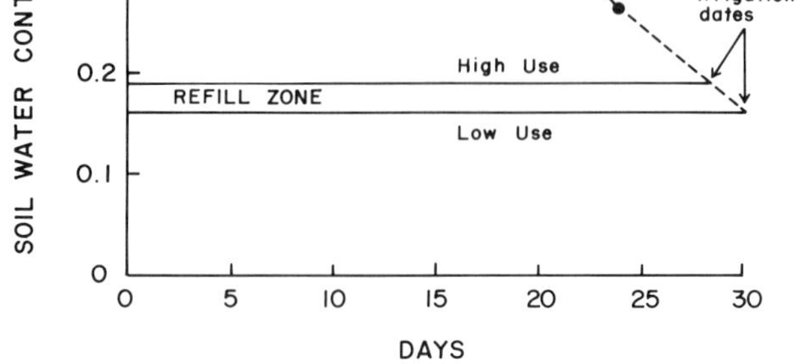

FIG. 5. Simulated record of field moisture depletion between irrigations showing projected irrigation dates for high- and low-demand conditions (after Gear *et al.*, 1977).

tion should be scheduled so that the soil water content stays between the refill and full values. If it goes above the full point, leaching will occur. If it goes below the refill point, production will be reduced. The record of water content or potential below the root zone shows whether the end points are correctly set and if monitoring in the root zone is adequate. An increase in water content of soil below the root zone indicates leaching.

E. Variations in Scheduling with Different Irrigation Methods

Surface irrigation methods require location of the measurement site about one-half to two-thirds of the distance from the head end of the field. This helps avoid placement in the zone of overirrigation. For furrow irrigation, measurements should be made at two-thirds of the distance from the head of the field and 10–20 rows from the field's edge. The sensors should be located in the row between normally spaced plants within the active root zone of vigorously growing plants on a normally shaped hill. If a fixed sequence is followed in irrigating a field, the monitoring site should be placed in the starting position.

Sprinkler systems require one sampling or monitoring site for each irrigation set or practice. Thus, if center pivots are operated uniformly on a field, only one site is required. However, if more than one practice is followed, a separate site will be needed for each. The monitoring site should not be located much above or below the elevation of the pivot point or inlet end of the sprinkler system unless variation in application due to pressure differences is taken into account. Otherwise, cumulative errors will result.

Some sprinkler systems may also require other provisions. For example, center-pivot systems require that the reading be taken nearer the soil surface than is required for surface irrigation because of the smaller amounts of water applied at a higher frequency (e.g., daily). Otherwise, the wetting or drying trend may not be observable for several days. Wheel lines and hand lines require a measurement site for each two sets within the first lateral traverse. This provides adequate accounting of variations due to weather and operations if the same sequence is always followed for all laterals on the field. If the same sequence is not followed, then it is necessary to provide a moisture measurement site at the beginning position of each lateral to show when the irrigation cycle must be repeated. Solid sets usually require one measurement site for each set. Measurement sites should not be located where they will intercept runoff water from any line or lateral or where one sprinkler will deliver unusual amounts of water.

Drip irrigation systems are easily monitored with the neutron probe or other soil moisture methods. The sampling site is placed within the drip area in the active water-uptake zone of the root system. Like solid set, drip irrigation systems require one access tube or measurement site for each drip system set.

Although it is common practice to irrigate so as to refill the entire root zone under surface irrigation systems and lateral-move sprinkler systems, it is unusual to refill the root zone completely, or to deplete the available soil water reserve completely, from drip or center-pivot systems. This fact leads to use of different irrigation-scheduling techniques. For example, instead of wanting to know when to turn water on and when to turn it off, the irrigator wants to know the fraction of time the system should run to provide adequate water for the crop while avoiding leaching. Drip and center-pivot systems are much more likely to be operated on a high-frequency basis and to maintain a more nearly steady water content than are other systems. They are, however, just as manageable as the other systems by direct soil moisture measurement.

F. Calibration of Neutron Probes

Calibration and interpretation of measurements made using tensiometers, gypsum blocks, or direct sampling are discussed in detail elsewhere (Black, 1965; Taylor *et al.*, 1961) and will not be covered here. The greatest difficulty encountered in the use of the neutron probe for irrigation scheduling is that of calibration. It therefore seems appropriate to discuss a simple method of calibration and standardization that may be adopted anywhere in support of reliable irrigation scheduling.

First, consideration must be given to the correct functioning of the neutron probe. The electronic detector must count neutron impulses produced in the detector tube without counting background impulses that come from other sources. The performance of the detector circuit is easy to test. Simply take a set of twenty 15-sec readings in the shield without changing the surroundings of the neutron tube, and compute the mean, square root of mean and standard deviation for the twenty-reading set. If the ratio of the standard deviation to the square root of the mean is between 0.75 and 1.25, the neutron probe is working acceptably well. If the ratio is smaller, there are impulses being counted that are not from neutrons. If the ratio is larger, there are impulses from neutrons that are not being counted. In either case the manufacturer should be consulted for the necessary corrective action. Once the proper functioning of the probe is assured, the probe should be calibrated for the soils in which it will be used.

Adequate calibration of the neutron probe consists of installing about 20 access tubes in the variety of soils to be monitored, determining gravimetric water content and bulk density of the soil by sampling at the tube locations, and taking neutron probe readings of ambient soil moisture. The calibration is obtained by a regression of volumetric water content (gravimetric water × bulk density) on neutron probe count. The calibration equation is

$$\theta \ (\mathrm{cm}^3/\mathrm{cm}^3) = mC + b \tag{22}$$

where C is the neutron count, and m and b are the slope and intercept from the regression analysis. The constant m may vary from probe to probe, and over time for any given probe, while b remains fairly constant. Once the regression line has been obtained, the calibration can be checked, and new calibrations obtained from measurements on a single reference standard. We chose a 30-gal barrel as the reference standard for our 50-mCi sources. The barrel is filled with water and has an access tube placed in its axial center. A metal spacer 8-cm long is placed at the bottom of the tube, on the bottom plug of the access tube, to hold the neutron probe off the floor of the barrel. The neutron probe is inserted into the access tube and lowered onto the spacer at the bottom. A set of eighty 15-sec readings is then taken and the apparent amount of water θ_{std} is calculated using Eq. (22). This value of apparent water content becomes the reference standard.

Recalibration consists of taking 20 readings in the reference barrel and calculating their mean, square root of mean, and standard deviation. If the ratio of standard deviation to square root of mean is between 0.75 and 1.25, the probe function is deemed acceptable and the value for m is calculated from

$$m = (\theta_{std} - b)/C_{std} \tag{23}$$

G. Integration of Soil Moisture Methods with Other Methods of Irrigation Scheduling

While soil moisture monitoring does provide a direct measure of soil moisture condition during a growing season, frequent direct monitoring of soil water can be time consuming and expensive. It may be possible to use other methods of scheduling to reduce sampling frequency and still maintain adequate control over soil moisture. The ET method of Jensen *et al.* (1971) uses a computed water budget from estimates of ET, drainage, and water applied to determine when irrigation should be applied. This calculation could be simplified using the simple model of Priestly and Taylor (1972) to estimate daily potential ET from solar radiation and temperature data. Estimates from pan evaporation could also be used. These estimates of daily water use could be used with occasional measurements of soil moisture to assure that cumulative errors in the climatological scheduling method did not lead to either crop stress or overirrigation.

References

Black, C. A., ed. (1965). "Methods of Soil Analysis," Monogr. No. 9. Am. Soc. Agron., Madison, Wisconsin.

Bloemen, G. W. (1980). Calculation of hydraulic conductivities of soils from texture and organic matter content. *Z. Pflanzenernaehr. Bodenkd.* **143,** 581–605.

Campbell, G. S. (1974). A simple method for determining unsaturated conductivity from moisture retention data. *Soil Sci.* **117,** 311-314.

Campbell, M. D., Campbell, G. S., Kunkel, R., and Papendick, R. I. (1976). A model describing soil-plant-water relations for potatoes. *Am. Potato J.* **53,** 431-441.

Clothier, B. E., Scotter, D. R., and Kerr, J. P. (1977). Water retention in soil underlain by a coarse textured layer: Theory and a field application. *Soil Sci.* **123,** 392-399.

Cowan, I. R. (1965). Transport of water in the soil-plant-atmosphere system. *J. Appl. Ecol.* **2,** 221-239.

Fischbach, P. E. (1971). Scheduling irrigations by electrical resistance blocks. *Nebr. Univ., Coll. Agric. Ext. Serv.* **EC 71-752.**

Gardner, W. R. (1960). Dynamic aspects of water availability to plants. *Soil Sci.* **89,** 63-73.

Gardner, W. R. (1970). Field measurement of soil water diffusivity. *Soil Sci. Soc. Am. Proc.* **34,** 832.

Gear, R. D., Dransfield, A. S., and Campbell, M. D. (1977). Irrigation scheduling with neutron probe. *J. Irrig. Drain. Div., Am. Soc. Civ. Eng.* **103** (IR3), 291-298.

Gupta, S. C., and Larsen, W. E. (1979). Estimating soil water retention characteristics from particle size distribution, organic matter percent, and bulk density. *Water Resour. Res.* **15,** 1633-1635.

Haise, H. R., and Hagan, R M. (1967). Soil, plant and evaporative measurements as criteria for scheduling irrigation. *Agronomy* **11,** 577-604.

Hillel, D. (1972). "Optimizing the Soil Physical Environment Toward Greater Crop Yields." Academic Press, New York.

Hillel, D., Talpaz, H., and van Keulen, H. (1976). A macroscopic-scale model of water uptake by a nonuniform root system and of water and salt movement in the soil profile. *Soil Sci.* **121,** 242-255.

Jensen, M. E., Wright, J. L., and Pratt, B. J. (1971). Estimating soil moisture depletion from climate, crop and soil data. *Trans. ASAE* **14,** 954-959.

Lesczynski, F. B., and Tanner, C. B. (1976). Seasonal variation of root distribution of irrigated field-grown Russett Burbank potato. *Am. Potato J.* **53,** 69-78.

Priestly, C. H. B., and Taylor, R. J. (1972). On the assessment of surface heat flux and evaporation using large-scale parameters. *Mon. Weather Rev.* **100,** 81-92.

Robertson, W. K., Hammond, L. C., Johnson, J. T., and Boote, K. J. (1980). Effect of plant-water stress on root distribution of corn, soybeans, and peanuts in sandy soil. *Agron. J.* **72,** 548-550.

Taylor, S. A. (1965). Managing irrigation water on the farm. *Trans. ASAE* **8,** 433-436.

Taylor, S. A., and Ashcroft, G. L. (1972). "Physical Edaphology: The Physics of Irrigated and Nonirrigated Soils." Freeman, San Francisco, California.

Taylor, S. A., Evans, D. D., and Kemper, W. D. (1961). Evaluating soil water. *Bull—Utah Agric. Exp. Stn.* **426.**

Warrick, A. W., and Nielsen, D. R. (1980). Spatial variability of soil physical properties in the field. *In* "Applications of Soil Physics" (D. Hillel, ed.), pp. 319-344. Academic Press, New York.

CANOPY TEMPERATURE AND CROP WATER STRESS

Ray D. Jackson

U.S. Department of Agriculture
Agricultural Research Service
U.S. Water Conservation Laboratory
Phoenix, Arizona

I. Introduction

Production of agricultural crops in the arid areas of the world is almost totally dependent on irrigation. In semiarid areas, irrigation is increasingly used to supplement rainfall. In normally humid areas, irrigation systems are being installed to offset the yield-reducing stress conditions that ensue after a few weeks without rain. Although the number of irrigation systems is growing, the supply of water available for irrigation is decreasing. This, coupled with the steeply rising costs of the energy required to pump and move water to the desired location, is causing an increased emphasis to be placed on means of attaining the maximum benefit from each unit quantity of water used for irrigation.

One aspect of irrigation science that has been of interest for some time is irrigation scheduling. Even today, many farmers irrigate by the calendar, while others receive water at scheduled times and therefore use the water whether their

ISBN 0-12-24301-6

crops need it or not. Another group, composed largely of farmers in high-energy-use areas who can control their water deliveries, is searching for ways to determine the most beneficial time to apply just the right amount of water.

In general, irrigation-scheduling techniques fall into three categories: soil based, meterologically based, and plant based. Combinations of the three categories are sometimes used. An example of a soil-based technique is the monitoring of soil water content within a field. Knowing the "field capacity" and the "wilting point" of the field soil, soil water content information allows estimation of the amount of water lost in evapotranspiration and drainage below the root zone. When the soil water content falls to certain value, the amount required to bring the soil profile back to field capacity is to be added by irrigation. Although in some places soil water content is still measured by the time-honored method of determining the weight of water in a sample, the neutron moisture meter (Van Bavel *et al.*, 1963; Bell, 1976; Nakayama and Reginato, 1982), has become commonplace. This method has been tried for a large portion of an irrigation district in western Arizona (Gear *et al.*, 1977). The technique assumes uniform soil properties throughout the field, since only one or two point measurements are made in each field. Furthermore, the only plant response parameter used in the technique is an indirect one through the wilting-point estimation.

Meteorological methods have been developed by a number of researchers (Jensen and Haise, 1963; Heermann *et al.*, 1976; Jensen and Wright, 1978; Wright and Jensen, 1978). Basically, these methods use air temperature, net radiation, vapor pressure, and wind speed as inputs to models which calculate the amount of water evapotranspired during a given time period. When the evapotranspiration reaches a certain value, irrigations are deemed necessary. The meteorological methods do not directly account for drainage below the root zone, and no direct plant information other than crop coefficients are used. These methods have been used with programmable calculators (Kincaid and Heermann, 1974; Kanemasu *et al.*, 1978).

Direct measurement of some plant parameter would appear to be a superior method for timing of irrigations in that the plant responds to both its aerial and soil environments. Such methods are generally limited to measurements on individual plant parts such as leaves and petioles. Examples are the pressure bomb (Scholander *et al.*, 1965), the leaf diffusion porometer (Kanemasu, 1975), and petiole and leaf water content measurements (Longenecker and Lyerly, 1969). These methods are time consuming and require numerous measurements in order to characterize a field.

Hiler and Clark (1971) and Hiler *et al.* (1974) developed a *stress day index* (SDI) based on plant parameters. The SDI was the product of a plant stress factor and a crop susceptibility factor. The former is an indicator of the crop's water deficiency and the latter is a function of the crop species and stage of growth.

After testing several plant measurements (including leaf temperatures), they selected the plant water potential as a stress indicator. Their concepts of stress quantification were expanded by Stegman *et al.* (1976), who related the xylem pressure of leaves to the ambient air temperature and the available soil moisture. Their results indicated a substantial decrease in the amount of water used without a reduction in yield for crops irrigated using scheduling techniques based on plant stress indicators.

Although leaf temperature did not prove sufficiently sensitive for Hiler *et al.* (1974) to use in their SDI, Ehrler (1973) found it to be a good indicator of stress, if vapor pressure effects were accounted for. He suggested that leaf temperature could be used for irrigation scheduling. Hsiao (1973), in a detailed review of plant water stress, stated that "... it would be reasonable to assume that elevation in leaf temperature does not play a general role in water stress effects." With such differing views on the usefulness of leaf temperatures for evaluating crop water stress, the issue is, to say the least, confusing.

Even if leaf temperature is, in principle, a good indicator of plant water stress, measurements on individual leaves suffer the same disadvantage as other point measurements, namely that many samples must be taken to adequately characterize a field. Recent advances in IR thermometry obviate that disadvantage by offering the possibility of rapidly surveying a large number of plants and integrating plant temperatures over entire fields or characteristic sections of fields.

This article progresses from the voluminous literature on leaf temperatures to the use of IR thermometry to determine canopy temperatures, and on to energy balance considerations. It develops a basis for the evaluation of plant stress from canopy temperature, vapor pressure, and net radiation data. A *crop water stress index,* as derived from theoretical considerations and empirical data, is discussed in detail. This index, similar in many respects to the SDI of Hiler *et al.* (1974), can be rapidly obtained by surveying a field with an IR thermometer, measuring wet- and dry-bulb air temperatures, and estimating net radiation. Its potential for use in the timing of irrigations appears promising.

II. Historical Perspective

The use of canopy temperatures to detect water stress in plants is based upon the assumption that transpired water evaporates and cools the leaves below the temperature of the surrounding air. As water becomes limiting, transpiration is reduced and the leaf temperature increases. If little water is transpired, leaves will warm above air temperature because of absorbed radiation.

There are reports in the literature that argue strongly that leaf temperatures are always warmer than the surrounding air. Others argue just as strongly that leaves can become much cooler than the air. The first argument has prevailed for years, and the numerous reports expounding that view are often cited as evidence that a measurement of leaf temperature is not a sensitive water stress indicator. These diametrically opposite points of view can perhaps be reconciled by examining them from an historical perspective. We shall therefore proceed to review the pertinent literature in chronological order. This review is not intended, however, to be exhaustive. For an annotated bibliography concerning applications of remote sensing for assessing crop water deficits, see David (1969). Hsiao (1973) compiled an excellent review of water stress in plants, and Cihlar (1976) prepared a bibliography on thermal IR remote sensing.

A. Leaf Temperatures

The literature concerning leaf temperatures dates back at least to the early part of the last century. Ehlers (1915) cited a paper by Rameaux (1843) who reported placing a number of leaves still attached to their stems on top of one another and wrapping the stack around a mercury thermometer. Askenasy (1875), as cited by Ehlers (1915) and by Ansari and Loomis (1959), measured the temperature of leaves of succulent plants by placing a mercury thermometer on the upper surface of a leaf and by inserting it in a cut made for that purpose. He observed that thin leaves in sunshine were 4–5°C warmer than air, and that thick leaves were in excess of 20°C over air temperature. Askenasy recognized the effects of reradiation, air movement, and transpiration on the cooling of leaves.

Ehlers (1915) was concerned with determining how evergreen leaves accumulated food in winter, his thesis being that the leaves were warmer than the air and could therefore photosynthesize at rates higher than would be calculated using air temperatures. His results (obtained in Michigan) showed that leaves of evergreen conifers were 2–10°C above air temperature.

Miller and Saunders (1923) measured the temperature of leaves of crop plants under natural field conditions in Kansas. They used a clamp made from brass tongs modified by enclosing the ends with cork. Thermocouples were embedded in the cork in such a manner that they would contact the leaf surface when the clamp was closed. About 15 sec were required to obtain a measurement. They concluded that turgid leaves of most crops were essentially at air temperature, but alfalfa consistently showed a temperature of about 1°C below the temperature of the air. These results were perhaps the first to show that leaf temperatures may be less than air temperatures under field conditions. They were later criticized by Clum (1926) and by Curtis (1936), who claimed that the leaf temperature was changed by the clamp shading the leaf.

Clum (1926) attempted to correlate leaf temperatures and transpiration. He measured leaf temperatures by threading fine thermocouples through the mesophyll. Air temperatures were measured by a thermocouple junction hanging freely in the air, usually in the shadow of one of the leaves containing one of the other junctions. Measurements were made on potted plants both in the greenhouse and in full sunshine. Some pots were watered more than others to achieve different transpiration rates. He found that the leaves were always warmer than the air by 5–10°C. In no case did he find a definite correlation between the transpiration rate and the leaf–air temperature difference. (Clum's paper is noteworthy for its excellent literature review.)

Eaton and Belden (1929) measured cotton leaves in Arizona and found leaf temperatures to be 2–4°C below that of the air. Curtis (1936) explained their results on the basis that the low humidity allowed long-wave energy to radiate to the clear, cold sky, thus cooling the leaves. Watson (1934) reported that radiative cooling and transpiration were equally important in controlling plant temperatures. Curtis discounted the effect of transpiration in cooling the leaves. His own experiments showed that leaf temperatures may be changed several degrees by allowing or preventing radiation to cold objects or to space. He also questioned data taken by other investigators that reported leaf temperatures to be less than the temperatures of the air on the basis that their readings for the latter were probably too high.

Wallace and Clum (1938) continuously recorded leaf and air temperatures of garden plants. They found leaf temperatures as much as 7°C below air temperature (as measured by a thermocouple suspended in the air directly behind and in the shade of the leaf). They showed that leaf orientation plays a major role in the temperature regime in that leaves were generally warmer than the air when facing the sun but cooler when at oblique angles. They concluded that the loss of energy to the sky did not explain the fact that the leaves were often below air temperature.

The ink was hardly dry on Wallace and Clum's paper when Curtis (1938) offered a strong rebuttal to their conclusions. He maintained that the air temperature, as measured by Wallace and Clum, was probably 2–7°C above true air temperature, at least part of the time. He concluded by stating that impossibly high transpiration rates would be required to lower the temperature of a leaf below that of air when in direct sunlight.

Waggoner and Shaw (1952) measured the temperature of potato and tomato leaves in Iowa. They found that on clear days, upper exposed leaves were 3.3–8°C warmer than the air temperature in an instrument shelter. Lower leaves were 0.5–0.8°C cooler than the air. They found no difference between similarly exposed potato and tomato leaves.

Ansari and Loomis (1959) measured leaf temperatures on a number of plants in a greenhouse, in the open, and in a dimly lit laboratory. The only leaves that they found to be below air temperature were cotton leaves in very dim light and 25% relative humidity. They concluded that leaves are warmed by air temperatures and radiation and cooled primarily by conduction to the air.

A theoretical heat transfer analysis by Wolpert (1962) showed that the temperature of a typical plant leaf was dominated by convection, assuming that the leaf remained motionless. If the leaf flaps due to wind, water evaporation controls leaf temperature. Gates (1964a) calculated the energy balance for a single leaf and concluded that transpiration was very important in the energy budget of plants. His calculations indicated that sunlit leaves would be warmer than air and the lower shaded leaves would be cooler than air.

Wiegand and Namken (1966) measured air temperatures at plant height with shielded thermocouples, and cotton-leaf temperatures with an IR thermometer. They observed that 30–45 sec were required to regain equilibrium temperature when insolation changes caused by clouds occurred. They showed that leaf temperatures increased linearly with increasing insolation, and decreased linearly with increasing relative turgidity of the leaves. The leaf–air temperature difference increased with increasing insolation (from −3.5° to +6°C for the "wet" treatment), but was different for each irrigation treatment. They concluded that leaf temperature interpretation requires simultaneously measured radiation data, and that the early afternoon is a good time of day for making leaf temperature measurements.

Stevenson and Shaw (1971) showed that upright leaves were cooler than horizontal leaves of soybean. For both orientations, leaf temperatures reached 4°C below air temperature. Carlson *et al*. (1972) found that leaf temperatures increased with decreasing values of the relative leaf water content and vapor pressure deficit. They observed that leaf temperatures of two soybean varieties were significantly different.

Clark and Hiler (1973) compared leaf water potential, leaf diffusion resistance, and leaf–air temperature differences (leaf temperatures measured with IR thermometers), of well-watered and water-stressed peas. Their air temperature reference was 1 m above the canopy. The data showed that leaves of well-watered peas were almost always cooler than the air and that leaves of plants on the stressed plot were 2–3°C warmer than leaves of well-watered plants. They concluded that leaf water potential was more responsive to changes in plant water status than either leaf diffusion resistance or the leaf–air temperature difference. These results were used by Hiler *et al.* (1974) in refining the SDI.

Ehrler (1973) directly addressed the possibility of using leaf–air temperature differences as a guide to irrigation scheduling. In field experiments with cotton he placed fine wire thermocouples in leaves and measured the air temperature and the vapor pressure at 1 m above the crop. His results showed a leaf–air temperature difference that ranged from −3° to +2°C, depending on the degree of soil water depletion. A significant result of this study was the demonstration of a linear relation between the leaf–air temperature difference and the vapor pressure deficit. This relation is shown in Fig. 1. The leaf–air temperature difference decreased about 1.3°C for each kPa increase in vapor pressure deficit. Ehrler's data indicated that the use of leaf temperatures for irrigation scheduling was

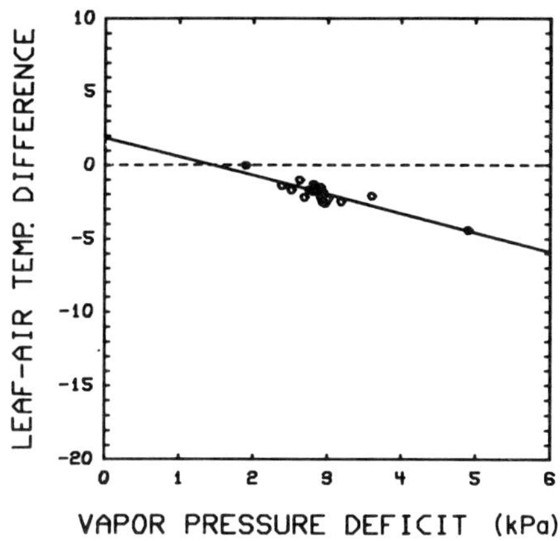

FIG. 1. The leaf–air temperature difference versus vapor pressure deficit for four varieties of cotton. Data were taken on predominantly sunny days when the crop was fully hydrated. Data points represent the mean of three replications (after Ehrler, 1973): slope, −1.29; intercept, 1.87; $R = 0.87$.

feasible. However, the measurement of individual leaf temperatures with thermocouples was cumbersome and tedious. Many leaves must be sampled in order to obtain a reasonable average for an entire field.

Sandhu and Horton (1978) found leaf temperatures of water-stressed oats to be 2.5–4.0°C warmer than well-watered oats. They attributed the temperature differences primarily to differences in transpiration. Sumayao *et al.* (1980) found that leaves of corn and sorghum, when well watered, were warmer than air when the air temperature was less than 33°C and cooler than air when the air temperature was greater than 33°C. Their results supported those of Linacre (1964, 1967) who first proposed 33°C as the "crossover" temperature. This constant value was disputed by Blad and Rosenberg (1976) who found that the crossover point for alfalfa occurred at temperatures between 23° and 30°C. Idso *et al.* (1981a) argued that a constant crossover temperature does not exist, but rather that it is dynamic and dependent on the vapor pressure deficit.

With the exception of Idso *et al.* (1981a), the papers reviewed in this section have been concerned with individual leaves. It is not surprising that the majority of the authors reported data that showed leaf temperatures greater than the air temperature. Many leaves were near the top of the plant and in direct sunlight. Air temperatures were measured in a variety of locations, ranging from the shade of the leaf to a meter or so above the plant, with no standard location suggested. Results were reported for single potted plants standing alone, without the canopy effects created by surrounding plants.

A number of experiments were conducted in glasshouses where the humidity was high. A majority of the field investigations were conducted in humid areas.

It will be shown in a later section that, theoretically, under conditions of high humidity, leaf temperatures will equal or exceed the air temperature. Field experiments conducted in semiarid areas, on the other hand, showed leaf temperatures to be as much as 7°C below that of the air.

Reradiation, convection, and transpiration were cited as the major means of cooling leaves exposed to radiation. For single leaves, transpiration was seldom indicated as a major cause of cooling. In fact, several studies discounted transpiration as an important cooling mechanism.

Part of the confusion about the use of leaf temperatures as plant water stress indicators arises, therefore, from the numerous studies on the temperature of single leaves on single plants; whereas in irrigation management, entire plant canopies are of concern. Gates (1964a) examined differences between single leaves and entire canopies. He measured Bowen's ratio (the ratio of sensible to latent heat exchange) for single leaves at several times during the morning hours. He found that the ratio increased throughout this period. This indicated that a single leaf exposed to sunlight would lose proportionally more heat by convection than by transpirational cooling as the temperature increases. However, lower leaves would have smaller values of the ratio, with the canopy as a whole having quite small values. These results were supported by the detailed calculations of Idso and Baker (1967). They concluded that for individual leaves reradiation was the major mode of heat transfer, but for canopies as a whole transpiration was the dominant mechanism. One can conclude that for a canopy (versus an individual leaf), transpirational cooling plays a major role in the energy balance, and therefore in determining plant temperature. Conversely, plant canopy temperature should be related to transpiration, and hence might serve as an indicator of plant water stress.

A problem encountered with leaf temperature measurements of field plants is that of variability. In water-stressed plots, leaf temperatures of plants 2 m apart were found to differ by 5.9°C, in contrast to relatively uniform temperatures for neighboring plants in well-watered plots (Blad *et al.*, 1982).

During the past 20 years, IR technology has developed to the point that canopy temperatures (both sunlit and shaded leaves, stems, and other plant parts) can readily be measured. With this development, the potential usefulness of plant temperature measurements to quantify plant water stress has greatly increased.

B. Canopy Temperatures

The extension of temperature measurements from individual leaves to canopies became possible with the development of instruments that measure emitted thermal radiation. Such instruments can be calibrated to read directly in terms of temperatures. Monteith and Szeicz (1962) and Tanner (1963) were among the first researchers to use IR radiometers to measure plant temperatures. Monteith and Szeicz (1962) presented a theoretical discussion concerning the relation between surface temperatures and stomatal resistance. From energy balance considerations, they derived an expression relating the canopy–air temperature difference to net radiation, wind speed, vapor pressure gradient, and the aerodynamic and canopy resistances. Their work formed the basis for the theoretical development of a crop water stress index that will be discussed in Section III. Their examples illustrated the dependence of surface temperature on stomatal resistance. They suggested that radiometrically determined surface temperatures could be used to evaluate the effective stomatal resistance of field crops.

Tanner (1963) used IR thermometry to explore the possibility of using plant temperature measurements to detect moisture stress differences between plants under different water regimes. Tanner recognized that the temperature sample would be weighted toward the upper part of the plant, but pointed out that it is these portions that participate most actively in transpiration, heat exchange, and assimilation. He concluded by stating ". . . plant temperatures may be a valuable qualitative index to differences in plant water regimes. Coupled with a better understanding of transfer processes at the plant surfaces, they may serve to provide quantitative data on plant-water status."

The instruments used by Monteith and Szeicz (1962) and by Tanner (1963) were essentially adaptations of laboratory apparatus. Shortly thereafter some aircraft-mounted IR scanning radiometers developed for military purposes were declassified and became available for civilian use. It would be nearly a decade before a truly portable IR radiometer was devised for field use. During the past few years, however, IR technology has developed rapidly, and IR thermometry is now used for many purposes. To provide a background for sections on airborne and ground-based thermal IR experiments, a brief review of some principles of IR thermometry is warranted.

1. *IR Radiation Thermometry*

Infrared thermometry is a noncontact method for estimating the surface temperature of a target. In principle, the measurement does not interfere with the surface and yields a temperature that is an integrated value over the field of view of the sensor. The instrument measures the radiation emitted from the target, and relates this radiation R to the surface temperature T_s by the Stefan–Boltzmann blackbody law,

$$R = \epsilon\sigma T_s^4 \tag{1}$$

where ϵ is the emissivity of the surface and σ is the Stefan–Boltzmann constant (5.674×10^{-8} W m^{-2} K^{-4}), with R in units of W m^{-2}. For discussions of basic principles of IR radiation thermomety, see Fuchs and Tanner (1966), Lorenz (1968), Gates (1968), and Perrier (1971).

Most IR thermometers have filters that allow only energy within the 8–14μm wave band to reach the sensor. This wave band includes the peak of blackbody emission at normal temperatures and has relatively low absorption by water vapor. The emissivity of natural surfaces within this wave band is usually high. However, water vapor absorption cannot always be neglected, and for precise measurements the emissivity must be known. Another complicating factor is that part of the radiation emitted by the surroundings and reflected by the surface in this wave band also may strike the sensor, adding to the radiation received from the target.

a. Atmospheric Water Vapor. Lorenz (1968) stated that the influence of water vapor (in the 8–14-μm wave band) on surface temperature measurements was neglible if the distance to the target was less than 154 m. His results suggest that ground-based thermometric measurements will generally not need to be corrected for water vapor absorption, but that aircraft- and spacecraft-based measurement will need correcting. Lorenz (1968) and Heilman *et al.* (1976) show how this correction can be made, if the water vapor distribution in the atmosphere is known. Computer models such as that of Selby *et al.* (1978) are frequently used to make this correction.

b. Emissivity and Sky Radiant Emittance. Fuchs and Tanner (1966) showed that the effects of radiation emitted by the surroundings and then reflected from the target to the sensor (termed sky radiant emittance by Conaway and Van Bavel, (1967a,b) can be accounted for if the emissivity is a constant over all wavelengths within the wave-band interval, and if the filter function of the instrument is known. The flux of radiation reaching the sensor is

$$R = \epsilon\sigma T^4 + (1 - \epsilon)B \qquad (2)$$

Where B is the reflected radiation from the surroundings. Fuchs and Tanner (1966), Conaway and Van Bavel (1967a,b), and Idso and Jackson (1968) have described how B could be evaluated. If $\epsilon = 1$, then B has no effect on R. If $\epsilon < 1$ and B is neglected [as in Eq. (1)], then T_s will be overestimated. If ϵ is erroneously assumed to be 1 when it is less than 1, then T_s will be underestimated. Perrier (1971) showed that the error involved by assuming $\epsilon = 1$ is

$$(T_s - T)/T = 1 - [\epsilon + (1 - \epsilon)B/\sigma T^4]^{1/4} \qquad (3)$$

where T is the calculated apparent temperature obtained by assuming $\epsilon = 1$ and using the relation $R = \sigma T^4$. If $B = R$ then $T = T_s$, regardless of the value of ϵ. The value of B is generally less than R, and varies with sky conditions. As B decreases the error increases. The maximum error $(1 - \epsilon^{1/4})$ occurs when $B = 0$ (i.e., the true surface temperature is always underestimated). The maximum difference between the true surface temperature T_s and the apparent temperature T is

$$T_s - T = T(1 - \epsilon^{1/4}) \qquad (4)$$

Figure 2 shows the maximum $T_s - T$ versus emissivity over the range from 0.9 to 1.0. Calculations were made for two temperatures.

There are two reasons why the values shown in Fig. 2 are the maximum to be expected. The first is that the sky radiant emittance B adds to the target radiance R, causing a higher apparent target temperature. Since B, although variable, is never 0, the actual $T_s - T$ difference will always be less than that shown in Fig. 2. The second reason involves the limited bandwidth of the sensor. Sutherland *et al.* (1979) calculated the detector response for an infinite bandpass using the Stefan–Boltzmann law [Eq. (1)], and the response for an infinitesimal bandwidth using the Planck equation. They showed, with the two extremes, that the apparent temperature is not well defined but that the error is less for the infinitesimal bandwidth. Since most IR thermometers do not exceed an 8–14-μm bandwidth, the apparent temperature is generally closer to the true temperature than the lines in Fig. 2 suggest.

c. Leaf and Canopy Emissivities. The thermal IR emittance of most plant surfaces ranges between 0.95 and 1.00. Gates (1964b) stated that all plant surfaces have a long-wave emittance of 0.95 or more, with most plant leaves being 0.97–0.98. This was substantiated by Idso *et al.* (1969) who found only two of 34 plant species studied that had single-leaf emissivities less than 0.95. The emittance of a canopy should be greater than for individual leaves because of cavities formed by canopy architecture. Fuchs and Tanner (1966) and Blad and Rosenberg (1976) measured canopy emissivities and reported values of 0.976, 0.976, and 0.971. Even with soil exposed between rows of trees, the composite emittance remains high (Sutherland and Bartholic, 1977). Assuming that the

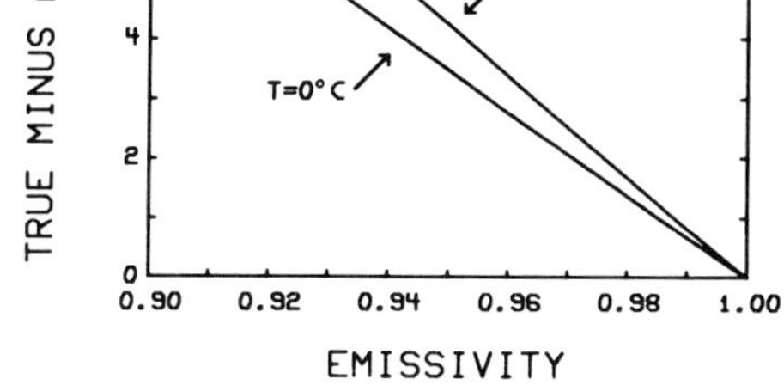

FIG. 2. The maximum increment to be added to the apparent temperature T obtained by assuming a unit emissivity, to obtain the actual surface temperature T_s as a function of emissivity for two values of the apparent temperature.

canopy emittance is on the low side, say 0.97, Fig. 2 shows that for an apparent temperature of about 30°C, the maximum error involved by assuming $\epsilon = 1$ would be about 2°C. Considering the various factors involved, it is doubtful that the error caused by assuming $\epsilon = 1$ for plant canopies would generally exceed 1°C.

d. Portable IR Thermometers. During the 1960s IR technology advanced rapidly, and instruments that could be used in the field became commercially available. The instruments used by Fuchs and Tanner (1966) and by Conaway and Van Bavel (1967a,b) consisted of an electronic console to which a relatively lightweight sensing head was connected by a cable. The instruments required external power sources. The sensing head contained a temperature-controlled blackbody to which, during operation, the sensor was exposed alternately with radiation coming from the target. The reference blackbody temperature was above 40°C. However, during the summer months in the Southwest, when the ambient air temperature was high and the sensing head was exposed to solar radiation, the reference temperature exceeded the preset control temperature, causing erroneous readings. Ambient temperature effects on these instruments were documented by Idso and Jackson (1968), and by Jackson and Idso (1969). Subsequent designs included a battery-operated console, making the instrument portable.

Gates (1968) described a small, lighweight IR thermometer specifically designed for hand-held use. The portability of the instrument was superb. However, it required frequent calibration and was sensitive to ambient temperature changes (McGinnies and Aronson, 1971).

During the past decade a number of lightweight, hand-held, battery operated IR thermometers have become available. For the most part, these instruments were designed for use in buildings or other enclosed areas where only small temperature changes would occur. When used in the field under conditions of changing insolation and air temperature, the instruments failed because components would be changing temperature, and erroneous readings of the target temperature would result. Because of the market for IR thermometers in agricultural research and the potential market as an operational irrigation scheduling tool, several companies have addressed this problem and have made significant progress toward its solution. It is highly recommended that a potential purchaser determine whether an adequate ambient temperature compensation circuit is an integral part of any instrument that is intended for field use.

Calibration and frequent checks of calibration are essential if reliable data are to be obtained from an IR thermometer. Jackson *et al.* (1980) discussed field-use calibration and the precautions to be observed when using these instruments. Sun azimuth angles and row direction can cause a small difference in canopy temperatures (Fuchs *et al.*, 1967). Clawson and Blad (1982) found negligible directional effects during measurements of corn canopy temperatures. Gardner *et al.* (1981b) reported a slight tendency for a southerly exposure to be warmer than plots with a northerly exposure. The effect can be minimized by taking readings from two directions 180° apart and averaging them.

2. *Airborne Thermal Scanner Experiments*

In a forward-looking science fiction type of scenario, Park *et al.* (1968) predicted the use of thermal IR imagery obtained from orbiting satellites for irrigation scheduling. Their scenario had an agronomist and a forester managing 800,000 acres of land devoted to intensive agriculture and forestry production. The two met frequently to review satellite data. Thermal imagery played a major role in their assessment of irrigated acreage. The fictional duo found "hotspots" in soybeans, but concluded that overall, their irrigation schedule had been correct. The projections of Park *et al.* (1968) were based on a few thermal images obtained from airborne scanners. The promise of remote sensing as a tool for resource management as outlined by Park *et al.* has not as yet been fulfilled. In the 13 years since they made their predictions, much has been accomplished yet much remains to be done.

Some of the data on which Park *et al.* (1968) based their predictions were reported by Wiegand *et al.* (1968). Their experiments included extensive ground measurements along with imagery from an airborne scanner. They showed that freshly irrigated crops were up to 20°C cooler than nonirrigated portions of the same fields. Myers and Allen (1968) presented similar data and stated that remotely sensed plant canopy temperatures appear to be a feasible means for assessing irrigation needs, as well as the extent and severity of drought. Later, Myers and Heilman (1969) demonstrated the effect of plant cover on remotely sensed temperatures. They showed that soil background greatly influences thermal images obtained from aircraft.

Bartholic *et al.* (1972) were among the first to use an airborne thermal scanner specifically to determine the temperature of soils and of crop canopies differing in water stress. They observed up to 6°C differences between least- and most-stressed plots planted to cotton. From airborne data taken over Kansas, Heilman *et al.* (1976) observed that soybean was 2.6°C cooler than sorghum. Blad and Rosenberg (1976) used an airborne scanner to compare surface temperatures of wheat, alfalfa, and pasture. The wheat and alfalfa were cooler than pasture. They also used portable radiation thermometers to measure surface temperatures. Alfalfa was found to be 5–7°C cooler than air (measured at 2 m) during mid- and late afternoon. Irrigated corn, on the other hand, was always warmer than alfalfa and was usually warmer than air (except for short periods in the late afternoon). This work pointed out the necessity of studying the temperature relations of all crops of interest, since not all are alike.

Millard *et al.* (1978) presented thermal imagery for six differently irrigated wheat plots. The stressed plot was 8°C above the air temperature measured at 1.5 m, whereas the well-watered plots were as much as 6°C below air temperature. The pseudocolored thermal imagery clearly showed the temperature differences and the temperature variations within plots. Their data showed that when full ground cover is achieved, airborne thermal imagery can readily distinguish different irrigation treatments, and that it could therefore be used as an irrigation-scheduling tool.

The soil background exposed by incomplete canopies poses a difficult problem for determining plant canopy temperatures from aircraft and spacecraft (David, 1969). A remote sensor above an incomplete canopy would see sunlit and shaded soil, and sunlit and shaded vegetation. These four surfaces would probably have somewhat different temperatures, and a composite of the various temperatures would be detected by the sensor. The problem is more serious for crops with wide rows than for crops that are broadcast or that are planted or drilled in narrow rows. Hatfield (1979) measured vertical and angular temperatures of wheat drilled in 0.18-m rows, using a 20°-field-of-view portable radiation thermometer. The angular–vertical temperature difference was near zero for low and full cover conditions, and reached a minimum of −2°C at about 40% plant cover. Heilman *et al.* (1981) measured temperatures from vertical and angular positions over barley. They found that the composite (vertical) temperatures were from 0.5° to 11.5°C higher than the canopy (angular) temperatures. The prediction of canopy temperatures from composite temperatures was improved by accounting for emissivity and sky radiance.

Jackson *et al.* (1979) modeled this problem for row crops. They assumed that plant rows could be represented by a rectangular block of specified height and width. The fractions of sunlit and shaded soils and vegetation could be calculated from sun elevation and azimuth angles for a particular location, hour, and date. Using ground-measured soil and plant temperatures for a cotton crop with 1-m row spacing, they calculated the composite temperature that a thermal scanner would see as a function of view angle. Results of these calculations are shown in Fig. 3. The maximum temperature occurred near the center of the scan because of the hot sunlit soil. As the scan moves to either side of nadir (0°), the temperature decreases. This occurs because as the scan angle increases, the amount of vegetation in the field of view of the instrument increases, and the amount of soil decreases. The scan line in Fig. 3 is not symmetrical because in one direction predominantly sunlit surfaces are seen, whereas in the other direction shaded surfaces form a significant portion of the scene. Jackson *et al.* (1979) suggested that concurrent measurement of reflected solar data may help in the extraction of plant temperatures from the composite scene. The problem is serious and far from solved (Gardner *et al.*, 1981c). Since the ultimate use of canopy temperature techniques for agricultural management decisions such as irrigation schedul-

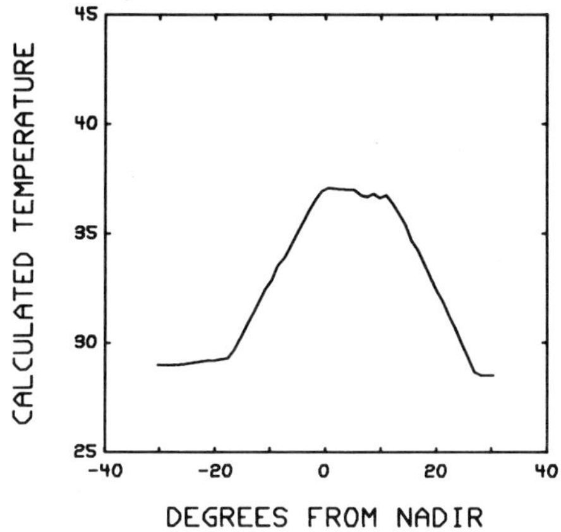

FIG. 3. Calculated composite temperatures of row cotton as a function of view angle at 1330 hr on Julian day 196 (for east–west rows). The temperatures used in the computation were sunlit soil, 55.3°C; shaded soil, 31.1°C; sunlit vegetation, 28.5°C; shaded vegetation, 29.5°C (from Jackson *et al.*, 1979).

ing will probably be from aircraft or spacecraft platforms, a solution to this problem is urgently needed.

3. *Ground-Based Canopy Temperature Experiments*

Aston and Van Bavel (1972) proposed that soil water depletion could be remotely detected by determining the increase in visible and thermal radiant heat loads upon plant leaves as the underlying soil surface dries. They cautioned that the temperature differences involved would be both small and variable. As plant cover increased the differences would become smaller. For full canopies, however, another approach could be taken. Aston and Van Bavel suggested that because of the inherent heterogeneity of soils, various locations in a field would become stressed before others, and the canopy temperature would show a greater variability than under well-watered conditions. They proposed that the variability of temperatures within a field be used to signal the onset of water deficits.

Stone and Horton (1974) installed an IR radiometer above a field of grain sorghum. They used the radiometrically measured canopy temperature along with the air temperature to calculate evapotranspiration (ET) using equations developed by Bartholic *et al.* (1970) and by Brown and Rosenberg (1973). One objective was to show that remotely sensed surface temperatures could be used to obtain regional ET estimates. Blad and Rosenberg (1976) mounted an IR radiometer over a crop of alfalfa and compared the resulting temperatures with leaf thermocouples and with aircraft scanner-derived temperatures. They made qualitative comparisons between ET of corn and alfalfa based on temperature measurements.

a. Stress Indices. Idso *et al.* (1977) and Jackson *et al.* (1977) measured canopy temperatures every day throughout a complete wheat-growing season. Their purpose was to develop techniques for evaluating crop water stress remotely with a minimum of measurements. They assumed that environmental factors (such as vapor pressure, net radiation, and wind) would be largely manifested in the canopy temperature, and they defined a *stress degree day* (SDD) as the difference between the canopy temperature T_c and the air temperature T_A. They reasoned that the SDD would increase with increasing plant water stress. Idso *et al.* (1977) used the SDD to predict final yields. They plotted the cumula-

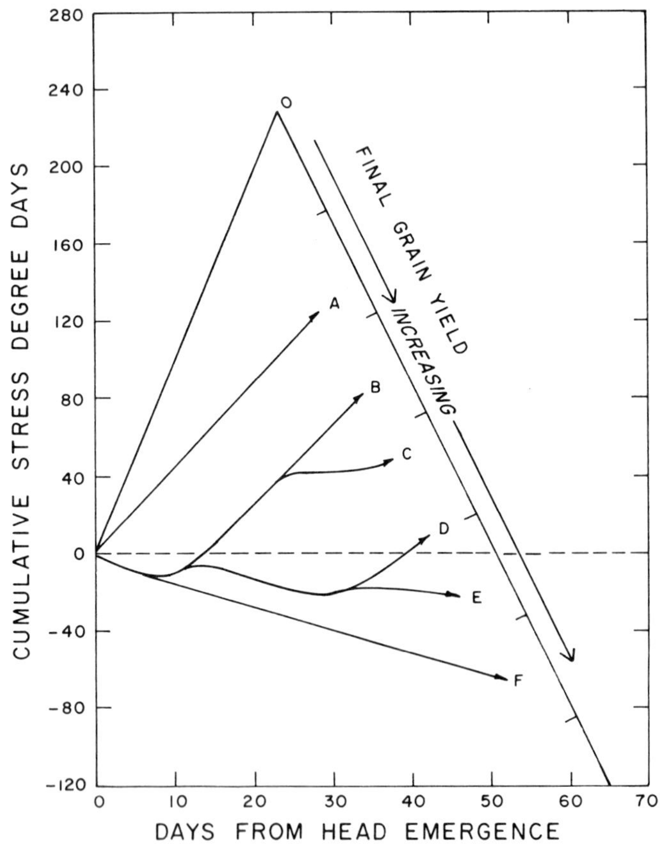

FIG. 4. Cumulative SDDs from the time of head emergence for Produra wheat. The lines A–F represent typical paths by which six differently irrigated plots approached final yield (after Idso *et al.*, 1977).

tive SDD from heading to maturity as a function of time for several wheat plots that had received different amounts of irrigation water. Figure 4 is a generalized plot of their results. Lines A–F indicate time courses for the different water treatments. All plots were irrigated immediately after planting. Line A resulted when no additional irrigations were given. The cumulated SDD increased continuously with time, and a low final grain yield was predicted. Curved-line B represents a plot that had received irrigations until the time of heading. About 10 days after heading, the water supply was depleted and the cumulative SDD steadily increased. Curve C followed B until another irrigation was given late in the season. Curves D and E had ample water until about 30 days after heading, curve E received another irrigation at that time. Curve F represents a plot that was not stressed for water. This figure demonstrated how canopy temperature data can be used to track crop water stress and the effect of stress on yield.

The SDD concept was applied to red kidney beans by Walker and Hatfield (1979). They found that final yield was inversely related to the SDD, in agreement with the results of Idso *et al.* (1977), and concluded that the SDD was a valid representation of the effect of moisture stress on yield. They plotted the total amount of water used (measured with a neutron moisture meter) as a function of accumulated SDDs. Results are shown in Fig. 5, which demonstrates that the less water available for transpiration, the higher the SDD. These data provide evidence that transpiration and canopy temperatures are related, a fact that was difficult to establish for leaf temperatures by short-term experiments (see Section II,A). The application of the SDD to irrigation scheduling was suggested, but not demonstrated.

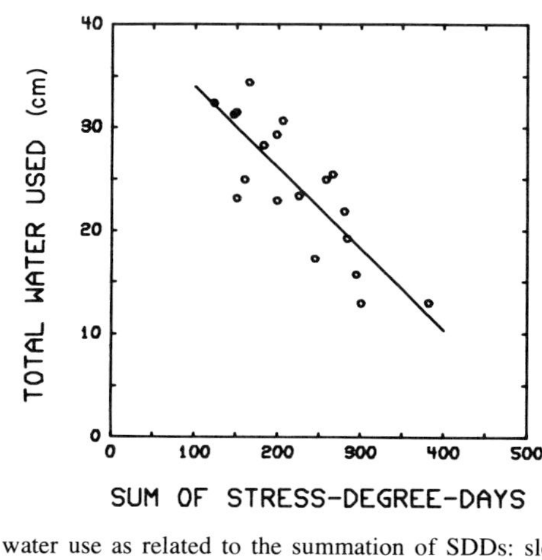

FIG. 5. Seasonal water use as related to the summation of SDDs: slope, −0.0788; intercept, 41.84; $R = 0.822$. [Redrawn from Fig. 5 of Walker and Hatfield (1979, p. 970) by permission of the authors and the American Society of Agronomy, Inc., 677 South Segoe Road, Madison, WI 53711.]

Jackson *et al.* (1977) evaluated the SDD as a possible irrigation-scheduling tool. They measured water depletion with a neutron moisture meter, and accumulated all positive values of the SDD. Whenever the SDD was negative, it was taken to be zero. Results for two plots are shown in Fig. 6. Numerical values on the ordinate refer both to the positive SDDs and centimeters of water depleted. It was concluded that irrigations should be given when or before the positive SDDs reached a value of 10.

Evapotranspiration (ET) calculated using canopy temperatures was compared with ET measured by weighing lysimeters and soil water depletion by Jackson *et al.* (1977). They found that the empirical relation

$$ET = R_n - 0.064(T_c - T_A) \qquad (5)$$

would describe ET for periods of a week or more, but not for individual days. In Eq. (5) R_n is the net radiation. The temperature measurements were taken at one time of day (postnoon), whereas the ET is given in cm day^{-1}. They found that net radiation could be estimated with sufficient accuracy to yield reliable values of ET. These results support the suggestions of Stone and Horton (1974): that net radiation could be estimated from incoming solar radiation and albedo, that air temperatures could be determined on the ground, and that airborne scanners

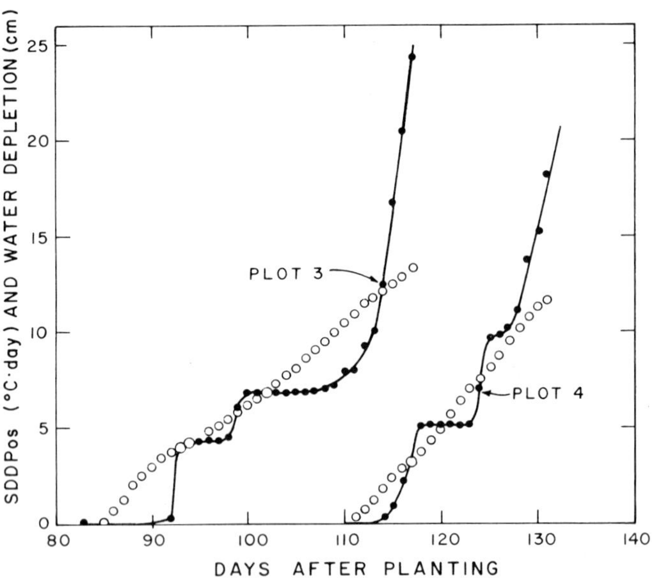

FIG. 6. Positive SDDs (-•-) and water depletion (o) as a function of days after planting for two wheat plots. The summation was set at zero at the last irrigation. Numerical values on the ordinate are the same for both factors (from Jackson *et al.*, 1977).

could be used to measure T_c, enabling water use by crops to be evaluated over large areas.

Sequin and Petit (1980) calculated ET for a dry and an irrigated zone in southern France using four methods, two of which used remotely sensed surface temperature as an input. One of the methods was the simplified approach of Jackson *et al.* (1977). They concluded that the temperature methods gave satisfactory results, with a precision of 10–15% compared to an energy balance reference method. The advantage of the simplified approach is that neither extensive ground measurements nor the evaluation of surface roughness is necessary.

Heermann and Duke (1978) used the average temperature difference between treatment plots and adjoining well-watered areas to study crop water stress under limited irrigation. Their plots were in a cornfield irrigated by sprinkler from a center-pivot system. The average temperature elevation was linearly related (with a negative slope) to applied water and to relative dry-matter yield. They concluded that an average temperature difference of greater than 1.5°C was significantly correlated with a reduction in yield.

The canopy–air temperature difference was combined with net radiation and vapor pressure data and used as an irrigation-scheduling tool by Geiser *et al.* (1982). They compared the temperature method with resistance blocks and a checkbook method and concluded that irrigation water use was less when the temperature difference was used as the scheduling criterion.

Blad *et al.* (1981), Clawson and Blad (1982), and Gardner *et al.* (1981a) built on the suggestion of Aston and Van Bavel (1972) and tested the deviation of midday canopy temperature for usefulness as an irrigation-scheduling tool. They found standard deviations of 0.3°C in fully irrigated plots of corn. In nonirrigated plots, the standard deviation was as great as 4.2°C. They concluded that plots which exhibited a standard deviation above 0.3°C were in need of irrigation. Figure 7, taken from Clawson and Blad (1982), shows the daily values of canopy temperature variability (CTV), defined as the range (maximum minus minimum) of all IR thermometer-sensed temperatures within a plot during a particular measurement period. For this plot, irrigations were given when the CTV reached a value of 0.8.

Figure 8, also taken from Clawson and Blad (1982), shows the difference in temperature between a stressed plot and a well-watered plot [called the *temperature stress day* (TSD) by Gardner *et al.* (1981b)]. Use of the well-watered plot as a reference compensates for environmental effects such as air temperature and vapor pressure deficit. For the data shown in Fig. 8, irrigations were given when the average of all canopy temperatures measured in the stressed plot during a particular time period were 1°C warmer than the average canopy temperatures of the well-watered plot. These experiments indicate that both methods, the CTV and the TSD, could be used as a viable irrigation-scheduling technique.

The SDD as used by Jackson *et al.* (1977) proved insufficient to assess water

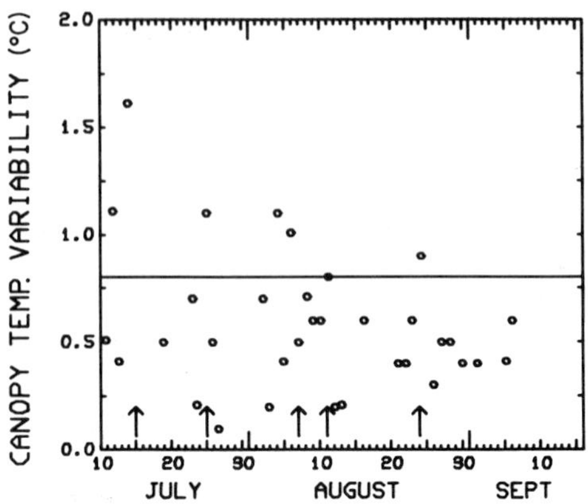

FIG. 7. Canopy temperature variability of corn during the growing season. The critical CTV was set at 0.8 for this treatment. Arrows indicate irrigations [after Clawson and Blad (1982), by permission of the authors and the American Society of Agronomy, Inc., 677 South Segoe Road, Madison, WI 53711].

stress in corn. Gardner *et al.* (1981a) showed that stressed corn plants were below air temperature much of the time and suggested that corn may be more sensitive to water stress than wheat. They also suggested that canopy–air temperature differences may be soil, crop, and climate specific.

Gardner *et al.* (1981b) plotted the fraction of the maximum yield as a function

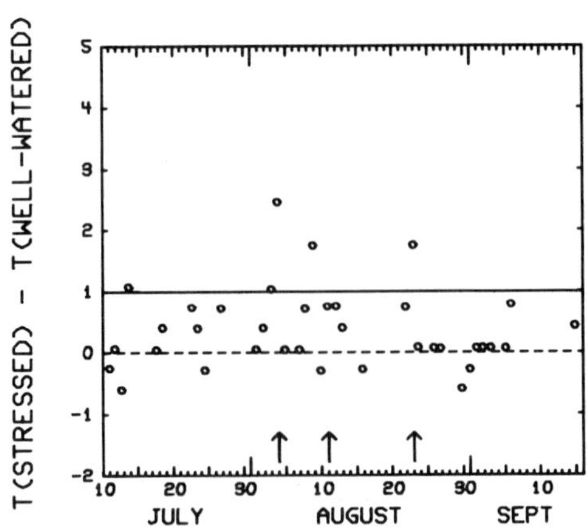

FIG. 8. Canopy temperature differences between a stressed and well-watered corn plot. The critical value of the TSD for this treatment was 1. Arrows indicate irrigations [after Clawson and Blad (1982), by permission of the authors and the American Society of Agronomy, Inc., 677 South Segoe Road, Madison, WI 53711.]

of the TSD for two varieties of sorghum. They found a quadratic relationship between yield and TSD, in contrast to the linear relation between yield and the SDD found by Idso *et al.* (1977). The possibility of using canopy temperatures to predict the growth stage of corn was investigated by Gardner *et al.* (1981c). They found that the summation with time of the canopy temperature was highly correlated with growth stage, but that indices developed as stress indicators (e.g., canopy–air temperature difference) were not well correlated.

b. Plant Water Potential. The assumption that canopy temperature is a measure of plant water stress is largely based on qualitative observations such as visual wilting. Physiological processes are often affected before wilting becomes apparent, and different species may wilt at different stress levels (Hsiao, 1973). Since plant water potential has gained wide acceptance as a fundamental measure of plant water status, it is of interest to compare such measurements with canopy temperature data.

The plant water potential can be measured by use of a pressure chamber such as that described by Scholander *et al.* (1965). Either whole plants or parts of plants can be measured, depending upon plant size and species and the chamber size. With plants such as cotton, a leaf with petiole is usually measured, whereas with grain, the entire above-ground portion of the plant may be placed in the chamber. In either case the sample measured is a small part of an entire canopy, and considerable variation is to be expected.

Plant water potentials were measured on whole (above-ground) wheat plants on 19 clear days (at 1400 hr) at Phoenix, Arizona, by Ehrler *et al.* (1978a). Canopy temperatures were measured concurrently, using IR thermometers. Measurements were made on six differently irrigated plots and the results are shown in Fig. 9. Even with a number of observations per plot, there was a considerable amount of scatter in the data. The line drawn through the points indicated that as stress increased (plant water potential decreased), the canopy–air temperature difference increased. The temperature difference reached a maximum of about 5°C and increased only slightly with further decrease in plant water potential. Data of Sumayao *et al.* (1980), if plotted on Fig. 9, would fall well within the range of values shown for wheat. One can conclude from these data that canopy temperatures and plant water potential are indeed correlated, but not linearly, and that variability due to sampling precludes a more quantitative comparison.

The diurnal change in plant water potential and canopy temperatures was documented for wheat by Ehrler *et al.* (1978b). The diurnal range of plant water potential was considerable, from about −0.1 to −1.6 MPa for well-watered plots, and from about −2.1 to −4.3 MPa for a water-deficient plot. They concluded that for canopy temperature measurement, 1400 hr was the best time of day to assess water stress, a conclusion also reached by Blad *et al.* (1981).

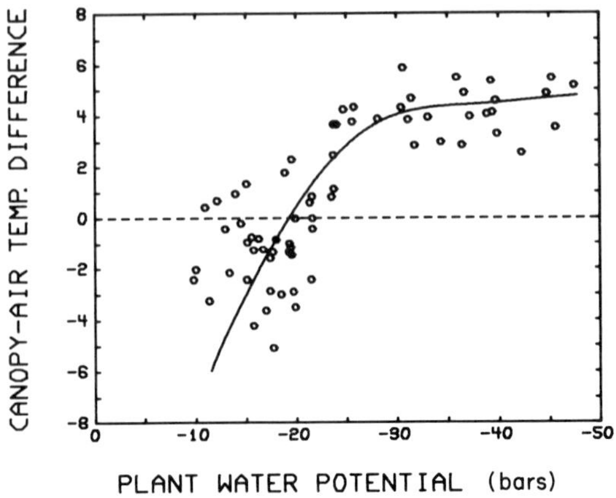

FIG. 9. Temperature of the canopy minus the air temperature at 1.5 m above the crop at 1400 as affected by the whole-plant water potential, the data consisting of 70 points from measurements taken on 19 clear days, with each point being the mean of 12–30 pressure chamber readings on separate plants, and 4–12 ΔT readings. The 70 points were derived from 220 individual sets of data (after Ehrler *et al.*, 1978a).

Heermann and Duke (1978) measured plant water potentials prior to dawn in conjunction with postnoon temperature measurements in corn. They found no trend in the plant water potential measurements for different water application rates, but observed significant yield reductions. Gardner *et al.* (1981b) compared the difference in plant water potential between stressed and nonstressed sorghum and the canopy temperature difference between the two plots. They found a parabolic relation between the two differences, i.e., the potential difference increased with an increase in the temperature difference to a maximum at about 4°C, then decreased. They recognized the effect of measurement variability and suggested that stomatal opening and closing may affect relationships between canopy temperature and plant water potential.

A general conclusion reached by considering the above results is that a quantitative measure of plant water stress under field conditions has not been achieved. The variability of point measurement of plant properties such as water potential is considerable. A canopy temperature measurement may be a better indicator of plant water stress, at least during the early stages of stress. As the canopy becomes several degrees warmer than the air, the temperature difference becomes insensitive to changes in plant water potential.

c. Environmental Factors. Changing cloud conditions can dramatically affect canopy temperatures. Measurements made under such conditions will increase data variability. Wiegand and Namken (1966) reported leaf-temperature changes of up to 6°C caused by changing solar insolation. Leaves required 30–45 sec to reach a new equilibrium after being shaded by a cloud. Blad *et al.* (1981) reported leaf temperature changes of 5.2°C in stressed corn, but only a 3.6°C change in nonstressed corn. They suggested that quantification of stress would be difficult under conditions of variable cloudiness.

Leaf temperature decreased with increased wind speed in wind-tunnel experiments of Barthakur (1975). He found the greatest variation to occur at wind speeds of 0–2 m sec^{-1}. Beyond 2 m sec^{-1} leaf temperatures were not significantly affected by increases in wind speed. Carlson *et al.* (1972) did not show a significant effect of wind speed on the temperature of soybean leaves under field conditions. Okuyama (1975) measured rice canopy temperatures with an IR thermometer. His data show no trend in canopy-air temperature differences with wind speeds ranging from 0.2 to 1 m sec^{-1}. From these results it can be assumed that wind will affect canopy temperatures, but in a minor way.

III. Quantification of Crop Water Stress

A review of the meterological, soil, and plant factors that are used to signal irrigation needs shows that whereas meterological and soil factors indicate when plants *may* be stressed, plant factors indicate when they *are* stressed. Plant factors, such as water potential, are point measurements that require numerous samples to characterize a field. Canopy temperature measurements can minimize this problem. Up till now, however, canopy temperature techniques have not been entirely satisfactory for quantifying plant water stress.

Three temperature indices have been proposed in the literature: the SDD, which is the canopy-air temperature difference measured postnoon near the time of maximum heating; the TSD, which is the difference in canopy temperatures between a stressed crop and a nonstressed (well-watered) reference crop; and the CTV, which is the range of temperatures encountered when measuring a plot during a particular measurement period.

With the SDD, plants were considered stressed if the value was positive and not stressed if negative. Experimental evidence has shown that the arbitrary division of SDD = 0 is not appropriate for all environmental conditions. The TSD requires a reference plot that is *not stressed* to be in close proximity to the field in question. The critical CTV value to be used as a signal to irrigate may be influenced by the degree of variability of soil properties inherent within a field.

The combined energy balance-aerodynamic relation used to predict evaporation from natural surfaces (Penman, 1948) can be written with the surface temperature as a function of net radiation and vapor pressure deficit (Monteith and Szeicz, 1962). This approach leads to a temperature-based stress index that may prove to be a reliable means of quantifying crop water stress. The following development is based on Jackson *et al.* (1981).

A. Energy Balance Considerations

The energy balance for a crop canopy can be written

$$R_n = G + H + \lambda E \tag{6}$$

where R_n is the net radiation (W m^{-2}), G is the heat flux below the canopy (W m^{-2}), H is the sensible heat flux (W m^{-2}) from the canopy to the air, λE is the latent heat flux to the air (W m^{-2}), and λ is the heat of vaporization. In their simplest forms, H and E can be expressed as

$$H = \rho c_p (T_c - T_A)/r_a \tag{7}$$
$$\lambda E = \rho c_p (e_c^* - e_A)/[\gamma(r_a + r_c)] \tag{8}$$

where ρ is the density of air (kg m^{-3}), c_p is the heat capacity of air (J kg^{-1} $°C^{-1}$), T_c is the surface temperature (°C), T_A is the air temperature (°C), e_c^* is the saturated vapor pressure (Pa) at T_c, e_A is the vapor pressure of the air (Pa), γ is the psychrometric constant (Pa $°C^{-1}$), r_a is the aerodynamic resistance (sec m^{-1}), and r_c is the canopy resistance (sec m^{-1}) to vapor transport. A detailed discussion of procedures and assumptions leading to Eqs. (6)–(8) was given by Monteith (1973).

Combining Eqs. (6)–(8), assuming that G is negligible and defining Δ as the slope of the saturated vapor pressure-temperature relation $(e_c^* - e_A^*)/(T_c - T_A)$ in units of Pa $°C^{-1}$, we obtain

$$T_c - T_A = \frac{r_a R_n}{\rho c_p} \cdot \frac{\gamma(1 + r_c/r_a)}{\Delta + \gamma(1 + r_c/r_a)} - \frac{e_A^* - e_A}{\Delta + \gamma(1 + r_c/r_a)} \tag{9}$$

which relates the difference between the canopy and the air temperatures to the vapor pressure deficit of the air $(e_A^* - e_A)$, the net radiation, and the aerodynamic and crop resistances. Equation (9) was given in a slightly different form by Monteith and Szeicz (1962). The upper limit of $T_c - T_A$ can be found from Eq. (9) by allowing the crop resistance r_c to increase without bound, i.e., as $r_c \to \infty$

$$T_c - T_A = r_a R_n/\rho c_p \tag{10}$$

The lower bound, found by setting $r_c = 0$ in Eq. (9) (the case of wet plants acting as a free water surface), is

$$T_c - T_A = \frac{r_a R_n}{\rho c_p} \cdot \frac{\gamma}{\Delta + \gamma} - \frac{e_A^* - e_A}{\Delta + \gamma} \tag{11}$$

Equations (9) and (11) describe a linear relation between $T_c - T_A$ and the vapor pressure deficit $e_A^* - e_A$. Thus, for a particular temperature, the lower bound is a line extending from the intercept at $e_A^* - e_A = 0$ (saturated air) to a value of $e_A^* - e_A = e_A^*$ (completely dry air). The upper bound is independent of vapor pressure and dependent only on r_a and R_n. The bounds are shown by the lines labeled 0 (lower) and ∞ (upper) in Fig. 10. Lines for intermediate values of r_c (5, 50, and 500 sec m^{-1}) are also shown. All lines in this figure were calculated for a temperature of 30°C, and net radiation of 600 W m^{-2}.

1. *Aerodynamic and Canopy Resistances*

In Eqs. (9)–(11) a value of r_a is needed. Reported values of r_a generally exceed 20 sec m^{-1} (Van Bavel and Ehrler, 1968; Szeicz and Long, 1969; Szeicz *et al.*, 1969). These values were obtained using aerodynamic methods. Using $r_a = 20$ sec m^{-1} and $R_n = 600$ W m^{-2} in Eq. (10) yields $T_c - T_A = 10$°C, a rather high average temperature for a crop canopy, even when transpiration is zero. Before further discussion of Fig. 10, it will be necessary to establish a reasonable value of r_a.

Jackson *et al.* (1981) evaluated r_a for wheat by measuring T_c, T_A, and R_n for a mature, fully senesced wheat crop (with no available soil water), and then solving Eq. (10) for r_a. Measurements involved monitoring one area with a fixed-position 15°-field-of-view (FOV) IR thermometer and a hand-held 3°-FOV instrument. The stationary instrument was automatically scanned every 6 sec, with these values averaged over a 6-min period. With the hand-held instrument, 6–8 instantaneous measurements were made over different parts of the field. The temperature differences ranged from 2 to 9°C, and averaged 5° in both cases. The

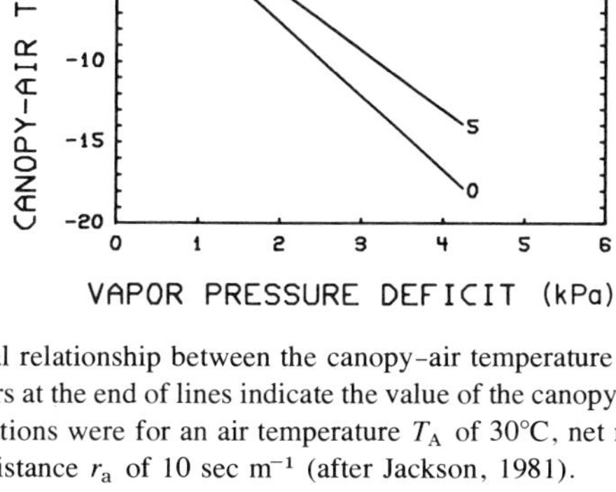

FIG. 10. Theoretical relationship between the canopy–air temperature difference and the vapor pressure deficit. Numbers at the end of lines indicate the value of the canopy resistance r_c used for the calculations. All calculations were for an air temperature T_A of 30°C, net radiation R_n of 600 m^{-2}, and an aerodynamic resistance r_a of 10 sec m^{-1} (after Jackson, 1981).

wide range of temperatures was apparently due to the instability of the air within the canopy. With the radiometer held stationary, $T_c - T_A$ values were observed to increase until a value of about 9°C was reached; then they dropped rapidly to about 2°C. During the measurements the wind speed was low (1 m sec^{-1} or less), the incoming radiation was high ($\simeq$800 W m^{-2}), and the air temperatures were >35°C, conditions conducive to high canopy temperatures. Since the average value of $T_c - T_A$ was 5°C under these conditions (and those of Ehrler *et al.*, 1978a), this value may be near the maximum that can be expected. With 5°C in Eq. (10), $r_a = 10$ sec m^{-1}.

Idso *et al.* (1981b) obtained $T_c - T_A$ and $e_A^* - e_A$ data for well-watered alfalfa fields located in Arizona, Kansas, Minnesota, and Nebraska, and for a stressed alfalfa field in Arizona (Fig. 11). The intercept of a $T_c - T_A$ versus $e_A^* - e_A$ plot for water-stressed alfalfa was about 1°C. Since r_c was finite, Eq. (10) does not apply, but the first term on the right-hand side of Eq. (11) can be used to estimate r_a. The resulting value is 6 sec m^{-1}. A likely value for r_a probably lies between 6 and 10 sec m^{-1}, and this value may be relatively constant.

In the above discussion, r_a was established for unstable conditions, i.e., the canopy was warmer than the air. When the canopy is cooler than the air, r_a may be greater (Szeicz and Long, 1969). A sensitivity analysis of Eq. (9) showed that $T_c - T_A$ was most affected by changes in r_a at low values of the vapor pressure deficit. For larger values of the deficit (3–5 kPa), increasing r_a by a factor of 3 had little effect on $T_c - T_A$.

The lines in Fig. 10 were calculated for several values of the canopy resistance

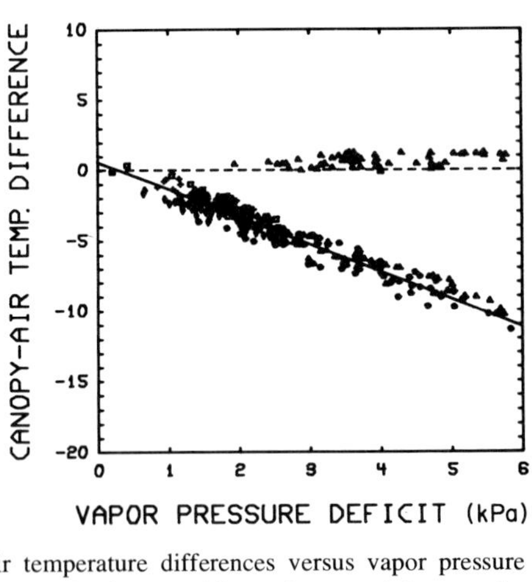

FIG. 11. Canopy–air temperature differences versus vapor pressure deficit for several well-watered plots of alfalfa assumed to be transpiring at the potential rate, and one severely water-stressed plot for which all temperature differences were positive: slope, −1.96; intercept, 0.58; $R = 0.95$. (After Idso *et al.*, 1981b.)

r_c. In this discussion, r_c is used to describe the overall resistance encountered within the soil–plant system by water moving to the leaf–air interface, from which it can evaporate. Components of this overall resistance are discussed by Van Bavel and Ehrler (1968). The lines shown in Fig. 10 extend from a completely saturated to a completely dry atmosphere, for an air temperature of 30°C. At 3 kPa, the temperature difference for a "wet" canopy would be about -12°C. With an increase of r_c to 5, 50, or 500 sec m^{-1}, the temperature difference would be, respectively, about $-9.2°$, $-1.7°$, or 3.8°C. These values indicate that, in a relatively dry environment, canopy–air temperature differences of 10–15°C could be expected, depending upon the degree of water stress of the crop. In a humid environment, canopy–air temperature differences would be much smaller, and would become positive for many situations. Thus, Eq. (9) and Fig. 10 partially explain the controversy that has raged for years over whether or not leaf temperatures should always be warmer than air.

2. *Experimental Verification*

The experimental results of Idso *et al.* (1981b), shown in Fig. 11, were obtained at five locations during the summer of 1980. All measurements were made under generally clear sky conditions, with canopy temperatures measured at intervals varying from 10–30 min at 2–3 hr before sunrise to 2–3 hr before sunset. At all sites the alfalfa was well watered and was assumed to be transpiring at the potential rate. Air temperatures ranged from about 20° to about 40°C during the course of the measurements.

Linear regression analysis of the data in Fig. 11 yielded the relation

$$T_c - T_A = 0.59 - 1.95(e_A^* - e_A) \qquad (12)$$

The slope (-1.95 °C kPa^{-1}) was somewhat smaller than for the leaf–air temperature, versus vapor pressure deficit relation (-1.29 °C kPa^{-1}) of Ehrler (1973), shown in Fig. 1. Since Ehrler (1973) measured leaf temperatures of cotton, whereas Idso *et al.* (1981b) measured canopy temperatures of alfalfa, we would not expect their slopes to be equal.

The scatter of data around the regression line in Fig. 11 may be due to net radiation changes, wind speed effects, and experimental errors involved in measuring the canopy, wet-bulb, and dry-bulb temperatures.

The linearity of the data lends validity to the assumptions underlying the development of Eq. (9) and to the assumption of a constant value of r_a. A comparison of Fig. 10 and 11 shows that a value of r_c between 10 and 20 sec m^{-1} would be necessary in Eq. (9) to predict the experimental temperatures at 3 kPa. However, the intercept of the experimental line (0.59) is less than any shown in Fig. 10. One explanation for this apparent discrepancy is that the lines in Fig. 10 were calculated assuming a constant air temperature of 30°C. The slope of the saturated vapor pressure–temperature relation (Δ) is a function of temperature.

Since Δ appears in both the slope and the intercept of Eq. (9), both terms are temperature dependent. Thus, the temperature-dependent lower bound is formed by a family of lines, one for each temperature. The result is a curved lower bound, as shown in Fig. 12. The symbols represent the end points of imaginary lines for temperatures incremented by 1° from 0° to 35°C. Lines for three of the temperatures (10°, 20°, and 30°C) are illustrated in the figure.

The lower limit shown by the plus signs in Fig. 12 is for a completely dry atmosphere, a situation not usually encountered. Also, in calculating this limit, various factors such as net radiation and resistance terms were held constant. In a natural situation the air is not completely dry, and the various environmental factors are all interrelated. Thus a curved lower limit may not be encountered in practice to the extent shown by Fig. 12, but a nearly linear relation, as observed experimentally by Idso *et al.* (1981c), may prevail over much of the vapor pressure deficit range. The situation for humid conditions is complicated by the fact that small temperature difference need to be measured.

3. *Potential Evapotranspiration*

Equation (11) represents the case of evaporation from a free water surface, which is not necessarily the case for potential evaporation from a crop. In irrigated areas the soil may be adequately supplied with water, yet the plant surfaces may appear dry. In this case the canopy resistance is probably not zero (Van Bavel and Ehrler, 1968), but has a value that will be called the *canopy resistance at potential evapotranspiration* r_{cp}. The value of r_{cp} will probably

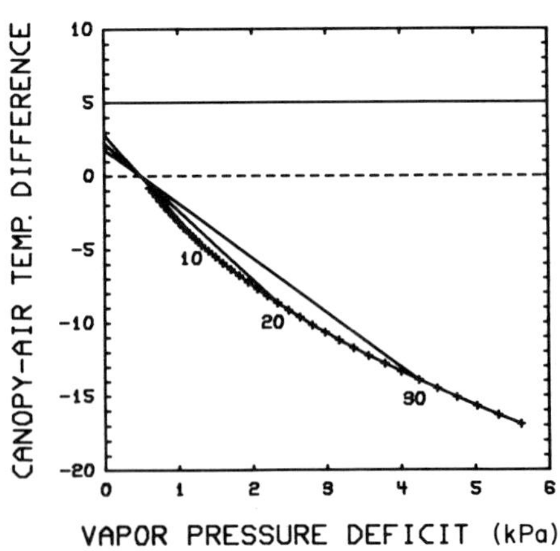

FIG. 12. The lower bound of the canopy-air temperature difference. The symbols represent the lower bound for a particular temperature at the maximum vapor pressure deficit for that temperature. Numbers identify the temperatures (°C) for the three straight lines. Other parameters were $r_a = 10$ sec m^{-1}, $r_c = 5$ sec m^{-1}, $R_n = 600$ W m^{-2} (after Jackson *et al.*, 1981).

be different for different crops, and may change with environmental conditions during a single day. Setting $r_c = r_{cp}$ in Eq. (9) we have

$$T_c - T_A = \frac{r_a R_n}{\rho c_p} \cdot \frac{\gamma^*}{\Delta + \gamma^*} - \frac{e_A^* - e_A}{\Delta + \gamma^*} \qquad (13)$$

where

$$\gamma^* = \gamma(1 + r_{cp}/r_a) \qquad (14)$$

4. *An Index of Crop Water Status*

A crop with adequate water will transpire at the potential rate for that crop. As water becomes limiting, the actual ET will fall below the potential rate. A measure of the ratio of actual to potential ET should, therefore, be an index of crop water status. Combining Eqs. (6)–(8) and solving for λE yields

$$\lambda E = \frac{\Delta R_n + \rho c_p (e_A^* - e_A)/r_a}{\Delta + \gamma(1 + r_c/r_a)} \qquad (15)$$

which is the Penman–Monteith equation for ET in terms of canopy and aerodynamic resistances (Monteith, 1973; Thom and Oliver, 1977). Taking the ratio of actual (λE *for any* r_c) to potential (λE_p for $r_c = r_{cp}$) gives

$$\frac{E}{E_p} = \frac{\Delta + \gamma^*}{\Delta + \gamma(1 + r_c/r_a)} \qquad (16)$$

with γ^* defined by Eq. (14). Jensen (1974) and Howell *et al.* (1979) discussed Eq. (16) for the case of $r_{cp} = 0$, i.e., $\gamma^* = \gamma$. Rearranging Eq. (16) will give r_c in terms of E/E_p, a result reported by Van Bavel (1967), Szeicz and Long (1969), and Russell (1980), again with $r_{cp} = 0$. Van Bavel measured E with lysimeters and calculated the canopy resistance.

The ratio E/E_p ranges from 1 (ample water, $r_c = r_{cp}$) to 0 (no available water, $r_c \to \infty$). In studying plant-water relations one thinks of a plant as going from a no-stress to a stressed condition. Therefore, it is esthetically pleasing for a stress index to go from 0 to 1. We consequently define a crop water stress index (CWSI) as

$$\text{CWSI} = 1 - \frac{E}{E_p} = \frac{\gamma(1 + r_c/r_a) - \gamma^*}{\Delta + \gamma(1 + r_c/r_a)} \qquad (17)$$

To calculate the CWSI or E/E_p using Eqs. (16) or (17) requires a value for the ratio r_c/r_a. This is obtained by rearranging Eq. (9) with the result

$$\frac{r_c}{r_a} = \frac{\gamma r_a R_n/(\rho c_p) - (T_c - T_A)(\Delta + \gamma) - (e_A^* - e_A)}{\gamma[(T_c - T_A) - r_a R_n/(\rho c_p)]} \tag{18}$$

giving the ratio r_c/r_a in terms of net radiation, canopy and air temperatures, vapor pressure deficit, and the aerodynamic resistance. In practice, r_c/r_a is evaluated using Eq. (18) and substituted into Eq. (17) to obtain the CWSI.

A graphic representation of the CWSI is given in Fig. 13. With $r_a = 10$ sec m^{-1} and $R_n = 600$ W m^{-2}, the upper bound at $T_c - T_A = 5$ is obtained. If $T_A = 30°C$ and $r_c = 5$ sec m^{-1}, the lower bound is obtained. The canopy temperature is found by experiment to be 27°C. The difference, $T_c - T_A = -3$, is plotted at point B. A vertical line is extended from point B to the upper (point A) and the lower (point C) bounds. The CWSI is BC/AC = $[-3 - (-9.3)]/[5 - (-9.3)] = 0.44$. The ratio of actual to potential ET [Eq. (16)] is AB/AC = 0.56.

The data shown in Fig. 11 indicate that a linear lower bound would obtain for well-watered conditions. Idso *et al.* (1981b) proposed that this lower bound (or baseline) be determined experimentally for a number of crops. This procedure is simpler in that a linear equation is used to determine $T_c - T_A$ at point C, instead of the more complicated calculations required by Eqs. (17) and (18). The empirical baseline, however, does not account for changes in net radiation, nor wind speed.

a. The Evaluation of Δ. The slope of the saturated vapor pressure–temperature relation Δ appears in most of the equations in the previous section. As a first approximation. Δ can be evaluated at the air temperature T_A. When the temperature difference $T_c - T_A$ is large (as for the case of well-watered crops at high vapor pressure deficits), a better approximation is to evaluate Δ at

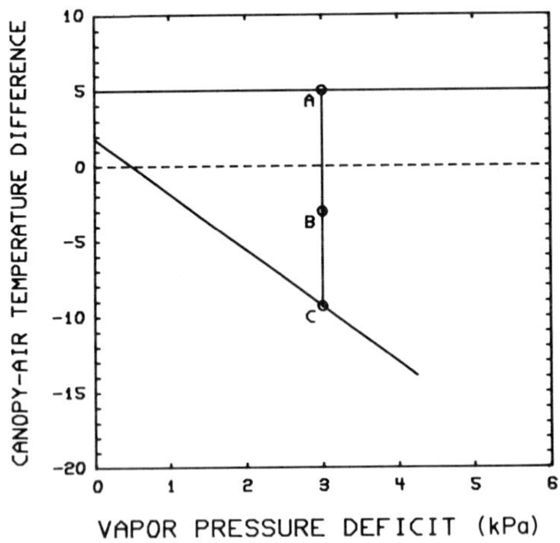

FIG. 13. A graphical representation of the CWSI calculation. Pertinent data are $r_a = 10$ sec m^{-1}, $R_n = 600$ W m^{-2}, $r_c = 5$ sec m^{-1}, $T_A = 30°C$, and $T_c = 27°C$.

$(T_c + T_A)/2$. Obviously, when T_c is near T_A the two approaches yield similar results. Taking the average of the canopy and air temperatures is sufficient for Eqs. (9), (11), (12), and (18), but Eqs. (16) and (17) pose a problem. Following closely the development of Eq. (16) from Eq. (15), we find that Δ in the numerator should be evaluated as the average of the air temperature and the canopy temperature that would obtain if the crop were evaporating at potential. The Δ in the denominator should be evaluated as the average of the actual measured canopy temperature and the air temperature. Keeping Δ as the value at the measured temperatures and Δ^* as the value using the calculated canopy temperature at potential, the numerator of Eq. (16) becomes $\Delta^* + \gamma^*$ and the numerator of Eq. (17) becomes $\gamma(1 + r_c/r_a - \gamma^* + \Delta - \Delta^*$.

The evaluation of Δ^* is complicated by the fact that T_c at potential may not be known. This can be estimated using an iterative procedure with Eq. (13) by evaluating Δ at T_A, calculating T_c, evaluating a new Δ at $(T_c + T_A)/2$, and recalculating T_c until an acceptable value is obtained. In practice, the use of Eqs. (16) and (17) with Δ evaluated at the average of the measured canopy temperature and the air temperature will yield similar results for both the low and high values of the indices, with the maximum error occurring near 0.5. At the midpoint, the difference between the two methods of calculation will be within 0.06, e.g., 0.48 for the one value of Δ and 0.54 when Δ and Δ^* are used. The results reported here are in terms of Δ evaluated at the average of the measured canopy and air temperatures.

b. The Effect of Wind. Wind is the only major environmental variable not explicitly appearing in Eq. (17). It is, however, implicit in that it affects the aerodynamic resistance r_a (Van Bavel and Ehrler, 1968). Thom and Oliver (1977) derived a wind function for Penman's evaporation equation. In terms of the aerodynamic resistance it is

$$r_a = \frac{4.72[\ln(z - d)/z_0]^2}{1 + 0.54U} \tag{19}$$

where z is the height above the surface, d the displacement, z_0 is the roughness parameter, and U is the wind speed in m $\sec^{-1}$. This relation shows that r_a is inversely related to a linear function of wind. With Eqs. (9)–(11) and (17), some qualitative observations concerning the effect of wind on $T_c - T_A$ can be made. For the case of a wet surface evaporating as a free water surface ($r_c = 0$), Eq. (11) shows that r_a appears only in the first term on the right-hand side. If wind speed were to increase, the intercept of the lines shown in Fig. 10 would decrease. The net result would be a slight lowering of $T_c - T_A$ with increasing wind for all values of the vapor pressure deficit. At the upper limit, Eq. (10) shows that $T_c - T_A$ would vary greatly with r_a and tend toward, but not reach, zero as the wind speed increases. For the intermediate region ($r_c > 0$), $T_c - T_A$

may increase or decrease with wind speed, depending on the values of r_a and r_c (with constant values for the net radiation and the vapor pressure deficit). This complex interaction shows the need for specific experiments to elucidate the effect of wind because, in a natural environment, an increase in wind speed changes not only r_a but also r_c, T_A, and possibly $e_A^* - e_A$ in a manner not easily calculable.

B. Experimental Aspects

Although the CWSI is based upon well-known principles, its very recent development means that the experimental data available for its evaluation are limited. A preliminary evaluation can be made with the data presented by Jackson *et al.*, (1981) and Jackson (1981). The following discussion is based largely on their work.

1. *Procedure*

Wheat (*Triticum durum* Desf. var. Produra) was planted in 11 × 13-m plots on 6 February 1980 (Julian day 37), and first irrigated on 8 February. Plot A received a second irrigation on 9 April (day 100). Plot B was irrigated on 2 April (day 93) and 23 April (day 114). Plot C, the wettest treatment, received a second irrigation on 19 March (79), a third on 15 April (106), a fourth on 2 May (123), and a fifth on 13 May (134). All irrigations were in the range of 0.10–0.12 m of water.

Canopy temperatures were measured with a portable IR thermometer held at an angle of about 30° from horizontal. By the time the plants were about 0.2-m tall (day 70), the instrument predominantly viewed plants. Plot canopy temperatures were taken as the average of eight measurments, four facing east and four facing west (to minimize sun angle effects). Wet- and dry-bulb temperatures were measured with a psychrometer held at a height of 1.5 m. Incoming solar radiation was recorded, from which net radiation was estimated.

Soil water contents were measured with a neutron soil moisture meter in each plot at 0.2-m intervals to a depth of 1.6 m, two to three times per week. Water contents for each depth were smoothed using a sliding cubic technique (an adaptation of the sliding parabola of DuChateau *et al.*, 1972). The smoothing procedure allowed the interpolation of water contents for each day.

The traditional way to determine the amount of "available" water is to calculate the amount held at "field capacity" and subtract the amount held at the "wilting point" (as estimated by a laboratory measurement that determines the water remaining in the sample after being subjected to 1.5 MPa of air pressure). Ritchie (1981) proposed that "extractable" water is a more precise measure of water availability to plants because the measurements are made *in situ*. This can be done by measuring the water content of a full soil profile (with an actively growing, fully developed crop) shortly after irrigation (taking drainage into account). This is called the drained upper limit. The lower limit is determined by withholding water from the crop after the root system has developed and then, when the plants have died, measuring the profile water content. The extractable water is the difference between the two profile measurements. Ritchie (1981) suggested that it be measured for each soil and each crop.

The drained upper limit and the lower limit for the wheat plots were measured. The total extractable water to a soil depth of 1.1 m was found to be 0.175 m. The fraction of extractable water used to that depth was calculated from the smoothed water content data.

2. *The Crop Water Stress Index and Soil Water*

Measured data for the CWSI are shown in Fig. 14. Lines were drawn through the points by eye. The data show the day-to-day change in CWSI in addition to the scatter that might be expected due to errors in measurements of canopy temperatures and wet- and dry-bulb air temperatures. The plus (+) symbols represent the fraction of extractable water used.

The fraction of extractable water used increased with time, dropping to a

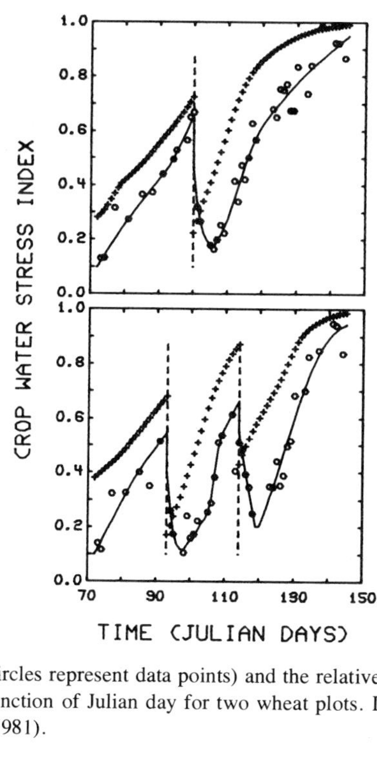

FIG. 14. The CWSI (circles represent data points) and the relative amount of extractable water used (plus symbols) as a function of Julian day for two wheat plots. Dashed vertical lines indicate irrigations (from Jackson, 1981).

minimum following irrigation (zero would indicate a full profile). The CWSI also increased with time, roughly in parallel with the extractable water used. At first glance, it appears that a reasonably good correlation exists between the two factors. However, it can be seen that the CWSI did not drop to its lowest value immediately after irrigation. Instead, the CWSI required 5–6 days to reach a minimum, implying that stressed wheat requires some time to recover (Hsiao, 1973). Some possible reasons for this are that leaves need to rehydrate and roots that were previously in dry soil need to develop new root hairs. The length of the recovery period depends largely upon the degree of stress experienced by the plants, but it may also vary with plant species and age. A similar recovery period has been documented for cotton by Ehrler (1973) and for sorghum by Idso and Ehrler (1976).

The existence of a recovery period for the temperature-based index is evidence that a unique relationship does not exist between plant temperatures and soil moisture. This is further demonstrated by plots of the CWSI versus extractable water used, as shown in Figs. 15 and 16. For plots A and B, the circles represent data after the first irrigation (at planting). These data are rather similar for both plots, the CWSI increasing in a linear manner with increasing amounts of ex-

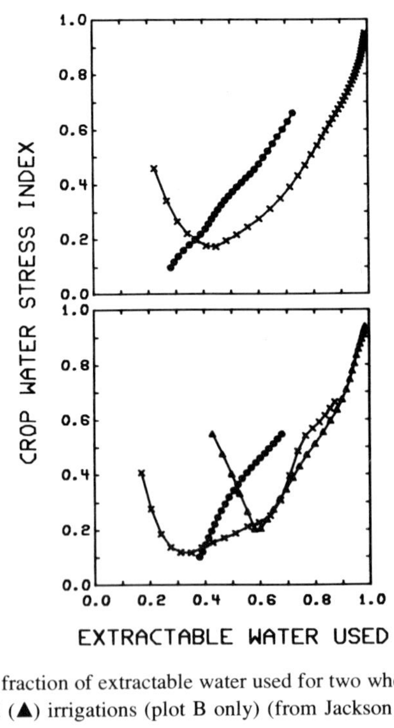

FIG. 15. CWSI versus fraction of extractable water used for two wheat plots. Data after the first (●), second (×), and third (▲) irrigations (plot B only) (from Jackson, 1981).

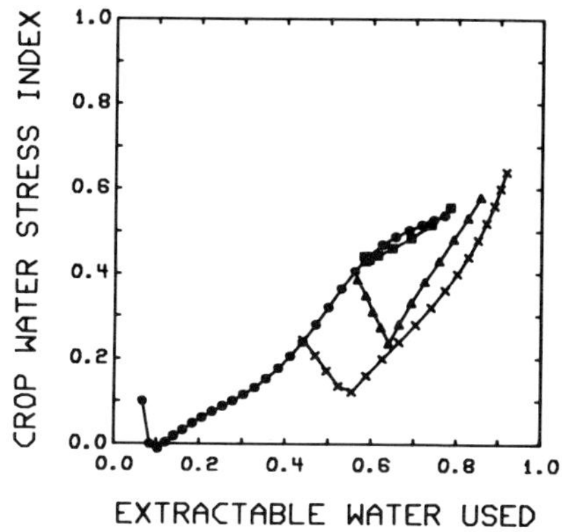

FIG. 16. CWSI versus relative amount of extractable water used for a wheat plot that received a total of five irrigations. Data after the second (●), third (×), fourth (▲), and fifth (■) irrigations (from Jackson, 1981).

tractable water used. The second irrigation was given to plot B seven days prior to the one given to plot A; thus plot A was the most stressed of the three plots. This point is also evident in the greater recovery time required for plot A as seen in Figs. 14–16. The wettest plot (Fig. 16) required nearly the same recovery time for the third (crosses) and fourth (triangles) irrigations as did plots A and B (Fig. 14). However, the second irrigation (circles) recovered in 1–2 days. This irrigation was given early in the season while the plants were actively growing and were not stressed. The fifth irrigation, given late in the season when much of the vegetation had senesced, showed no recovery period but also no decrease in the CWSI after irrigation.

The CWSI lines for each irrigation do not coincide with the values for a prior irrigation, even after the plants have recovered. This is further evidence that the relation between CWSI and soil moisture is not unique. This probably results, in part, from changes in rooting volume with time due to plant growth and the location of available water. The complexity of the situation becomes evident when one considers that soil water availability depends on root distribution, which, in turn, is determined predominantly by irrigation history (Chaudhary and Bhatnager, 1980) as well as by soil and aerial factors such as nutrient availability and evaporative demand. Since the precise rooting volume cannot be determined, exact correspondence of CWSI and extractable water should not be expected.

Another factor of importance that is evident in Fig. 16 (data for the fifth irrigation—squares) is the effect of plant senescence on the CWSI-extractable water relation. As the wheat matured, green leaves began to die, causing transpiration (with its consequent evaporative cooling) to decrease. Thus, after an irrigation, plant temperatures remained high (causing a high CWSI) even though the fraction of extractable water used was low.

3. *The Crop Water Stress Index and Crop Water Stress*

In order for the CWSI to be a useful irrigation-scheduling tool, its relation to crop water stress must be established. Attempts to relate other temperature indices to plant water stress have not been entirely satisfactory because of data variability (Section II,B,3b).

Recently, Pinter (1982) reviewed how reflected light in certain wave bands can be used to assess green biomass. As green plants begin to cover the soil, the reflection of visible red light is reduced because of absorption by plants. Near-IR light generally is reflected more by green plants than by soils. The ratio of the reflected near-IR to the visible red is therefore a sensitive indicator of green biomass.

Jackson and Pinter (1981) measured the reflectance of the near IR (0.8–1.1 μm) and the visible red (0.6–0.7 μm) over the same wheat plots (at the same time of day) as were used for the CWSI measurements, using a hand-held spectral radiometer (Jackson *et al.*, 1980). The near-IR : red ratio was formed, and the data are shown in Fig. 17 for the driest (plot A) and the wettest (plot C) treatments. From shortly after planting until about day 80, the ratios are essentially the same for the two plots. On day 79, plot C was irrigated. Within two days the ratio for plot C was markedly higher than for plot A. The data show that an irrigation at about day 80 (indicated by the arrow labeled INF) was critical if growth was to be maintained. Figure 14 shows that the CWSI was about 0.28 on day 80. The ratio reached a maximum for plot A on day 94, and for plot C on day 99. The maximum of the ratio is the point where new green growth is offset by senescence (indicated by the arrow labeled MAX). At the maximum ratio for

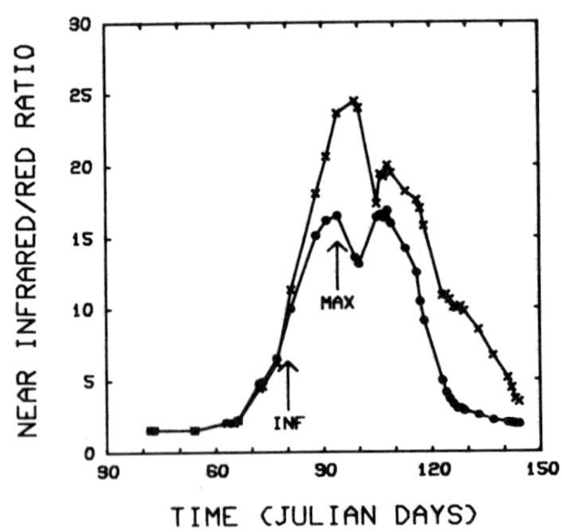

FIG. 17. The ratio of near IR (0.8–1.1 μm) to red (0.6–0.7 μm) for plot C (crossed symbols) as a function of Julian data (from Jackson and Pinter, 1981).

both plots the extractable water used was about 0.67 and the CWSI was about 0.5.

These data suggest that if the CWSI becomes greater than 0.3, a reduction in growth rate is imminent. If it reaches 0.5, net growth will cease and may decrease. Irrigations should therefore be given when the CWSI is within the range 0.3–0.5, the precise value being determined by water availability and other management factors.

The above criteria were determined for wheat in the vegetative stage, using specific values of r_{cp} (5 sec m^{-1}) and r_a (10 sec m^{-1}). There is some indication that r_{cp} may increase with growth stage. Obviously, much research needs to be done before an operational system of irrigation scheduling using the CWSI is perfected.

IV. Concluding Remarks

"Leaf temperatures are always warmer than the air" is a statement that is still heard today. This concept stems from research done during the first half of this century, for the most part in humid regions and glasshouses. Reports of leaves being cooler than the air, such as the 7°C difference measured by Wallace and Clum (1938), were soon challenged by Curtis (1938). It is entirely possible that Curtis never measured a leaf that was cooler than the air. One only needs to look at Eq. (9) and Figs. 10 and 11 to see that in humid climates leaves are likely to be warmer than the air, especially when the leaf is sunlit and near the top of the plant.

The concept of leaves being warmer than air has persisted, and the suggestion that leaf or canopy temperatures could be used as an irrigation-scheduling tool is sometimes received with skepticism. However, it is now rather obvious that leaf and canopy temperatures may be either warmer or cooler than the air, depending upon environmental factors that can for the most part be specified.

Equation (9) and Figs. 10 and 11 also show the limitations of any temperature-based stress index. In humid climates canopy temperatures will be near to or higher than air temperatures, with only a small range of tamperatures. In arid areas, however, canopy temperatures may be more than 10°C below air temperature and have a range of perhaps of 15°C. It is in the arid areas where irrigation is practiced that temperature techniques can work best and are most needed. This does not preclude their use during dry periods in normally humid areas. With Eq. (9), and pertinent meterologoical data, the potential usefulness for temperature techniques can be ascertained.

All temperature-based stress indices assume that only vegetation temperatures are measured. With IR thermometry, care must be taken to avoid the complication of the soil background, especially when sparse vegetation is being measured.

This is a particular problem when airborne scanners are used over row crops that do not completely cover the soil. Some plant species will orient their leaves to be normal to incident sunlight if well watered and to be parallel to the light when stressed, with an apparent reduction of ground cover from 100% to about 20% (Byrne *et al.*, 1979). We need to know much more about how plant and canopy architecture affects canopy temperature measurements.

Critical values of the temperature indices need to be established for various crops. For example, a CWSI of 0.3 may be the critical value for wheat but not for cotton. Some evidence indicates that water stress prior to the bloom stage in cotton may reduce vegetative growth and save water, without reducing yields (Guinn *et al.*, 1981). Temperature measurements may not be a sufficient indiator of irrigation needs for potatoes grown for tuber quality on certain soils. In this case irrigations are given to keep the soil moist, facilitating tuber expansion, with the plants never stressed. These are but a few management factors that should be considered as the development of temperature-based indices progresses.

The tremendous advance in IR technology during the past few years has allowed the production of lightweight hand-held IR thermometers that can be used to measure plant canopy temperatures rapidly. Many of these instruments have just recently become available to agricultural researchers. Some have already been purchased by farmers and farm consultants. The next few years should see an explosion of knowledge about canopy temperatures and about how such information can be used in irrigation management. The timing of irrigations is the likely area for the maximum use of these instruments. The data accumulated so far strongly suggest success.

Even the casual reader of this review will easily discern the personal bias of the author. For this I do not apologize, but hope that enough information is presented so that the reader can evaluate the material objectively. In my opinion, agricultural application of remotely sensed temperatures (and other remote measurements) is just beginning. Much exciting research is in store.

References

Ansari, A. Q., and Lommis, W. E. (1959). Leaf temperatures. *Am. J. Bot.* **46,** 713–717.

Askenasy, E. (1875). Ueber die temperature welche pflanzen im sonnenlicht annehmen. *Bot. Zg.* **33,** 441–444.

Aston, A. R., and Van Bavel, C. H. M. (1972). Soil surface water depletion and leaf temperature. *Agron. J.* **64,** 368–373.

Barthakur, N. (1975). Leaf temperatures under controlled environmental conditions. *Int. J. Biometeorol.* **19,** 21–27.

Bartholic, J. F., Namken, L. N., and Wiegand, C. L. (1970). Combination equations used to calculate evaporation and potential evaporation. *U.S. Agric. Res. Serv., ARS* **USDA-ARS-41-170,** 1–14.

Bartholic, J. F., Namken, L. N., and Wiegand, C. L. (1972). Aerial thermal scanner to determine temperatures of soils and of crop canopies differing in water stress. *Agron. J.* **64,** 603–608.

Bell, J. P. (1976). "Neutron Probe Practice," Inst. Hydrol. Rep. No. 19. Natural Environmental Research Council, Wallingford, England.

Blad, B. L., and Rosenberg, N. J. (1976). Measurement of crop temperature by leaf thermocouple, infrared thermometry, and remotely sensed thermal imagery. *Agron. J.* **68,** 635–641.

Blad, B. L., Gardner, B. R., Steinmetz, S., and Rosenberg, N. J. (1982). Plant and air temperature patterns in alfalfa, corn, grass, sorghum and soybeans as measured with thermocouples and infrared thermometers. *Agron. J.* (in press).

Blad, B. L., Gardner, B. R., Watts, D. G., and Rosenberg, N. J. (1981). Remote sensing of crop moisture status. *Remote Sens. Q.* **3,** 4–20.

Brown, K. W., and Rosenberg, N. J. (1973). A resistance model to predict evapotranspiration and its application to a sugar beet field. *Agron. J.* **65,** 341–347.

Byrne, G. F., Begg, J. E., Fleming, P. M., and Dunin, F. X. (1979). Remotely sensed land cover temperature and soil water status—A brief review. *Remote Sens. Environ.* **8,** 291–305.

Carlson, R. E., Yarger, D. N., and Shaw, R. H. (1972). Environmental influences on the leaf temperatures of two soybean varieties grown under controlled irrigation. *Agron. J.* **64,** 224–229.

Chaudhary, T. N., and Bhatnagar, V. K. (1980). Wheat root distribution, water extraction pattern and grain yield as influenced by time and rate of irrigation. *Agric. Water Manage.* **3,** 115–124.

Cihlar, J. (1976). "Thermal Infrared Remote Sensing: A Bibliography," Res. Rep. 76-1. Canada Center for Remote Sensing, Ottawa.

Clark, R. N., and Hiler, E. A. (1973). Plant measurements as indicators of crop water deficit. *Crop Sci.* **13,** 466–469.

Clawson, K. L., and Blad, B. L. (1982). Infrared thermometry for scheduling irrigation of corn. *Agron J.* **74** (in press).

Clum, H. H. (1926). The effect of transpiration and environmental factors on leaf temperatures. I. Transpiration. *Am. J. Bot.* **13,** 194–230.

Conaway, J., and Van Bavel, C. H. M. (1967a). Radiometric surface temperature measurements and fluctuations in sky radiant emittance in the 600 to 1300 cm^{-1} waveband. *Agron. J.* **59,** 389–390.

Conaway, J., and Van Bavel, C. H. M. (1967b). Evaporation from a wet soil surface calculated from radiometrically determined surface temperatures. *J. Appl. Meteorol.* **6,** 650–655.

Curtis, O. F. (1936). Leaf temperatures and the cooling of leaves by radiation. *Plant Physiol.* **11,** 343–364.

Curtis, O. F. (1938). Wallace and Clum, "Leaf temperatures": A critical analysis with additional data. *Am. J. Bot.* **25,** 761–771.

David, W. P. (1969). "Remote Sensing of Crop Water Deficits and its Potential Applications," Remote Sens. Cent. Tech. Rep. RSC-06. Texas A&M University, College Station.

DuChateau, P. C., Nofziger, D. L., Ahuja, L. R., and Swartzendruber, D. (1972). Experimental curves and rates of change from piecewise parabolic fits. *Agron. J.* **64,** 538–542.

Eaton, F. M., and Belden, G. O. (1929). Leaf temperatures of cotton and their relation to transpiration, varietal differences, and yields. *U.S., Dep. Agric., Tech. Bull.* **91.**

Ehlers, J. H. (1915). The temperature of leaves of pinus in winter. *Am. J. Bot.* **2,** 32–70.

Ehrler, W. L. (1973). Cotton leaf temperatures as related to soil water depletion and meteorlogical factors. *Agron. J.* **65,** 404–409.

Ehrler, W. L., Idso, S. B., Jackson, R. D., and Reginato, R. J. (1978a). Wheat canopy temperature: Relation to plant water potential. *Agron. J.* **70,** 251–256.

Ehrler, W. L., Idso, S. B., Jackson, R. D., and Reginato, R. J. (1978b). Diurnal changes in plant water potential and canopy temperature of wheat as affected by drought. *Agron. J.* **70,** 999–1004.

Fuchs, M., and Tanner, C. B. (1966). Infrared thermometry of vegetation. *Agron. J.* **58,** 597–601.

Fuchs, M., Kanemasu, E. T., Kerr, J. P., and Tanner, C. B. (1967). Effect of viewing angle on canopy temperature measurements with infrared thermometers. *Agron. J.* **59,** 494–496.

Gardner, B. R., Blad, B. L., and Watts, D. G. (1981a). Plant and air temperatures in differentially irrigated corn. *Agric. Meteorol.* **25,** 207–217.

Gardner, B. R., Blad, B. L., Garrity, D. P., and Watts, D. G. (1981b). Relationships between crop temperature, grain yield, evaptranspiration and phenological development in two hybrids of moisture stressed sorghum. *Irrig. Sci.* **2,** 213–224.

Gardner, B. R., Blad, B. L., Maurer, R. E., and Watts, D. G. (1981c). Relationship between crop temperature and the physiological and phenological development of differentially irrigated corn. *Agron. J.* **73,** 743–747.

Gates, D. M. (1964a). Leaf temperature and transpiration. *Agron. J.* **56,** 273–277.

Gates, D. M. (1964b). Characteristics of soil and vegetated surfaces to reflected and emitted radiation. *Proc. Int. Symp. Remote Sens. Environ., 3rd, 1964,* pp. 573–600.

Gates, D. M. (1968). Sensing biological environments with a portable radiation thermometer. *Appl. Opt.* **7,** 1803–1809.

Gear, R. D., Dransfield, A. S., and Campbell, M.D. (1977). Irrigation scheduling with neutron probe. *J. Irrig. Drain. Div., Am. Soc. Civ. Eng.* **103** (IR3), 291–298.

Geiser, K. M., Slack, D. C., Allred, E. R., and Stange, K. W. (1982). Irrigation scheduling using crop canopy-air temperature difference. *Trans. ASAE* (in press).

Guinn, G., Mauney, J. R., and Fry, K. E. (1981). Irrigation scheduling and plant population effects on growth, bloom rates, boll abscission, and yield of cotton. *Agron. J.* **73,** 529–534.

Hatfield, J. L. (1979). Canopy temperatures: The usefulness and reliability of remote measurements. *Agron. J.* **71,** 889–892.

Heermann, D. F., and Duke, H. R. (1978). Evaluation of crop water stress under limited irrigation. *Pap.—Am. Soc. Agric. Eng.* **78-2556,** 1–12.

Heermann, D. F., Haise, H. R., and Mickelson, R. H. (1976). Scheduling center pivot sprinkler irrigation systems for corn production in eastern Colorado. *Trans. ASAE* **19,** 284–287, 293.

Heilman, J. L., Kanemasu, E. T., Rosenberg, N. J., and Blad, B. L. (1976). Thermal scanner measurement of canopy temperatures to estimate evapotranspiration. *Remote Sens. Environ.* **5,** 137–145.

Heilman, J. L., Heilman, W. E., and Moore, D. G. (1981). Remote sensing of canopy temperature at incomplete cover. *Agron. J.* **73,** 403–406.

Hiler, E. A., and Clark, R.N. (1971). Stress day index to characterize effects of water stress on crop yields. *Trans. ASAE* **14,** 757–761.

Hiler, E. A., Howell, T. A., Lewis, R. B., and Boos, R. P. (1974). Irrigation timing by the stress day index method. *Trans. ASAE* **17,** 393–398.

Howell, T. A., Jordan, W. R., and Hiler, E. A. (1979). Evaporative demand as a plant stress. *In* "Modification of the Aerial Environment of Plants" (B. J. Barfield and J. F. Gerber, eds.), Monogr. No. 2, pp. 97–113. Am. Soc. Agric. Eng., St. Joseph, Missouri.

Hsiao, T. C. (1973). Plant response to water stress. *Annu. Rev. Plant Physiol.* **24,** 519–570.

Idso, S. B., and Baker, D. G. (1967). Relative importance of reradiation, convection, and transpiration in heat transfer from plants. *Plant Physiol.* **42,** 631–640.

Idso, S. B., and Ehrler, W. L. (1976). Estimating soil moisture in the root zone of crops: A technique adaptable to remote sensing. *Geophys. Res. Lett.* **3,** 23–25.

Idso, S. B., and Jackson, R. D. (1968). Significance of fluctuations in sky radiant emittance for infrared thermometry. *Agron. J.* **60,** 388-392.

Idso, S. B., Jackson, R. D., Ehrler, W. L., and Mitchell, S. T. (1969). A method for determination of infrared emittance of leaves. *Ecology* **50,** 899-902.

Idso, S. B., Jackson, R. D., and Reginato, R. J. (1977). Remote sensing of crop yields. *Science* **196,** 19-25.

Idso, S. B., Reginato, R. J., Jackson, R. D., and Pinter, P. J., Jr. (1981a). Foliage and air temperatures: Evidence for a dynamic "equivalence point." *Agric. Meteorol.* **24,** 223-226.

Idso, S. B., Reginato, R. J., Reicosky, D. C., and Hatfield, J. L. (1981b). Determining soil-induced plant water potential depressions in alfalfa by means of infrared thermometry. *Agron. J.* **73,** 826-830.

Idso, S. B., Jackson, R. D., Pinter, P. J., Jr., Reginato, R. J., and Hatfield, J. L. (1981c). Normalizing the stress degree day for environmental variability. *Agric. Meteorol.* **24,** 45-55.

Jackson, R. D. (1981). Soil moisture inferences from thermal infrared measurements of vegetation temperatures. *Dig., Int. Geosci. Remote Sens. Symp., 1981* Vol. I, pp. 364-374.

Jackson, R. D., and Idso, S. B. (1969). Ambient temperature effects in infrared thermometry. *Agron. J.* **61,** 324-325.

Jackson, R. D., and Pinter, P. J., Jr. (1981). Detection of water stress in wheat by measurement of reflected solar and emitted thermal IR radiation. *Int. Colloq. Spectral Signatures Objects Remote Sens., 1981,* pp. 399-406.

Jackson, R. D., Reginato, R. J., and Idso, S. B. (1977). Wheat canopy temperature: A practical tool for evaluating water requirements. *Water Resour. Res.* **13,** 651-656.

Jackson, R. D., Reginato, R. J., Pinter, P. J., Jr., and Idso, S. B. (1979). Plant canopy information extraction from composite scene reflectance of row crops. *Appl. Opt.* **18,** 3775-3782.

Jackson, R. D., Pinter, P. J., Jr., Reginato, R. J., and Idso, S. B. (1980). Hand-held radiometry. *Agric. Rev. Man., West. Ser. (U.S. Dep. Agric., Sci. Educ. Adm.)* **ARM-W-19,** 1-66.

Jackson, R. D., Idso, S. B., Reginato, R. J., and Pinter, P. J., Jr. (1981). Canopy temperature as a crop water stress indicator. *Water Resour. Res.* **17,** 1133-1138.

Jensen, M. E., ed. (1974). "Consumptive Use of Water and Irrigation Water Requirements," Irrig. Drain. Div., Am. Soc. Civ. Eng., New York.

Jensen, M. E., and Haise, H. R. (1963). Estimating evapotranspiration from solar radiation. *J. Irrig. Drain. Div., Am. Soc. Civ. Eng.* **89,** 15-41.

Jensen, M. E., and Wright, J. L. (1978). The role of evapotranspiration models in irrigation scheduling. *Trans. ASAE* **21,** 82-87.

Kanemasu, E. T., ed. (1975). "Measurement of Stomatal Aperature and Diffusive Resistance," Bull. No. 809. Washington State University, Pullman. pp. 39.

Kanemasu, E. T., Rasmussen, V. P., and Bagley, J. (1978). Estimating water requirements for corn with a pocket calculator. *Kans., Agric. Exp. Stn., Bull.* **615,** 1-24.

Kincaid, D. C., and Heermann, D. F. (1974). Scheduling irrigations using a programmable calculator. *U.S., Agric. Res. Serv. [Rep.]* **ARS-NC-12,** 1-55.

Linacre, E. T. (1964). A note on a feature of leaf and air temperatures. *Agric. Meteorol.* **1,** 66-72.

Linacre, E. T. (1967). Further notes on a feature of leaf and air temperature. *Arch. Meteorol., Geophys. Bioklimatol., Ser. B* **15,** 422-426.

Longnecker, D. E., and Lyerly, P. J. (1969). Moisture content of cotton leaves and petioles as related to environmental moisture stress. *Agron. J.* **61,** 687-690.

Lorenz, D. (1968). Temperature measurements of natural surfaces using infrared radiometers. *Appl. Opt.* **7,** 1705-1710.

McGinnies, W. J., and Aronson, R. C. W. (1971). Effects of environment on an infrared field thermometer. *Agron. J.* **63,** 813-814.

Millard, J. P., Jackson, R. D., Goettelman, R. C., Reginato, R. J., and Idso, S. B. (1978). Crop water-stress assessment using an airborne thermal scanner. *Photogramm. Eng. Remote Sens.* **44,** 77–85.

Miller, E. C., and Saunders, A. R. (1923). Some observations on the temperature of the leaves of crop plants. *J. Agric. Res.* **26,** 15–43.

Monteith, J. L. (1973). "Principles of Environmental Physics." Arnold, London.

Monteith, J. L., and Szeicz, G. (1962). Radiative temperature in the heat balance of natural surfaces. *Q. J. R. Meteorol. Soc.* **88,** 496–507.

Myers, V. I., and Allen, W. A. (1968). Electrooptical remote sensing methods as nondestructive testing and measuring techniques in agriculture. *Appl. Opt.* **7,** 1819–1838.

Myers, V. I., and Heilman, M. D. (1969). Thermal infrared for soil temperature studies. *Photogramm. Eng.* **35,** 1024–1032.

Nakayama, F. S., and Reginato, R. J. (1982). Simplifying neutron moisture meter calibration. *Soil Sci.* (in press).

Okuyama, T. (1975). Measurements of active surface temperature of rice plant grown in a small model of the paddy field using an infrared thermometer. *J. Agric. Met.* **30,** 191–194 (in Japanese, translated by M. Nakayama).

Park, A. B., Colwell, R. N., and Myers, V. I. (1968). Resource survey by satellite: Science fiction coming true. *USDA Yearb. Agric.* pp. 13–19.

Penman, H. L. (1948). Natural evaporation from open water, bare soil and grass. *Proc. Soc. London, Ser. A* **193,** 120–145.

Perrier, A. (1971). Leaf temperature measurement. *In* "Plant Photosynthetic Production Manual and Methods" (Z. Sestak, J. Gatsky, and P. G. Jarvis, eds.), pp. 632–671. Junk, The Hague.

Pinter, P. J., Jr. (1982). Remote sensing of microclimatic stress. *In* "Biometerology in integrated Pest Management" (J. L. Hatfield and I. J. Thompson, eds.), pp. 101–146. Academic Press, New York.

Rameaux (1843). Des températures végétales. *Ann. Sci. Nat., Bot. Biol. Veg.* **19**[2], 10.

Ritchie, J. T. (1981). Soil water availability. *Plant and Soil* **58,** 327–338.

Russell, G. (1980). Crop evaporation, surface resistance and soil water status. *Agric. Meteorol.* **21,** 213–266.

Sandhu, B. S., and Horton, M. L. (1978). Temperature response of oats to water stress in the field. *Agric. Meteorol.* **19,** 329–336.

Scholander, P. F., Hammel, H. T., Bradstreet, E. D., and Hemmingsen, E. A. (1965). Sap pressure in vascular plants. *Science* **148,** 339–346.

Seguin, B., and Petit, V. (1980). "Utilisation de la thermographie infrarouge pour l'estimation de l'évaporation. Analyse critique des résultats des mésures sur le site de la Crau," TELLUS Newsl. No 16. Comm. Eur. Commun.

Selby, J. E. A., Kneizys, F. X., Chetwynd, J. H., Jr., and McClatchey, R. A. (1978). "Atmospheric Transmittance/Radiance: Computer Code LOWTRAN IV," Air Force Geophys. Lab. Rep. No. AFGL-TR-78-0053. Hanscom AFB, Massachusetts.

Stegman, E. C., Schiele, L. H., and Bauer, A. (1976). Plant water stress criteria for irrigation scheduling. *Trans. ASAE* **19,** 850–855.

Stevenson, K. R., and Shaw, R. H. (1971). Effects of leaf orientation on leaf resistance to water vapor diffusion in soybean (Glycine max L. Merr) leaves. *Agron. J.* **63,** 327–329.

Stone, L. R., and Horton, M. L. (1974). Estimating evapotranspiration using canopy temperatures: Field evaluation. *Agron. J.* **66,** 540-545.

Sumayao, C. R., Kanemasu, E. T., and Brakke, T. W. (1980). Using leaf temperature to assess evapotranspiration and advection. *Agric. Meteorol.* **22,** 153–166.

Sutherland, R. A., and Bartholic, J. F. (1977). Significance of vegetation in interpreting thermal radiation from a terrestrial surface. *J. Appl. Meteorol.* **16,** 759–763.

Sutherland, R. A., Bartholic, J. F., and Gerber, J. F. (1979). Emissivity correction for interpreting thermal radiation from a terrestrial surface. *J. Appl. Meteorol.* **18,** 1165–1171.

Szeicz, G., and Long, I. F. (1969). Surface resistance of crop canopies. *Water Resour. Res.* **5,** 622–633.

Szeicz, G., Enrodi, G., and Tajchman, S. (1969). Aerodynamic and surface factors in evaporation. *Water Resour. Res.* **5,** 380–394.

Tanner, C. B. (1963). Plant temperatures. *Agron. J.* **55,** 210–211.

Thom, A. S., and Oliver, H. R. (1977). On Penman's equation for estimating regional evaporation. *Q. J. Roy. Meteorol. Soc.* **103,** 345–357.

Van Bavel, C. H. M. (1967). Changes in canopy resistance to water loss from alfalfa induced by soil water depletion. *Agric. Meteorol.* **4,** 165–176.

Van Bavel, C. H. M., and Ehrler, W. L. (1968). Water loss from a sorghum field and stomatal control. *Agron. J.* **60,** 84–86.

Van Bavel, C. H. M., Nixon, P. R., and Hauser, V. L. (1963). Soil moisture measurement with the neutron method. **USDA ARS-41-70,** 1–39.

Waggoner, P. E., and Shaw, R. H. (1952). Temperature of potato and tomato leaves. *Plant Physiol.* **27,** 710–724.

Walker, G. K., and Hatfield, J. L. (1979). Test of the stress-degree-day concept using multiple planting dates of red kidney beans. *Agron. J.* **71,** 967–971.

Wallace, R. H., and Clum, H. H. (1938). Leaf temperatures. *Am. J. Bot.* **25,** 83–97.

Watson, A. N. (1934). Further studies on the relation between thermal emissivity and plant temperatures. *Am. J. Bot.* **21,** 605–609.

Wiegand, C. L., and Namken, L. N. (1966). Influences of plant moisture stress, solar radiation, and air temperature on cotton leaf temperature. *Agron. J.* **58,** 552–556.

Wiegand, C. L., Heilman, M. D., and Gerbermann, A. H. (1968). Detailed plant and soil thermal regime in agronomy. *Proc. Int. Symp. Remote Sens. Environment, 5th, 1968,* pp. 325–342.

Wolpert, A. (1962). Heat transfer analysis of factors affecting plant leaf temperature. Significance of leaf hair. *Plant Physiol.* **37,** 113–120.

Wright, J. L., and Jensen, M. E. (1978). Development and evaluation of evapotranspiration models for irrigation scheduling. *Trans. ASAE* **21,** 88–96.

USE OF SOLUTE TRANSPORT MODELS TO ESTIMATE SALT BALANCE BELOW IRRIGATED CROPLAND

William A. Jury

Department of Soil and Environmental Sciences
University of California
Riverside, California

I. Introduction

Modern society places multiple and often contradictory demands on land-use projects. For example, irrigated agriculture is urged to maximize crop yields to

ADVANCES IN IRRIGATION, VOLUME 1

ISBN 0-12-24301-6

provide food for an expanding population at the same time that it is constrained to minimize the pollution of drainage water. Similarly, use of reclaimed municipal wastewater is a high priority for cities in arid areas with a restricted water supply, but this use is constrained by severe discharge regulations which may make the reuse operation unfeasible. In fact, many new water reuse projects and land waste-disposal operations require an environmental impact report prior to project approval. Because of the long time required for salt to travel through the soil, *in situ* field experiments are not a feasible tool for satisfying the requirements of such an environmental impact report.

Much of the research being conducted today in soil physics and soil chemistry is directed toward developing tools for dealing with the complex problems discussed above. One of the major tools under development is the solute transport model. Such a model, if reasonably accurate, could be used to design optimum irrigation water and salt management strategies to minimize drainage salt emissions while maintaining a given salinity range in the root zone. It may also be useful for estimating the potential mobility of various dissolved hazardous species held in a land disposal pond over soil with a low, but finite, permeability. Finally, it may be useful for estimating surface water management strategies to reclaim salt-affected soil.

In its simplest form a solute transport model may be nothing more than a guide for interpreting measurements, while in the most complex form it may seek to describe the space and time dependence of all phases of every chemical species in the soil–water system. The purpose of this article is to review the state of the art of solute transport modeling and to discuss in some detail the uses and limitations of models of varying degrees of complexity.

II. Literature Review

A. Solute Transport without Chemical Reactions

The first quantitative model of salt movement in soil water was presented in a series of papers by Nielsen and Biggar (Nielsen and Biggar, 1961, 1962, 1963; Biggar and Nielsen, 1962) who considered the principal mechanisms of transport of dissolved species to be convection within flowing soil water, dispersion of the solutes by virtue of the variations in local soil water velocity, and diffusion of the dissolved solutes within the soil solution. Their model has received many tests in the laboratory and seems to be valid beyond question when the porous medium is isotropic and homogeneous. A thorough discussion of many of these studies is given in several reviews of soil solute transport (Gardner, 1965; Biggar and Nielsen, 1967; Boast, 1973).

The model presented by Nielsen and Biggar was solved by these authors only for a specialized set of surface conditions and for steady water flow without plant water uptake. More complex mathematical models were presented by Bresler (1967), who used a mixing cell model for infiltration and redistribution, and by Bresler and Hanks (1969), who solved the general one-dimensional water and salt flow equations on a computer. These latter approaches require substantial data input on soil water properties to characterize the water regime. However, Wierenga and co-workers (Wierenga, 1977; Beese and Wierenga, 1980) have shown that only a small error in the prediction of solute movement results from using a steady-state water flux approximation even when the water regime has wetting and drying cycles, provided that the output is plotted against cumulative drainage. This approximation greatly simplifies the calculation of solute transport.

Studies of solute transport through aggregated soils with preferential flow pathways and blocked channels showed that the simple convection–dispersion model was inadequate to accurately predict the breakthrough curves resulting from a change in solute concentration at the inlet end. The theory has been modified in different ways to account for this phenomenon, by dividing the water volume into a mobile region and an exclusion volume where water is stagnant and salt is excluded (El Prince and Day, 1977) or by treating the stagnant region of the porous medium as a region accessible to salts by diffusion from the mobile water region (van Genuchten and Wierenga, 1976, 1977). The difficulty with such approaches is that they introduce new parameters into the model which cannot be independently measured or estimated.

Field studies of solute movement have been less successful than laboratory studies because of the large variation in soil water flow properties in both the horizontal and the vertical directions. However, basic principles developed in the laboratory have been successfully generalized to the field. For example, Miller *et al.* (1965) showed that chloride could be leached more effectively if the field soil was kept unsaturated during water infiltration than if the surface was ponded.

Until recently, most field tests of solute flow models have been on small plots with controlled inputs. Warrick *et al.* (1971) conducted a field test of the solute model assuming both constant and variable water flow rates. Good agreement between predicted and measured chloride concentrations could be achieved only when a value for the solute dispersion coefficient was chosen which was greatly in excess of laboratory values measured at similar water velocities. Jury *et al.* (1976) found a similar amount of apparent dispersion when the one-dimensional solute transport model was used to simulate measured nitrate concentrations in lysimeter drainage, but found that only a modest amount of apparent dispersion was present when the model was generalized to take horizontal variations in water input into account.

Solute transport through tile-drained fields has been studied by Jury (1975a,b) who used a piston flow model along the curved streamlines in the saturated zone. Much of the apparent dispersion in drainage concentration was found to be attributable to variable travel times to the drain.

The extent of field variability of solute transport was illustrated in a paper by Biggar and Nielsen (1976) who showed that water velocities and solute dispersion coefficients were log-normally distributed over a 150-ha field where solutes were leached by ponding. They calculated that hundreds of measurements of solute transport coefficients would have to be taken on their field to calibrate a salt transport model designed to estimate mean behavior of the field. In a separate field experiment, van der Pol *et al.* (1977) found log-normal variation in salt concentration at a given depth even when water input flux was controlled by drip irrigation across a field. These results and others (Luxmoore and Sharma, 1980; Sharma *et al.*, 1980) have resulted in proposals that solute transport be modeled stochastically, either as a random Monte Carlo problem (Warrick and Amoozegar-Fard, 1979), or in terms of a frequency reponse function (Dagan and Bresler, 1979). To date, these approaches have not been tested in the field, except on the data used to calibrate them.

B. Solute Transport with Cation Exchange

Cation transport through soil with negatively charged mineral surfaces is more complex than anion transport because of exchange adsorption. Transport of the principal cations in irrigation water, Ca^{2+}, Mg^{2+}, and Na^{+}, has long been of interest to soil scientists because of the importance of these ions in soil reclamation by leaching. A general transport-exchange model for Ca^{2+}–Mg^{2+} and for Na^{+}-divalent systems based on chromatography theory was proposed and tested by scientists at the U.S. Salinity Laboratory (Bower *et al.*, 1957; Brooks *et al.*, 1958) with good agreement found when the rate constants and equilibrium constants were fitted to data. Other studies used the equilibrium relation between sodium adsorption ration (SAR) and exchangeable sodium ratio to predict the effect of a reclamation operation (Reeve, 1957). Under certain conditions this may result in a simple graphical relation between applied Ca^{2+} and reduction in exchangeable sodium (Jury *et al.*, 1979).

C. Chemical Equilibrium Models

A model suitable for calculating the equilibrium speciation of a soil solution containing the major ions found in most irrigation waters was presented by Oster and McNeal (1971) based on a seawater model by Garrels and Christ (1965). The model includes the effects of ion pairing and precipitation–dissolution reactions. Many of the features of the model of Oster and McNeal were based on earlier work by Dutt and co-workers (Dutt, 1962; Dutt and Tanji, 1962; Tanji *et al.*, 1967).

A recent model by Sposito and Mattigod (1980) called GEOCHEM is capable of determining the equilibrium speciation of any chemical system for which the thermodynamic data are stored (currently, 36 metals and 69 ligands). This model, which is applicable to a wide variety of soil and environmental chemistry problems, is based on a geochemistry model REDEQ2 (Morel and Morgan, 1972) which Sposito and Mattigod generalized to soil solutions.

D. Coupled Solute Transport with Chemical Equilibrium

The models which represent chemical reactions as well as solute transport either assume that the reactions are rate limited and thus depend on the residence time, or that the reactions proceed instantaneously to equilibrium. Dutt and co-workers have done much of the early work in equilibrium–transport modeling, beginning with simple gypsum solutions (Dutt, 1962) to gypsum–cation exchange systems (Tanji *et al.*, 1965) and finally to a full representation of ion pairing, precipitation–dissolution, and cation exchange (Dutt *et al.*, 1972). These models used a mixing-cell concept where regions of the soil profile were periodically equilibrated with the moving soil solution. A recent version of this model, using the chemical subroutine of Oster and McNeal (1971) has been used by Jury *et al.* (1978a,b) to study the transient effects of soil salinization and to calculate the salt burden of irrigation waters (Jury and Pratt, 1980). A model with a fully nonlinear water flow subprogram but similar chemical subroutine has been recently published by Robbins *et al.* (1980).

E. Theoretical Foundations of Solute Transport Models

Many of the current theories for macroscopic solute transport in soils have been reviewed and criticized by Sposito *et al.* (1979) for their lack of a realistic dynamic model for solute molecular movement. These authors and co-workers derived the solute flux equation from a stochastic molecular model (Gupta *et al.*, 1980) and the flux equation and mass balance by statistical mechanics (Chu and Sposito, 1980). These efforts provide a firm physical foundation for transport of solutes in isotropic porous mediums.

III. Model Description

Three types of solute transport models will be used to predict drainage water composition in the calculations to follow.

A. Proportional Model

This model, which may only be used to predict drainage in steady state, assumes that all salts in solution move through the root zone without undergoing chemical reactions. Thus, the predicted drainage concentration C_d is given by

$$C_d = C_i/\mathrm{LF} \tag{1}$$

where C_i is the irrigation concentration and LF is the leaching fraction (drainage rate/irrigation rate). Similarly, the salt flux J_s below the root zone is given by

$$J_s = C_d D = C_i \, \mathrm{ET}/(1 - \mathrm{LF}) \tag{2}$$

where D is the drainage rate and ET the evapotranspiration rate.

B. Steady-State Chemistry Model

This model, described below in greater detail, allows precipitation or dissolution reactions to occur as the salts are concentrated according to Eq. (1). In the examples below, the soil is assumed to be calcareous with no native gypsum present. The CO_2 concentration at the bottom of the root zone is assumed to be 0.03 atm.

C. Dynamic Solute Transport Model with Chemical Reactions

The model to be described and used in subsequent sections is that of Jury *et al.* (1978a,b) and Jury and Pratt (1980). In the calculations it is assumed that the soil is initially unsalinized and 90% saturated with Ca^{2+} ions on an exchange complex of cation-exchange capacity (CEC) 0.45 meq/cm^3 soil. Any differences between this model and others in the literature will be noted.

1. *Water Transport Equation*

The water transport equation is a gravity flow model, assuming a sequence of layers of uniform water content. Each layer has the mass balance

$$L_i(d\theta_i/dt) = K(\theta_{i-1}) - K(\theta_i) - S_i \tag{3}$$

where L_i is the layer thickness, θ_i is the layer water content, $K(\theta)$ is the hydraulic conductivity, and S_i is the water extraction from layer i. The model may be run with time-dependent boundary inputs or may be set to a steady-state value. As noted by Wierenga (1977), the use of steady-state rather than transient water flow will not significantly affect solute flow as long as the flow is always in the same direction. The nonlinear water flow model of Robbins *et al.* (1980) might be needed to describe a situation where the water flow was upward as well as downward (i.e., infrequent irrigations).

2. *Solute Transport Equation*

When transport is occurring, all dissolved species are assumed to be convected and dispersed in the same way, irrespective of any possible differences in self-diffusion. Thus, a single equation (4) is used to describe ion transport:

$$\frac{\partial C\theta}{\partial t} = \frac{\partial}{\partial z}\left(D\frac{\partial C}{\partial z}\right) - \frac{\partial}{\partial z}(J_{\mathrm{w}} C) \tag{4}$$

where C is total solute concentration, $J_{\mathrm{w}} \simeq K(\theta)$ is water flux, and D is the dispersion coefficient. This equation is solved by finite differences.

3. *Chemical Equilibrium Subroutine*

The chemistry subroutine calculates equilibrium speciation for a water solution containing Ca^{2+}, Na^{+}, K^{+}, Mg^{2+}, Cl^{-}, SO_4^{2-}, HCO_3^{-}, and CO_3^{2-}, the major ions present in irrigation water. The reactions permitted are ion pair formation, precipitation–dissolution, and cation exchange. The concentration of CO_2 in the soil atmosphere is assumed to be known and provided, while pH is calculated by tbe requirement of charge neutrality and soil carbonate equilibrium. The GEOCHEM program (Sposito and Mattigod, 1980) does not require charge neutrality but instead specifies CO_2 and pH. The program of Robbins *et al.* (1980) requires charge balance and fixed pH.

Activity Coefficients. The activity coefficients γ are calculated by the extended Debye–Hückel equation, as explained by Oster and Rhoades (1975):

$$\log \gamma_i = -0.509 z_i^2 \sqrt{I}/(I + b a_i \sqrt{I}) + C_i I \tag{5}$$

where z_i is the valence, $b = 3.28 \times 10^7$, a_i and C_i are constants characteristic of each ion, and I is the ionic strength

$$I = \frac{1}{2} \sum_{i=1}^{N} m_i z_i^2 \tag{6}$$

where m_i is the molality of the ith ionic species. Table I gives the values of a_i and C_i used in the program.

Ion Pairing. Only ion pairs involving a single cation and anion were considered in the model, and associations involving chloride were considered neg-

TABLE I

EMPIRICAL PARAMETERS USED IN EXTENDED DEBYE–HÜCKEL MODEL OF ION ACTIVITY COEFFICIENT

Ion	a_i (cm × 10^8)	c_i	Ion	a_i (cm × 10^8)	c_i
Na^+	4.0	0.075	Cl^-	3.0	0.015
K^+	3.0	0.015	SO_4^{2-}	4.0	−0.040
Ca^{2+}	6.0	0.165	HCO_3^-	4.0	0.000
Mg^{2+}	8.0	0.200			

ligible. Dissociation constants for the important species are given in Table II, as reported in Oster and McNeal (1971).

Solubility Products. The two solid phases considered in the model are gypsum ($CaSO_4 \cdot 2H_2O$) and lime ($CaCO_3$) with solubility products 2.4×10^{-5} and 4.82×10^{-9}, respectively.

Cation Exchange. The cation-exchange equations solved in the program represent Na^+–divalent exchange and Ca^{2+}–Mg^{2+} exchange. The Na^+–divalent equation is taken from the empirical relation (U.S. Salinity Laboratory Handbook 60)

$$\mathrm{ESR} = E_N/(\mathrm{CEC} - E_N) = K_1 \cdot \mathrm{SAR} \tag{7}$$

where ESR is the exchangeable sodium ratio, E_N is the exchangeable sodium concentration (meq/cm^3), $\mathrm{SAR} = Na^+/[(Ca^{2+} + Mg^{2+})/2]^{1/2}$ is the sodium adsorption ratio [(mmol/cm^3)$^{-1/2}$], and $K_1 = 0.466$ (mmol/cm^3)$^{-1/2}$ (saturation extract).

Calcium–magnesium exchange is represented by the mass-action relation

$$Ca^{2+}/Mg^{2+} = K_2 E_C/E_M \tag{8}$$

where E_C and E_M are exchange concentrations of calcium and magnesium (meq/cm^3) and $K_2 = 0.7$ is taken from Tanji *et al.* (1967).

TABLE II

DISSOCIATION CONSTANTS K_D FOR ION PAIR SPECIES

Ion pair	K_D	Ion pair	K_D	Ion pair	K_D
$CaSO_4^0$	4.0×10^{-3}	$CaHCO_3^+$	5.53×10^{-2}	$CaCO_3^0$	6.3×10^{-4}
$MgSO_4^0$	6.3×10^{-3}	$MgHCO_3^+$	6.01×10^{-2}	$MgCO_3^0$	3.98×10^{-4}
$NaSO_4^-$	1.0×10^{-1}	$NaHCO_3^0$	1.78	$NaCO_3^-$	5.39×10^{-2}
KSO_4^-	1.1×10^{-1}	$H_2CO_3^0$	4.16×10^{-7}	HCO_3^-	4.87×10^{-11}

4. *Simultaneous Chemical Reactions and Solute Transport*

The calculation of ion transport and reactions proceeds as follows:

(*i*) Ions are transported passively according to Eq. (4), for a time small compared to the time required to traverse a mixing cell.
(*ii*) The soil profile is divided into mixing cells (e.g., 20-cm thick) and the concentrations of each ion within a cell are averaged.
(*iii*) The total average ion concentrations of a given mixing cell, along with the exchange concentrations and average water content of the cell, are sent to the chemical equilibrium subroutine.
(*iv*) The series of equations describing carbonate equilibrium, charge balance, ion pairing, precipitation–dissolution, and cation exchange are solved iteratively by the method described by Oster and McNeal (1971). The new total ion concentrations are obtained by summing over the various species in which the ion appears.
(*v*) The new ion concentrations within a mixing cell are placed in the same relative profile they were in prior to averaging in the main program.
(*vi*) When all cells have been averaged, transport begins again.

IV. Illustrative Calculations

Table III gives the chemical composition of four prototype waters used in the simulations: Colorado River water, Feather River water, saline power-plant cooling water, and a typical reclaimed municipal wastewater. The waters were cho-

TABLE III

SAMPLE WATER USED IN THE CALCULATIONS

Component[a]	Colorado River	Feather River	Power-plant cooling water	Municipal wastewater
Ca^{2+}	3.9	0.45	26.2	4.35
Mg^{2+}	2.5	0.36	18.3	2.50
Na^{+}	4.5	0.25	49.2	6.09
K^{+}	0.1	0.04	—	0.36
Cl^{-}	2.6	0.08	26.2	4.51
SO_4^{2-}	5.9	0.16	62.7	5.21
HCO_3^{-}	2.4	0.86	4.8	3.15
TDS (mg/liter)	737	83.6	6108	860
EC (mho/cm)	1	0.1	8.1	1.3
SAR (meq/liter)	2.5	0.4	10.4	3.3

[a] Amounts of ions given in meq/liter.

sen to represent extremes of waters which might be used for irrigation. The waters are assumed to be applied at a fixed LF. The ET is taken constant at 90 cm/yr. The calculations of drainage concentration and salt flux are either for steady state, when concentrations have reached their final value, or transient state, when salt flux and concentrations are changing with time.

A. Steady-State Calculations

Figure 1 shows predicted steady-state drainage concentrations as a function of LF for each of the four waters. The solid line gives the prediction of the proportional model [Eq. (1)], whereas the dashed curve gives the prediction of the steady-state chemistry model, which includes precipitation–dissolution reactions. The proportional model will underestimate the drainage concentration of high-quality waters—such as that of the Feather River—which dissolve $CaCO_3$ from the soil at all values of LF. At high LF this underestimation may be substantial. On the other hand, for waters high in gypsum or lime saturation, the proportional model will overestimate drainage TDS by neglecting precipitation. For example, at an LF of 0.1, the chemistry model predicts that power-plant cooling water will precipitate 27% of applied salt, a salt burden reduction which

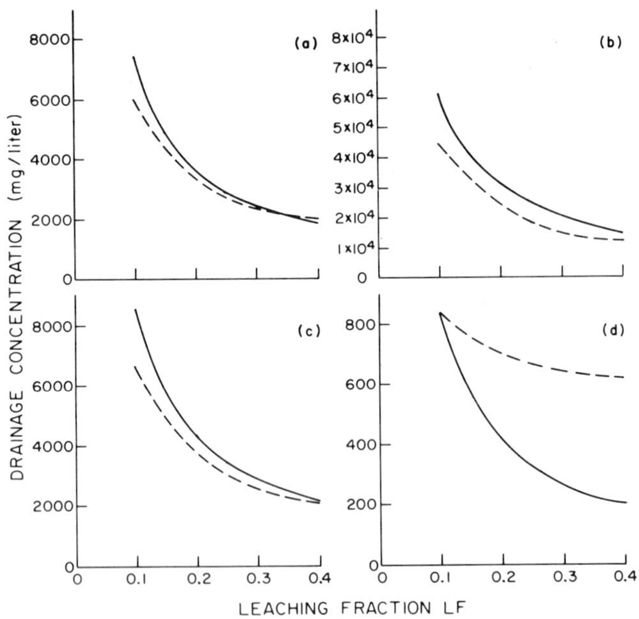

FIG. 1. Predicted steady-state drainage concentrations for four waters versus LF using the proportional model (solid line) and the chemistry model (dashed line): (a) Colorado River; (b) power-plant cooling water; (c) municipal wastewater; (d) Feather River.

could play an important role in management design strategies for this water (Jury *et al.*, 1980). For waters intermediate in saturation with lime, such as Colorado River water, the steady-state chemistry model predicts precipitation at low LF and dissolution at high LF.

When drainage concentration is used as an index of salinity, all models predict an increase with decreasing LF. However, when mass flow of salt in drainage is used as a salinity index, the result decreases with decreasing LF, both because less salt is applied and because increasing drainage concentration tends to decrease mineral dissolution (Rhoades *et al.*, 1973, 1974). Figure 2 shows salt flux (metric tons/ha yr^{-1}) plotted against LF for the four models. Here the effect of chemical reactions on salt balance is quite pronounced. Whereas, for example, the proportional model predicts a ratio of 1.5 between salt flux at LF = 0.4 and 0.1 [see Eq. (2)], the chemistry model predicts ratios of 1.7, 1.9, 1.9, and 4.4 for the power plant, Colorado, wastewater and Feather River irrigation waters, re-

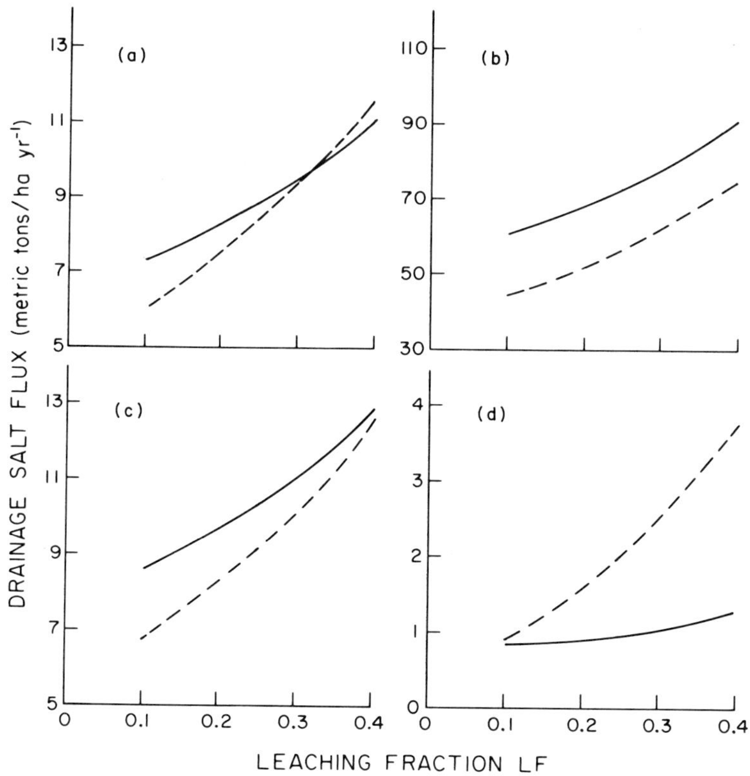

FIG. 2. Predicted steady-state salt flux for four waters versus LF using the proportional model (solid line) and the chemistry model (dashed line): (a) Colorado River; (b) power-plant cooling water; (c) municipal wastewater; (d) Feather River.

spectively. The predictions of the chemistry model shown in Fig. 2 are very close to the salt balance measurements made by Rhoades *et al.* (1974) for the Colorado and Feather Rivers.

B. Transient-State Calculations

The calculations given in the previous section apply to the steady state, which is reached only after the profile has had sufficient solution passed through it to reach equilibrium between the final solution composition and the composition of the exchange complex. During the transitional period, which may last for many years, precipitation reactions are enhanced by addition of Ca^{2+} ions into solution by exchange reactions (Jury *et al.*, 1978a,b). Because the transitional period can last for such a long time, steady-state models may give an unrealistic estimate of salt balance, and thus the more complex dynamic model may have to be used.

Figure 3 is a plot of the fraction of applied salt precipitated per year in the top 2.5 m for the municipal wastewater at three leaching fractions. After 10 years, only the soil irrigated at the high LF 0.4 is near steady state. Even after 27 years, a time in excess of the lifetime of many projects, the soil receiving waters at LF 0.1 is still far from steady state. For soils with many meters of exchange surfaces above the groundwater table, steady state would never be reached and a transient model of salt balance would be needed to make realistic estimates. That such transient effects may be substantially in excess of steady-state values was shown by Jury *et al.* (1978c), who measured up to 50% precipitation of applied power-plant cooling water over three years of irrigation. This result was compatible with

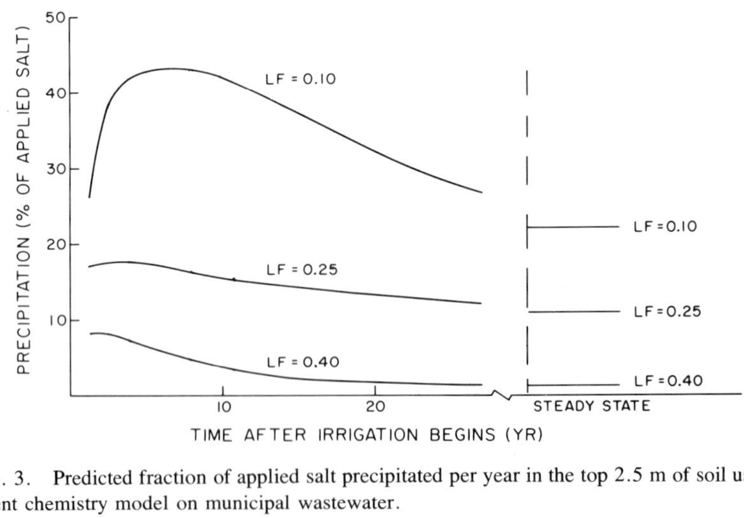

FIG. 3. Predicted fraction of applied salt precipitated per year in the top 2.5 m of soil using the transient chemistry model on municipal wastewater.

the transient model, but greatly in excess of the predictions of the steady-state models (Jury *et al.*, 1978b).

The enhanced precipitation during the transitional stage is caused by Na^+–Ca^{2+} exchange and by Mg^{2+}–Ca^{2+} exchange. When both Na^+ and Mg^{2+} are present in the water applied to a Ca^{2+}-saturated soil, Na^+ exchange proceeds nearly to equilibrium before Mg^{2+} exchange reaches its maximum rate (Jury and Pratt, 1980). Thus, different ions in solution proceed to equilibrium at different rates, and each has an effect on the salt balance of the profile. Figure 4, taken from Jury and Pratt (1980), shows a simultaneous plot of precipitation, solution ion concentration, and change in Na^+ and Mg^{2+} exchange concentration versus time for a 30-cm layer below the root zone of a soil irrigated by the saline power-

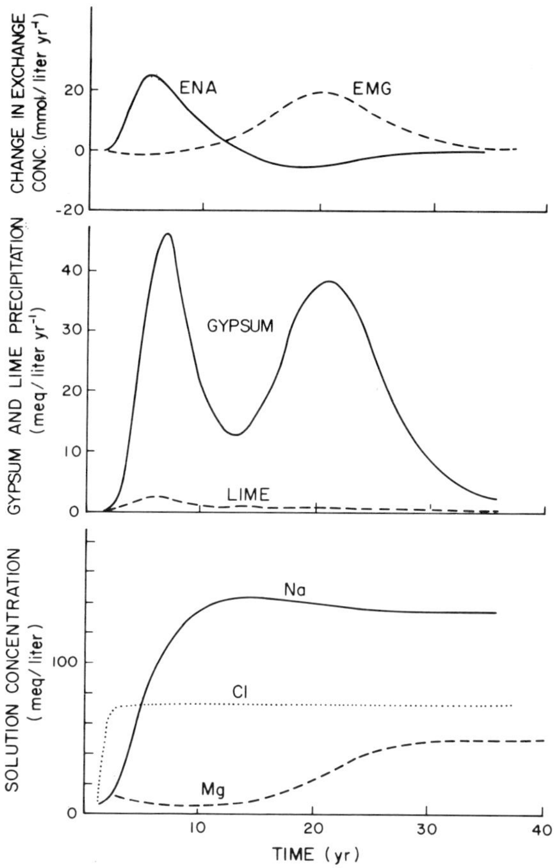

FIG. 4. Predicted solution concentrations, precipitation, and change in exchangeable sodium (ENA) and magnesium (EMG) in the 270–300-cm layer for the power-plant cooling water irrigated at LF = 0.4 and ET = 90 cm/yr (from Jury and Pratt, 1980).

plant cooling water (SCW) at LF = 0.4. Several conclusions may be drawn from this figure. First, Cl^-, Na^+, and Mg^{2+} all proceed to their final values at different rates. Second, Na^+ exchange and Mg^{2+} exchange occur at different times. Third, each exchange process causes an enhancement in precipitation. Fourth, Mg^{2+} exchange is the limiting process delaying the onset of steady state.

C. Transition Times

The calculated transition times to reach steady state for Cl^-, Na^+, and Mg^{2+} for the power-plant cooling water (irrigated at constant LF) in years per meter of soil assuming ET = 90 cm/yr and CEC = 0.45 meq/cm³ soil are given in the accompanying tabulation.

LF	Cl^-	Na^+	Mg^{2+}
0.25	1.33	4.08	12.05
0.40	0.66	2.22	8.20

The time required for Mg^{2+} equilibration is substantial, even at the high LF = 0.4. The exact times depend on the composition of the irrigation water, the exchange capacity of the soil, and the steady-state exchange concentrations (Jury and Pratt, 1980).

There are several reasons why Mg^{2+} requires so long to equilibrate relative to the other ions. First of all, the ratio of Mg^{2+} to Ca^{2+} in solution increases as the solution is concentrated, because Ca^{2+} concentrations are limited by $CaCO_3$ solubility. Thus, the ratio of Mg^{2+} to Ca^{2+} on the exchange sites will increase as the solution is concentrated, requiring substantial transfer of incoming Mg^{2+} to exchange [see Eq. (8)]. Second, since the concentration of Mg^{2+} in irrigation waters is usually low, a large amount of solution must be passed through the soil in order to add enough Mg^{2+} to reach equilibrium.

The extent to which exchange effects alter the salt balance depends on the type of water used, the kind of soil, and the LF of the water application. A detailed discussion of the salt balance of several different waters during the transitional stage is given by Jury *et al.* (1978a,b) and Jury and Pratt (1980).

V. Limitations of the Modeling Approach

A. Spatial Variability

The large variation found in water transmission rates of natural field soils has made the prediction of solute transport very difficult. The one-dimensional models discussed above are most useful in making comparisons between waters of different concentrations and compositions, and between different application rates (LF) of a given water. Furthermore, the chemical equilibrium subroutine may, in principle, be interfaced with any kind of water transport model, so that as methods are developed for modeling water transport in the field, they may be used for solute transport modeling as well.

B. Kinetic Effects

All of the chemical processes discussed above are assumed to proceed to equilibrium. If any of the important processes are rate limited, then the model will overpredict the extent of the reactions occurring at a given place and time. The factors which could contribute to kinetic limitations of reactions are many and complex, such as water velocity, pore bypass, soil aggregation, etc., and would be difficult to account for in a model. However, the major reactions affecting irrigation waters (cation exchange, gypsum and lime precipitation–dissolution) proceed rapidly to equilibrium under normal soil water velocities, so that the assumption of equilibrium is not likely to create too large an error.

C. Limitations Caused by Lack of Data

The major limitation to solute transport modeling is uncertainty in modeling water movement. The data required to calibrate a multidimensional water flow model in the field seem to be unobtainable, thus requiring the use of approximate one-dimensional models of water flow (Nielsen *et al.*, 1973). Wben these water flow models are used to represent solute flow, a large dispersion coefficient is required to model one-dimensional solute movement. This coefficient, which arises from use of an inadequate model, is very difficult to measure or predict (Biggar and Nielsen, 1976).

The many constants used in the chemistry program have been obtained for the most part in laboratory programs. Their utility in the field has only been checked by comparing predicted and measured ion concentrations (Wierenga *et al.*, 1975; Jury *et al.*, 1978b; Robbins *et al.*, 1980).

VI. Summary and Conclusions

Solute transport models differ greatly in complexity and utility, ranging from simple steady-state models which ignore chemical reactions, to complex dynamic models which include all major soil–solution reactions. In many cases the steady-state assumption will be in error because of the persistence of exchange reactions. During the transitional period or even in steady state, precipitation or dissolution reactions may have an appreciable influence on the salt balance of the drainage water, and calculations based on simple models may be greatly in error.

The usefulness of solute transport models to make predictions in field conditions is limited by the inadequacy of water transport models in describing the spatial and temporal distribution of water flow. The solute transport model has its greatest utility as a tool for predicting relative behavior of different irrigation waters, different soil types, or different water management strategies when a scenario for drainage ion composition is produced.

References

Beese, F., and Wierenga, P. J. (1980). Solute transport through soil with adsorption and root uptake computed with a transient and a constant-flux model. *Soil Sci.* **129,** 245–252.

Biggar, J. W., and Nielsen, D. R. (1962). Miscible displacement. II. Behavior of tracers. *Soil Sci. Soc. Am. Proc.* **26,** 125–128.

Biggar, J. W., and Nielsen, D. R. (1967). Miscible displacement and leaching phenomenon. *Agron. Monogr.* **11,** 254–274.

Biggar, J. W., and Nielsen, D. R. (1976). Spatial variability of the leaching characteristics of a field soil. *Water Resour. Res.* **12,** 78–84.

Boast, C. W. (1973). Modeling the movement of chemicals in soil by water. *Soil Sci.* **115,** 224–230.

Bower, C. A., Gardner, W. R., and Goertzen, J. O. (1957). Dynamics of cation exchange in soil columns. *Soil Sci. Soc. Am. Proc.* **21,** 20–24.

Bresler, E. (1967). A model for tracing salt distribution in the soil profile. *Soil Sci.* **104,** 227–233.

Bresler, E., and Hanks, R. J. (1969). Numerical method of estimating simultaneous flow of water and salt in unsaturated soil. *Soil Sci. Soc. Am. Proc.* **33,** 827–832.

Brooks, R. H., Goertzen, J. O., and Bower, C. A. (1958). Prediction of changes in the compositions of the dissolved and exchangeable cations in soils upon irrigation with high sodium waters. *Soil Sci. Soc. Am. Proc.* **22,** 122–124.

Chu, S.-Y., and Sposito, G. (1980). A derivation of the macroscopic solute transport equation for homogeneous saturated porous media. *Water Resour. Res.* **16,** 542–546.

Dagan, G., and Bresler, E. (1979). Solute dispersion in unsaturated soil at field scale. *Soil Sci. Soc. Am. J.* **43,** 461–466.

Dutt, G. R. (1962). Prediction of the concentration of solutes in the soil solution for systems containing gypsum and exchangeable CA and MG. *Soil Sci. Soc. Am. Proc.* **26,** 341–343.

Dutt, G. R., and Tanji, K. K. (1962). Predicting concentrations of solutes in water percolated through a column of soil. *J. Geophys. Res.* **69,** 3437–3493.

Dutt, G. R., Shaffer, M. J., and Moore, W. J. (1972). Computer simulation model of dynamic biophysicochemical processes in soils. *Ariz., Agric. Exp. Stn., Bull.* **196.**

El Prince, A. M., and Day, P. R. (1977). Fitting solute breakthrough equations to data. *Soil Sci. Soc. Am. J.* 41, 39–41.

Gardner, W. R. (1965). Movement of nitrogen in soil. *Agron. Monogr.* **10,** 550–572.

Garrels, R. M., and Christ, C. L. (1965). "Solutions, Minerals, and Equilibria," Harper, New York.

Gupta, V. K., Bhattacharya, R. N., and Sposito, G. (1980). A molecular approach to the foundation of the theory of solute transport in porous media. *J. Hydrol.* **50,** 333–353.

Jury, W. A. (1975a). Solute travel-time estimates for tile-drained fields. I. Theory. *Soil Sci. Soc. Am. Proc.* **39,** 1020–1024.

Jury, W. A. (1975b). Solute travel-time estimates for tile-drained fields. II. Applications to experimental studies. *Soil Sci. Soc. Am. Proc.* **39,** 1025-1028.

Jury, W. A., and Pratt, P. F. (1980). Estimation of the salt burden of irrigation drainage waters. *J. Environ. Qual.* **9,** 141-146.

Jury, W. A., Gardner, W. R., Saffigna, P. G., and Tanner, C. B. (1976). Model for predicting movement of nitrate and water through a loamy sand. *Soil Sci.* **122,** 36-43.

Jury, W. A., Frenkel, H., and Stolzy, L. H. (1978a). Transient changes in the soil water system from irrigation with saline water. I. Theory. *Soil Sci. Soc. Am. J.* **42,** 579-585.

Jury, W. A., Frenkel, H., Devitt, D., and Stolzy, L. H. (1978b). Transient changes in the soil water system from irrigation with saline water. II. Analysis of experimental data. *Soil Sci. Soc. Am. J.* **43,** 585-590.

Jury, W. A., Frenkel, H. Fluhler, H., Devitt, D., and Stolzy, L. H. (1978c). Use of saline water and minimum leaching for crop production. *Hilgardia* **46,** 169-192.

Jury, W. A., Jarrell, W. M., and Devitt, D. (1979). Reclamation of saline-sodic soils by leaching. *Soil Sci. Soc. Am. J.* **43,** 1100-1106.

Jury, W. A., Vaux, H. J., and Stolzy, L. H. (1980). Reuse of power plant cooling water for irrigation. *Water Resour. Bull.* **16,** 830-836.

Luxmoore, R. J., and Sharma, M. L. (1980). Runoff responses to soil heterogeneity. *Water Resour. Res.* **16,** 675-684.

Miller, R. J., Biggar, J. W., and Nielsen, D. R. (1965). Chloride displacement in Panoche clay loam in relation to water movement and distribution. *Water Resour. Res.* **1,** 63-73.

Morel, F., and Morgan, J. (1972). A numerical method for computing equilibria in aqueous chemical system. *Environ. Sci. Technol.* **6,** 58-67.

Nielsen, D. R., and Biggar, J. W. (1961). Miscible displacement in soils. I. Experimental information. *Soil Sci. Soc. Am. Proc.* **25,** 1-5.

Nielsen, D. R., and Biggar, J. W. (1962). Miscible displacement in soils. III. Theoretical considerations. *Soil Sci. Soc. Am. Proc.* **26,** 216-221.

Nielsen, D. R., and Biggar, J. W. (1963). Miscible displacement in soils. IV. Mixing in glass beads. *Soil Sci. Soc. Am. Proc.* **27,** 10-13.

Nielsen, D. R., Biggar, J. W., and Erh, K. T. (1973). Spatial variability of field-measured soil-water properties. *Hilgardia* **42,** 215-260.

Oster, J. D., and McNeal, B. L. (1971). Computation of soil solution composition variation with water content for desaturated soils. *Soil Sci. Soc. Am. Proc.* **35,** 436-442.

Oster, J. D., and Rhoades, J. D. (1975). Calculated drainage water compostions and salt burdens resulting from irrigation with river waters in the Western United States. *J. Environ. Qual.* **4,** 73-79.

Reeve, R. C. (1957). The relation of salinity to irrigation and drainage requirements. *Congr. Irrigation Drainage* **5,** 10.175-10.187.

Rhoades, J. D., Ingvalson, R. D., Tucker, J. M., and Clark, M. (1973). Salts in irrigation drainage waters. *Soil Sci. Soc. Am. Proc.* **37,** 770-774.

Rhoades, J. D., Oster, J. D., Ingvalson, R. D., Tucker, J. M., and Clark, M. (1974). Minimizing the salt burdens of irrigation draingage waters. *J. Environ. Qual.* **3,** 311-316.

Robbins, C. W., Wagenet, R. J., and Jurinak, J. J. (1980). A combined salt transport-chemical equilibrium model for calcareous and gypsiferous soils. *Soil Sci. Soc. Am. J.* **44,** 1191-1194.

Sharma, M. L., Gander, G. A., and Hunt, C. G. (1980). Spatial variability of infiltration in a watershed. *J. Hydrol.* **45,** 101-122.

Sposito, G., and Mattigod, S. V. (1980). "Geochem." University of California, Riverside.

Sposito, G., Gupta, V. K., and Battacharya, R. N. (1979). Foundational theories of solute transport in porous media: A critical review. *Adv. Water Resour.* **2,** 54-68.

Tanji, K. K., Dutt, G. R., Paul, J. L., and Doneen, L. D. (1967). Quality of percolating waters. *Hilgardia* **38,** 307-318.

van der Pol, R. M., Wierenga, P. J., and Nielsen, D. R. (1977). Solute movement in a field soil. *Soil Sci. Soc. Am. J.* **41,** 10–13.

van Genuchten, M. T. H., and Wierenga, P. J. (1976). Mass transfer studies in sorbing porous media. I. Analytical solutions. *Soil Sci. Soc. Am. J.* **40,** 473–480.

van Genuchten, M. T. H., and Wierenga, P. J. (1977). Mass transfer studies in sorbing porous media. II. Experimental evaluation with tritium. *Soil Sci. Soc. Am. J.* **41,** 272–278.

Warrick, A. W., and Amoozegar-Fard, A. (1979). Infiltration and drainage calculations using spatially scaled hydraulic properties. *Water Resour. Res.* **15,** 1116–1120.

Warrick, A. W., Biggar, J. W., and Nielsen, D. R. (1971). Simultaneous water and solute transfer for an unsaturated soil. *Water Resour. Res.* **7,** 1216–1225.

Wierenga, P. J. (1977). Solute distribution profiles computed with steady state and transient water flow models. *Soil Sci. Soc. Am. J.* **41,** 1050–1055.

Wierenga, P. J., Shaffer, M. J., Gomez, S. P., and O'Connor, G. A. (1975). Ionic distributions in large soil columns. *Soil Sci. Soc. Am. Proc.* **34,** 1080–1084.

LEVEL-BASIN IRRIGATION

A. R. Dedrick, L. J. Erie, and A. J. Clemmens

U.S. Department of Agriculture
Agricultural Research Service
U.S. Water Conservation Laboratory
Phoenix, Arizona

I. Introduction

Level-basin irrigation is a surface irrigation method whereby a controlled amount of water is applied to level soil surfaces, generally over a short time period. The basin may be of any configuration and is surrounded by a control barrier such as a dike. A key difference between level basins and other traditional surface irrigation systems commonly used in the United States is that much of the infiltration into the soil occurs after the water advance has been completed. Hence this is more of a static process rather than the dynamic process commonly associated with surface irrigation such as graded borders. Design of level-basin

ISBN 0-12-24301-6

systems depends upon the water supply flow rate and the soil infiltration characteristics. Larger flow rates and/or low soil infiltration rates may permit larger basins, which lead to improved machine efficiency. Level basins can, however, be used with higher intake soils and/or lower flow rates if the user is willing to reduce basin size. Level-basin irrigation can generally be used for any crops grown.

Irrigation techniques somewhat similar to the level basins used in the United States (Dedrick *et al.*, 1978; Erie and Dedrick, 1979) have been referenced by others. Commonly used names include *check flooding* and *basin flooding* (Israelsen, 1950), *level borders* (Zimmerman, 1966; Merriam, 1968), *check irrigation* (Schwab *et al.*, 1966), and *check-basin irrigation* (Michael, 1978). Most of these systems refer to relatively small parcels of land within the confining area, which generally differs from the large, precision-leveled basins discussed herein.

This concept of irrigation is certainly the most common method of irrigation used in many parts of the world today. Historically, it is likely to have been the first irrigation technique used. The check-basin irrigation system as described by Michael generally involves dividing a field into small unit areas so that each is nearly a level surface. The size of the unit is mostly dependent upon the general terrain on which the system is installed or used. This irrigation method is particularly well suited for rice production. The use of level basins in the United States is relatively new compared to the more traditional worldwide usage. The first report on level basins in the United States was by Ross and Swanson (1957), and it outlined some of the general ideas of level basins as used in the Pecos area of Texas. Phelan (1960) discussed bench leveling for surface irrigation and erosion control. Jensen and Howe (1965) reported on performance and presented empirical information for design of border checks. Generally, the fields used in the United States are larger than those that have been used in other parts of the world and involve considerable land grading in order to attain basin sizes that fit the machinery being used.

Irrigation water can be applied uniformly and efficiently to level basins. These features, along with other advantageous characteristics and potentially limiting factors, are now discussed.

High water application efficiencies (E_a) can be attained since no water runs off. In addition to the high application efficiency (the ratio of water required for an irrigation and the amount of water applied), the distribution uniformity (DU) of water applied can be very high.

Deep percolation water should be minimal for a properly designed and managed level-basin system. This reduces water table buildup and minimizes drainage requirements. Water-soluble fertilizers are commonly applied to level basins by injecting the chemicals directly into the irrigation supply stream. Conventional application techniques are unnecessary and application is uniform. Precise water application also maintains these fertilizers in the root zone and minimizes the degrading effect of these fertilizers on the groundwater.

The key to attaining high efficiency and uniformity is to get the water over the entire basin as rapidly as possible so that the difference in infiltration opportunity time is small for all areas of the basin. Variable soils within a basin create water distribution problems and efforts to eliminate such variations may be costly.

Large streams of water can be used with level-basin irrigation, thereby reducing the irrigation time. Changing irrigation sets on a predetermined time schedule eliminates the need for continual attention by the irrigator, with regard to where the water is in the field. Depending on the size of the basins, however, the application time may be quite short if the field is small and the flow rate is large. Hence careful attention is required on the part of the irrigator, but only for a short period of time. Regardless of the stream size, single or multiple outlets requiring little time to open and close can be used to turn water into the field. There is no need to direct water into individual furrows; hence labor requirements within the basin will be relatively low. A large stream size is generally desirable to achieve best results with level-basin irrigation, but a rather elaborate erosion-prevention structure may be required. As a way of reducing erosion, the number of outlets may be increased, but this eliminates some of the advantages associated with the low labor requirements of level-basin irrigation.

Precision land leveling is essential to attain uniform water distribution. After the total amount of water has been applied, the standing water is allowed to penetrate. If high or low areas exist, uneven penetration will occur and the uniformity of application, which is potentially high for level-basin systems, may be low. Releveling, or maintenance, of the level-basin system will probably be necessary to correct for nonuniform soil settlement and the surface irregularities resulting from tillage operations. With the recent commercial availability of laser-controlled land-leveling equipment, precise land leveling is now attainable (the procedure and results are reported in Section II). More land grading will generally be necessary for level-basin irrigation than for other types of surface irrigation systems. The amount of land grading will be limited by the depth of topsoil and the leveling costs. Costs of developing land for level-basin irrigation depend on the size of the basins and the topography of the land. From an irrigation design standpoint, basin size is dependent on the flow rate and intake characteristics of the soil. Large flows will result in large basins with fewer turnout structures and associated equipment. But large basins may mean excessive earth moving if the topography is not relatively level. Hence, topography rather than flow rate may limit the basin size.

Furrows, as well as flat-bed crops, are well adapted to this method of irrigation (Fig. 1). All furrows are connected at both ends of the field by a secondary ditch, so differences in the rate of advance among furrows are of minor importance. Crops grown using furrows can be germinated evenly by maintaining a uniform water depth throughout the length of the furrow to prevent inundation and minimize soil crusting over seed beds. Secondary ditches along the edge of basins, required to distribute water into bed or row crops, may use appreciable

FIG. 1. Lettuce being germinated on level beds. Precision leveling and planting result in essentially the same water elevation throughout the basin. All furrows are connected at both ends of the basin by secondary ditches.

field area. However, this area is usually necessary as a turn row for equipment. Temporary dikes may also be required to prevent overtopping of beds directly in front of turnouts. Many crops which are typically grown on beds can be flat planted in level basins since the beds are no longer needed for water control.

If single outlets are used at the corners of the field nearest the water supply, the supply ditch is not needed along the edge of the last two basins. Less ditch means reduced waste area and reduced cost of construction and maintenance.

Automatic controls can be conveniently applied to this method of irrigation. Since the water is applied volumetrically, only a few outlet structures are needed, and the gate is either opened or closed (no regulation required). The guesswork in applying the right amount of water is eliminated since there is no surface runoff, and all water applied to the basin is used within that basin. The time of application is a relatively straightforward calculation once the area of the basin is known, the depth of water to be applied is determined, and the flow rate coming into the basin is measured. Some allowance for deep percolation loses (inefficiency) may need to be made when applying the water. This deep percolation loss is, essentially, the only inefficiency of the system. Thus the main obligation of the irrigator is to change the water from basin to basin at predetermined times.

As with all methods of irrigation, the correct amount of water must be applied. This requirement is perhaps more important for the level-basin method because overapplication may lead to excessive inundation times and result in crop damage. In addition, excessive deep percolation losses may result. Conversely, underapplication will likely result in incomplete basin coverage.

Leaching of salts is accomplished easily with a level-basin irrigation system. Since water covers the entire surface uniformly, there is even penetration of the water. Since rainfall will not run off, it too may be useful for leaching purposes.

All rainfall will be confined within the diked area for plant growth and leaching. If rainfall intensity is high, surface drainage of the level-basin system may be necessary or even required. Such drainage systems could also be used as protection against irrigator error (overirrigation). Unlike sloping irrigated land, erosion from rainfall will be negligible.

Benches and dikes will likely require maintenance. The amount of maintenance will increase as the terrain steepens, since the benches are susceptible to dike breakage caused by rodent burrowing, overapplication of water, or cracking soils.

Relatively light applications of water are possible. Success is dependent upon proper design: balancing flow rate with the size of the field and the water intake characteristics of the soil. This advantage is especially important for vegetable crops, fertilizer application, frost control, and germination. Such management and cultural requirements can increase the seasonal water use beyond the typical consumptive use requirements of the crop.

II. Land Preparation

Since there is no runoff water with level-basin irrigation systems, the only inefficiencies occur either in the distribution of water within a field or by overapplication by the irrigator. As with other surface methods of irrigation, irregularities in soil infiltration rates, or general soil type and land elevation differences, result in uneven distribution of water.

A. Soil Variability

The topsoil should be as uniform as possible within a basin to assure even water distribution. If the soil is not uniform over the entire area the following measures can be taken to correct the situation:

(1) For broad soil variability across the field, select basin size and shape to provide similar soils within a basin. (A field with surface soils changing from clay loam to fine sandy loam with significantly different infiltration characteristics is illustrated in Fig. 2. The concrete-lined canal was located to provide soil

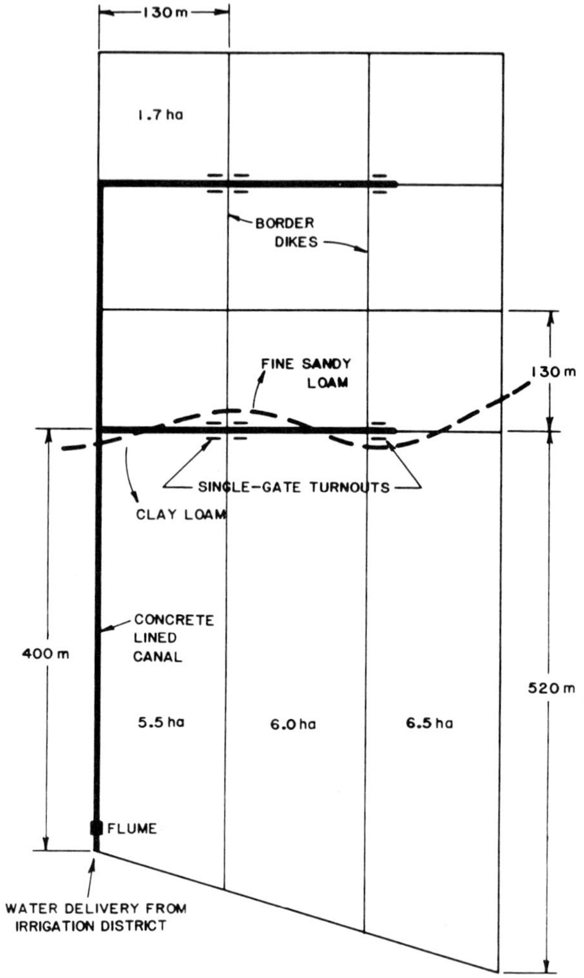

FIG. 2. Field illustrating canal placement to provide nearly similar soils within basins. Surface soils (root zone) changed from clay loam to fine sandy loam with significantly different infiltration characteristics and water-holding capacities.

similarity within the basins. Details of the system's design are given in the Appendix.)

(2) For spotty or random soil variability within a basin, undercut dissimilar streaks of soil occurring within a basin and replace with soil similar to the rest of the basin.

(3) For nonuniform soil within a root zone (e.g., lenses dissimilar to most of the profile), deep plow or rip the soil to attain mixing.

B. Land Grading

Land grading must be accurate for successful level-basin irrigation. The principle of level-basin irrigation is to allow all parts of the field to receive water for nearly equal periods of time; hence, high and low spots must be minimized. In addition to unequal distribution, low spots are subject to excessive inundation and high spots may become salty and dry prematurely.

1. *Rough Grading*

The primary earth-moving process may use traditional surveying procedures of staking and marking cuts and fills or more modern laser-controlled equipment. The amount of cut that the land can be subjected to will depend on the type of soil and the depth of the productive soil layers.

2. *Precision Finishing with Laser-Controlled Equipment*

In the past, conventional methods of surveying and equipment were used in an effort to achieve the leveling precision needed. The field was staked as many as three times on progressively smaller grids. Each staking was followed by soil being carefully moved by a highly skilled operator using some type of scraping equipment or a conventional land plane. The degree of precision attained varied with the ability of the equipment operator. Laser-controlled scrapers, used as finishing tools, were first introduced in the southwestern part of the United States in 1975. This method can indeed provide the precision needed to distribute water uniformly over level basins.

a. Laser Equipment and Field Procedures. The equipment necessary includes a tractor, scraper, and laser-control equipment (see Fig. 3). The tractor horsepower requirement is dependent upon the type of scraper used and the operating conditions. Four-wheel drive equipment is helpful but not essential. The finishing tools were initally confined to drag scrapers but have recently been expanded to include combination carryall scraper units. Both types of equipment work satisfactorily. The type selected will depend upon the particular farming unit.

The laser equipment includes four components: transmitter, receiver, control station, and interface unit. Mounted on a tall tripod, the transmitter generates a thin red laser beam which is rotated rapidly to form a virtual plane of light over the working area. The plane can either be level (for level basin) or sloping. Power is provided by a 12 V dc automotive battery, which is generally adequate for about 12 hr of continuous operation. The maximum range of the laser is about 300 m, provided that dust, wind, or heat do not become limiting.

The receiving unit mounted on the scraper references the scraper blade height from the laser plane of light. The receiver consists of a series of silicone photocells to sense the laser light position. The receiver then supplies elevation correc-

FIG. 3. Laser-controlled scrapers used for moderate grading and precision surface finishing. The laser transmitter is shown in the foreground and receivers are shown mounted on the scrapers. (Courtesy of Laserplane Corporation, Dayton, Ohio.)

tional signals to an electric valve which is connected to the control station mounted in the tractor cab, where the operator can monitor system functions. Generally the systems can be used to both level and survey. The mode of operation can be selected at the control station. The operator uses the survey mode to actually survey the field to be finished by observing the displayed elevation while driving the tractor in a systematic grid across the field with the scraper grazing the field's surface. Generally, periodic readings about 30 m apart are recorded on a field sketch (see Fig. 4). The elevation readings are averaged to obtain a zero cut and fill elevation, and this average—or one slightly lower to meet cut and fill requirements—is entered into the control station. The operator will rely on his experience with a particular soil and grading job when determining the adjustment. Operation is then switched to the level mode and the grading process is started.

The field sketch is also used as a guide in locating major cut–fill areas. Shading all areas above the average elevation reading can help the operator determine the most efficient way to grade the field (Fig. 4). The level mode strives to fix the distance from the beam of laser light to the finished field elevation. All electrical signals are interfaced to the tractor hydraulic system through the interface unit.

b. Precision of Finishing Operation. Topographical surveys of basins that had been conventionally leveled and others that had been leveled using laser-controlled equipment were presented by Dedrick (1979). Deviations from the mean elevation for all surveyed points within two 4.3-ha basins, one that was conventionally leveled (Fig. 5) and one that was leveled using laser-controlled equipment (Fig. 6) help illustrate the performance of the laser equipment. In the conventionally leveled basin, which was cropped to alfalfa, the edges were found to be high and the center low. This concave surface shape generally occurs when

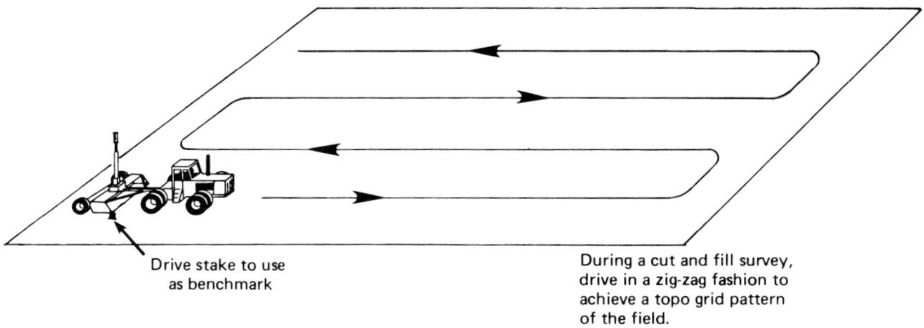

5.55	**5.52**	**5.54**	5.47	5.47	5.38	5.46	**5.54**	**5.57**	5.46	5.47	5.48	**5.57**
5.62	**5.50**	5.47	**5.52**	**5.53**	5.39	5.49	5.52	**5.51**	**5.53**	5.41	5.49	**5.52**
5.57	**5.59**	**5.66**	**5.51**	5.48	5.45	**5.57**	5.49	5.43	5.42	5.46	5.47	5.48
5.47	5.45	**5.58**	**5.50**	**5.53**	**5.57**	**5.54**	**5.59**	5.39	5.35	5.39	5.44	5.45
5.59	**5.54**	5.49	5.47	5.42	5.48	**5.59**	**5.67**	5.38	5.44	**5.50**	**5.56**	**5.57**
5.51	**5.56**	5.49	5.44	5.42	**5.50**	**5.59**	**5.64**	5.38	5.42	**5.51**	**5.54**	**5.57**
5.49	5.43	**5.52**	5.39	5.44	5.43	**5.55**	5.45	5.46	**5.54**	**5.56**	5.48	5.47
5.47	5.48	**5.54**	5.40	5.48	5.49	**5.61**	5.39	5.40	**5.53**	**5.51**	**5.61**	**5.54**
44.27	44.07	44.29	43.70	43.77	43.69	44.40	44.29	43.52	43.69	43.81	44.07	44.17

FIG. 4. The tractor/scraper unit is driven in a grid pattern to obtain a topographical map of the basin. Readings are recorded in a notebook by the operator, generally on a 30 m grid. The average elevation reading is then determined for the basin. The total number of readings was 104; the sum of all the readings was 571.74; thus the average elevation reading (AER), defined as the latter quantity divided by the former, was 571.74/104 = 5.50. (Courtesy of Laserplane Corporation, Dayton, Ohio.)

conventional land planes are used. With the laser-controlled equipment, the edges were not consistently the high areas of the basin.

The precision obtained by subsequent passes when grading basins using laser-controlled equipment is illustrated in Table I. These basins had been graded with laser-controlled equipment two years earlier and were producing sudan grass when surveyed prior to the leveling operation. In this particular case, once the rough leveling was completed (soil moved from the high spots to the low spots as determined from the initial survey map), an acceptable leveling job was attained. Scraping the basins again yielded little improvement.

Dedrick (1979) also presented numerical comparisons of the elevation deviations within basins graded conventionally and with laser-controlled equipment (Table II). Four deviation levels were considered. Leveling was consistently better with laser-controlled equipment. The Soil Conservation Service in southwestern Arizona requires that 80% of the field be within ±1.5 cm (±0.05 ft) of the mean before the grading job can be accepted. Neither of the conventionally

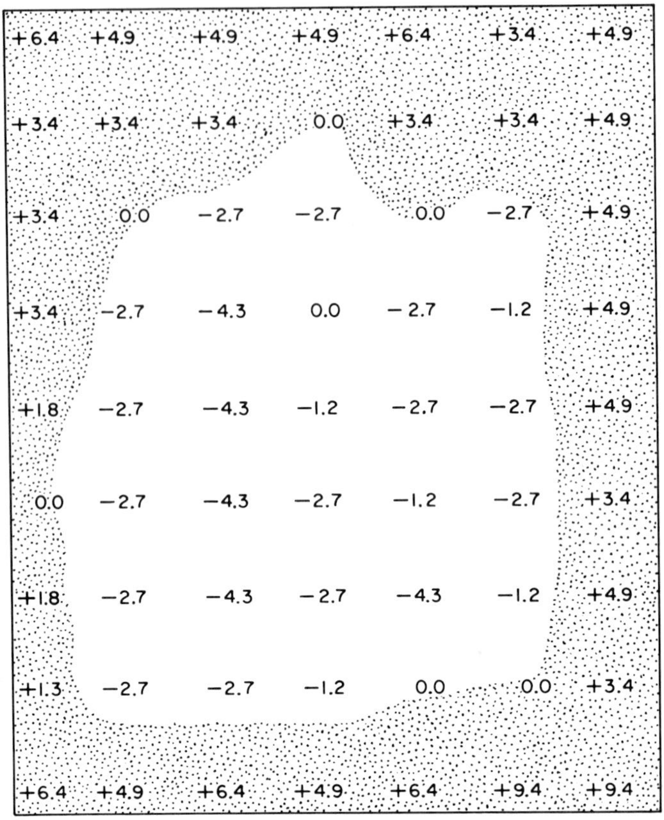

FIG. 5. Elevation deviations (cm) from the mean of a conventionally graded basin (nonlasered). Grid centers are approximately 30 m apart. Contour line represents zero deviation.

leveled basins met this requirement, whereas in nearly all instances the standard was met with the laser-controlled equipment. The leveling precision remained unchanged after the leveling operation even when the fields were tandem-disked, beds were prepared, and a crop of lettuce was planted and germinated on the beds. The average basin size studied was about 4 ha. These fields were considered to be typical of the general area and were selected by chance without prior coordination with the farmer.

c. Maintenance. The soil surfaces of the level basins require refinishing periodically to compensate for soil settlement associated with major cutting and filling and surface irregularities created by tillage operations. In many instances laser-controlled finishing after major land grading may not be justified until the land has been irrigated. The extent of these maintenance procedures will depend

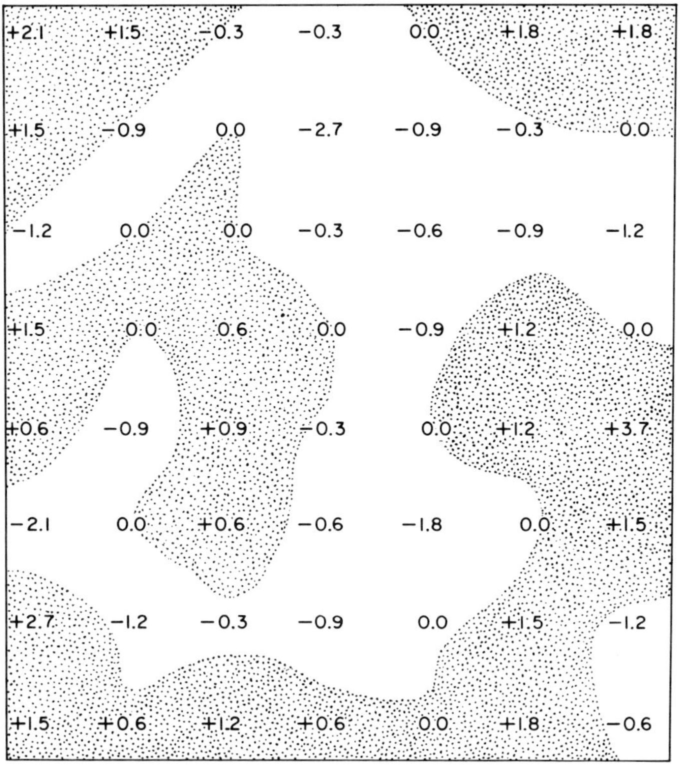

FIG. 6. Elevation deviations (cm) from the mean of a basin that had been graded using a laser-controlled drag scraper. Grid centers are approximately 30 m apart. Contour line represents zero deviation.

on several factors such as the type of tillage operation, time since major land-grading processes were completed, type of crop to be grown, value of crop to be grown, and susceptibility of the crop to damage caused by non-uniform water application. The time required to complete each leveling job depends on how level the field was initially.

3. *Cost of Land Grading*

Leveling costs are dependent on the amount of soil that must be moved, which in turn is related to field slope and basin size. Hence, level basins should be confined to generally flat terrain. Grading costs decrease with basin size, but other management factors must be considered. What the farmer can afford depends on the economics of the entire development. Hinz and Halderman (1978) prepared a pamphlet which illustrated the costs of leveling land with laser-

TABLE I

PORTION (%) OF 4-ha BASINS WITHIN VARIOUS ELEVATION DEVIATIONS FROM THE MEAN FOR SUBSEQUENT PASSES WITH A LASER-CONTROLLED DRAG SCRAPER

Basin	Deviation from mean elevation (cm)			
	±0.75	±1.5[a]	±2.25	±3
Basins leveled with laser-controlled equipment two years earlier:				
1	45	63	78	86
2	43	62	75	86
Mean	44	62	76	86
Rough grading, mainly moving high areas to low areas:				
1	55	78	90	86
2	64	86	94	98
Mean	60	82	92	97
Final grading, scraping entire basin in systematic fashion:				
1	65	86	94	98
2	66	84	95	99
Mean	65	85	94	98

[a] Soil Conservation Service (SCS) standards require that 80% of a basin be within ±1.5 cm of the mean.

controlled equipment. The volume of earth to be moved is the important factor, with the final cost depending upon the time and the location at which the leveling must be done. Finishing costs using laser-controlled equipment also depend on how level the field is when the leveling is started as well as on prevailing prices in the particular region. Rate of completion has ranged from about 1/2 to 1 ha/hr on level basins in the southwest part of the United States. The higher rates are associated with fields that had previously been graded with laser-controlled equipment, whereas the lower rates occur when basins have not been leveled with laser-controlled equipment before.

4. *Reclamation of Deeply Cut Areas*

If soil cutting has been exessive, the exposed surface material may be non-productive and hence must be reclaimed. Crop response and productivity can also be affected in fill areas due to soil dispersion and/or compaction from the grading operation. Reclamation procedures used have included the use of green manure crops, heavy manuring, addition of various fertilizers, chiseling or ripping, deep plowing, and subsequent leaching. The seriousness of the problem and the length of the required reclamation period depend upon the particular site and soil conditions, and should be considered before grading begins.

TABLE II

PORTION (%) OF BASIN WITHIN VARIOUS ELEVATION DEVIATIONS FROM THE MEAN FOR DIFFERENT CROPS GROWN ON BASINS GRADED EITHER BY CONVENTIONAL OR LASER-CONTROLLED TECHNIQUES

Grading technique	Crop	Basin	Deviation from mean elevation (cm) ±0.75	±1.5[a]	±2.25	±3.0
Conventional	Wheat	1	34	64	70	84
		2	45	70	79	95
		3	47	58	80	88
		4	52	63	82	85
		Mean	44	64	78	88
	Alfalfa	1	40	55	71	77
		2	34	50	70	77
		3	44	58	78	88
		4	18	27	48	73
		Mean	34	48	67	79
Laser/drag scraper	None, irrigated	1	66	84	91	96
		2	64	87	95	100
		Mean	65	86	93	98
	Tandem-disked twice	1	62	81	90	98
		2	62	85	95	98
		Mean	62	83	92	98
	Lettuce beds, after plant emergence	1	67	89	95	98
		2	62	83	95	98
		Mean	64	86	95	98

[a] SCS standards require that 80% of a basin be within ±1.5 cm of the mean.

III. Design and Evaluation

A. INTRODUCTION

In a level-basin irrigation system, water first moves over the basin and then becomes static. To attain uniformity of penetration, the water is generally applied so as to cover the field relatively quickly, and it is applied on a volume basis. A major portion of the water infiltrates during the static flooding period. This is different from the pattern in sloping border irrigation, in which a balancing of flow rate and application time is necessary to provide uniform and adequate coverage. For sloping irrigation techniques without static ponding, all infiltration occurs while the water is moving (i.e., dynamic).

Design of level basins involves sizing the basins so that a desired distribution uniformity is met for the particular volume of water to be applied (depth of application). This requires balancing soil infiltration characteristics, available flow rate, and basin size to yield the design limitations on uniformity. The tendency is to use the largest stream available, but beyond certain limiting conditions (to be discussed later) increased flow rates no longer help to increase uniformity or to allow longer basins. In addition, these high flow rates may cause excessive surface water depths requiring larger control barriers (dikes).

The application efficiency of a level-basin system is primarily a measure of the amount of water that percolates beyond the root zone of the crop. The amount of water infiltrated in a level basin is greatest where the water is turned in and least at the far end of the basin. Recession is the time when all water applied infiltrates into the soil. For level basins, it is reasonable to assume that recession occurs at the same time throughout a basin. Sample advance and recession curves for level basins are shown in Fig. 7. The time for water to advance to the end of the basin is designated t_t, and the time of recession t_r. The difference between t_r and t_t is the infiltration opportunity time at the end of the basin t_n.

The application efficiency E_a is defined as the average depth of water stored in the root zone divided by the average or gross depth Z_g of water applied (Kruse, 1978). The depth infiltrated Z_n at the end of the basin divided by the average or gross depth applied was defined as the distribution uniformity (DU) by Clemmens and Dedrick (1981a) and is similar in form to that defined by Kruse (1978). Erie and Dedrick (1979) referred to the term as distribution efficiency. Different situations of interest, comparing application efficiency and distribution uniformity, are shown in Fig. 8, where Z_d is the soil moisture deficit. When the entire field has been overirrigated, the value for distribution uniformity is larger than that for application efficiency (Fig. 8a). When the minimum depth infiltrated is equal to the desired depth, the application efficiency has the same value as the distribution uniformity (Fig. 8b). When part of the field has been underirrigated, the value for application efficiency is greater than that for the distribution uniformity but less than the desired efficiency Z_d/Z_g by the relative amount of underirrigation (Fig. 8c). For level-basin design, the conditions of Fig. 8b are assumed, i.e., the desired depth of application is the amount of water

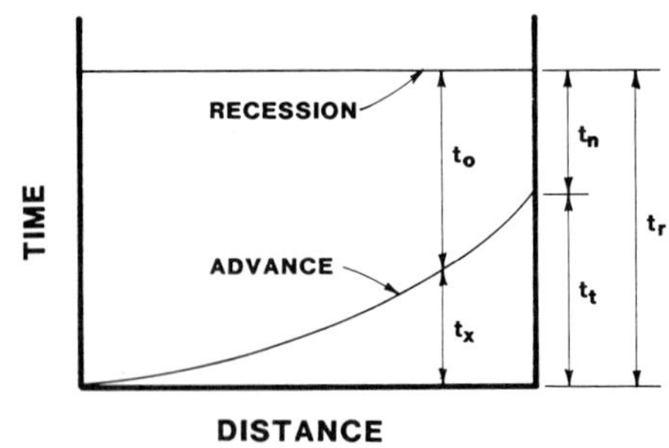

FIG. 7. Definition sketch of advance and recession curves for a level basin.

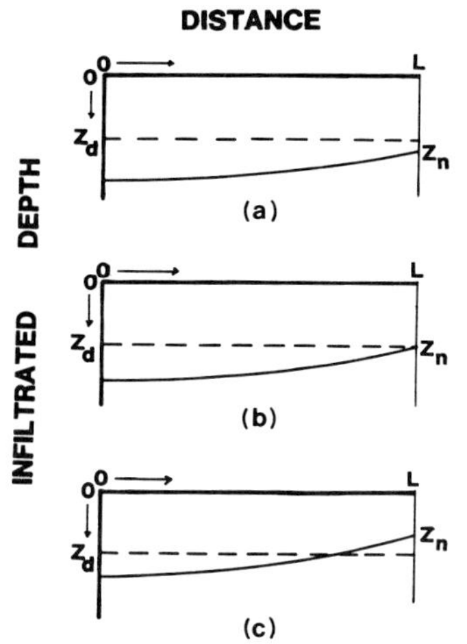

FIG. 8. Profiles of infiltrated water: (a) for overirrigation throughout a basin, $Z_n > Z_d$, DU = Z_n/Z_g, $E_a = Z_d/Z_g$, DU > E_a; (b) for adequate irrigation with $Z_n = Z_d$, DU = $E_a = Z_n/Z_g$; and (c) for slight underirrigation over part of the field, $Z_n < Z_d$, DU = Z_n/Z_g, DU < E_a < Z_d/Z_g.

needed to fill the root zone. If overirrigation or underirrigation occur, this is generally the result of improper management. Herein distribution uniformity will be used for design purposes. None of the design procedures consider spatial variability in soil infiltration conditions, soil water holding capacities, or soil surface elevations.

B. DESIGN

1. *Infiltration*

Distribution uniformity (DU) of a level basin can be functionally described as:

$$\text{DU} = f(q_{\text{in}}, t_{\text{co}}, L, n, I) \tag{1}$$

where q_{in} is the inflow rate per unit width, t_{co} is the application time, L is the basin length, n is the flow resistance (Manning n), and I is the infiltration characteristic. In most instances the infiltration characteristic of the soil [relation between depth infiltrated and time, symbolized by I in Eq. (1)] can be reasonably described by a power function:

$$Z = k\tau^a \quad \text{(cumulative)} \qquad \text{or} \qquad i = ak\tau^{a-1} \quad \text{(instantaneous rate)} \tag{2}$$

where Z is the cumulative infiltration, τ is the infiltration opportunity time, k is a constant, and a is the exponent. Occasionally the infiltration does not follow a power function throughout the irrigation period but approaches a constant infiltration rate. Infiltration could be described in these cases by:

$$Z = k\tau^a + i_f\tau, \qquad i = ak\tau^{a-1} + i_f \tag{3}$$

or by a branch function where the infiltration follows a power function until it reaches some time value τ_f, after which time it remains constant at that value:

$$Z = (1 - \backslash a)k\tau_f^a + i_f\tau \qquad \text{with} \quad i = i_f \quad \text{when} \quad t \geqslant \tau_f \tag{4}$$

where i_f is the "final" constant infiltration rate and τ_f is the time at which infiltration rate becomes constant. All other terms are as defined earlier. For design purposes Eq. (2) is generally used.

The Soil Conservation Service (SCS) of the U.S. Department of Agriculture has published a handbook for the design of border irrigation systems (U.S. Department of Agriculture, 1974). Therein they introduce the concept of intake families, which can be described by an equation of the form

$$Z = k\tau^a + b, \qquad i = ak\tau^{a-1} \tag{5}$$

where b is a constant and all other terms are as defined previously. They discuss some eight intake families which are assigned numbers such as 0.1 and 0.3 through 4.0. These approximately represent the basic intake rate of the soil in inches per hour.

Infiltration characteristics for level basins are usually measured with ring infiltrometers. Procedures for using ring infiltrometers have been presented by Erie (1962). Merriam (1971) and Clemmens (1981) have presented procedures for adjusting ring infiltration data to account for the effects of the ring. The procedures require that the infiltration data be collected in conjunction with an irrigation. The basic procedure provides for adjusting the cumulative infiltration data to reflect the quantity of water actually infiltrated over the area of the basin (i.e., an adjustment of the cumulative infiltration versus time curve to match a known infiltrated quantity of water after a particular time). The interested reader is directed to these publications for specific details of using ring infiltrometers.

2. *Resistance to Flow*

When designing level basins a form of the Manning's equation has commonly been used. An important factor is the roughness coefficient n. This coefficient expresses the flow-resistance effects of different hydraulic boundary conditions. Soil surface conditions, cultivation practices, and various crop factors cause flow resistance differences. The SCS (U.S. Department of the Agriculture, 1974) has recommended values of 0.04 for smooth, bare soil surfaces and row crops irrigated by the level-basin method; 0.10 for drilled small-grain crops if the drill rows run in the direction of water flow; 0.15 for alfalfa, mint, broadcast small grains; and 0.25 for dense sod crops and small grains drilled perpendicular to the flow. The n value selected will depend on the operational requirements of the system. Clemmens (1979) in several instances used the SCS recommended values of n and found them to be satisfactory when used to verify the zero-inertia model for border irrigation systems.

3. *SCS Design*

In the SCS design manual for border irrigation (U.S. Department of Agriculture, 1974), a series of design charts are presented for level basins. The design charts relate unit flow rate, basin length, depth of flow, and time of application to application efficiency (distribution uniformity). A range of Manning roughnesses (0.04, 0.15, and 0.25); soil infiltration families (0.1, 0.3, 0.5, 1.0, 1.5, 2.0, 3.0, and 4.0); and net depths of application (F_n in their charts is Z_n in this article) are presented in the charts. The design chart for $n = 0.15$, an intake family $= 0.5$, and $Z_n = 7.6$ cm is shown in Fig. 9. The design charts have been successfully used to design level-basin systems in the Wellton–Mohawk Irrigation and Drainage District in Arizona for most conditions. In some instances the design did not yield the predicted results, and these discrepancies may be due to a number of factors and can best be highlighted by considering some of the basic assumptions used in development of the design charts:

(1) The time of application equals or exceeds the time of water advance. Generally high efficiencies are attained if the incoming stream is cut off before the entire basin has been covered.

(2) The hydraulic slope equals the maximum flow depth divided by the basin length. This is an unrealistic assumption.

(3) Infiltration characteristics of all soils can be fitted into the intake family concept. Values of a found in the literature on surface irrigation range from less than 0.10 to greater than 0.80 (Clemmens, 1979). When the intake family curves are fitted with a power function through infiltration depths of 5.08 and 10.16 cm, the exponent a ranges from 0.595 to 0.740 (Clemmens, 1980), a much narrower range than that found in practice.

The charts are easy to use. Referring to Fig. 9 for a basin length of 200 m to attain an application efficiency (distribution uniformity) of 80%, the required unit flow rate q_{in} would be 3.3 liters/sec m^{-1}, the application time t_{co}, would be 96 min, and depth of flow would be 10.4 cm. If the total flow were 150 liters/sec, the basin width would be about 45 m. To increase the basin length to 400 m with the same distributon uniformity, the corresponding values would be $q_{in} = 11.2$ liters/sec m^{-1}, $t_{co} = 55$ min, and depth of flow would be 18.8 cm. For the same flow rate to the basin, the basin width would be decreased from 45 to 13 m. Depth of flow may become a limiting factor depending on how the basin borders are to be constructed.

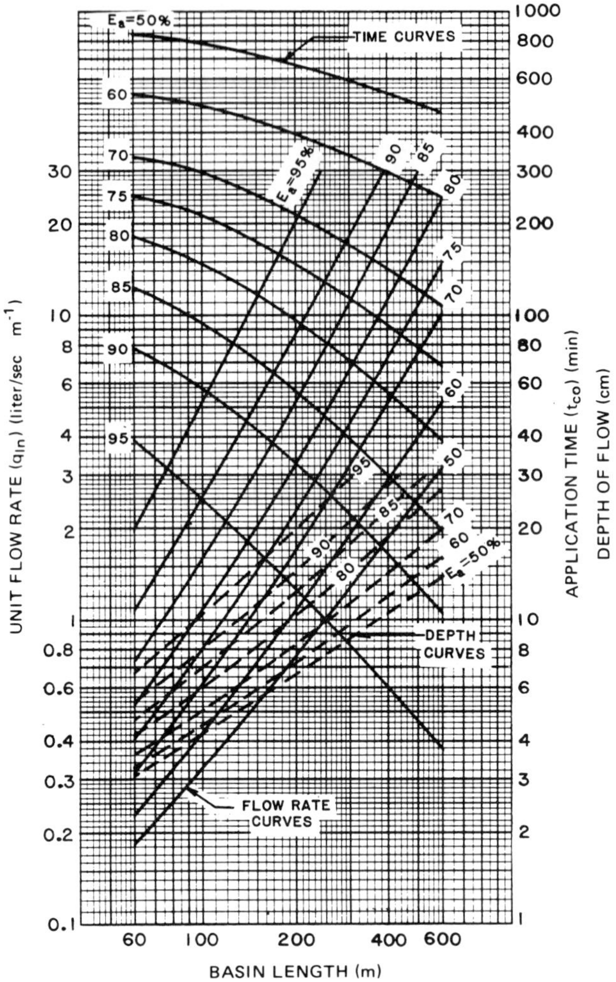

FIG. 9. SCS design chart for a 7.6 cm net depth of application, roughness coefficient n of 0.15 and an intake family of 0.5.

4. *Theoretical Design*

Theoretically based level-basin design procedures have been developed and verified since the SCS design manual was published in 1974. The ability to predict border irrigation advance and recession with the zero-inertia model (Strelkoff and Katopodes, 1977) has been the key to this improved design development. The model is based on the fact that flow velocities in surface irrigation are relatively low, and hence inertial terms in the mathematical expressions for continuity and momentum can be neglected with essentially no loss of accuracy (Katopodes and Strelkoff, 1977). The zero-inertia model was verified by Clemmens (1979) using several advance–recession data sets obtained by different researchers.

Dimensional analysis techniques have been used to display advance (Clemmens and Strelkoff, 1979), depth of flow (Clemmens, 1980), and design information (Clemmens *et al.* 1981; Clemmens and Dedrick, 1982) for level basins. Reference terms for flow rate Q, distance X, and depth of flow Y were used to relate dimensionless variables to actual dimensional variables for use in design. The dimensionless to dimensional relations are: $q^*_{in} = q_{in}/Q$, $L^* = L/X$, and $Y_{max} = Y^*_{max}/Y$.

When water flows into a level basin, the flow depth rises, quickly at first and then at a decreasing rate until the stream is cut off. The maximum depth of flow occurs at the time of cutoff. This maximum flow depth was discussed by Clemmens (1980) and is presented here in Fig. 10. The reference value for Y to use with this figure is found from

$$Y = (n/C_u)^{3/8} q_{in}^{9/16} t_{co}^{3/16} \tag{6}$$

where C_u is the unit coefficient ($1.0\ m^{1/2}$/sec in the SI system and $1.486\ ft^{1/2}$/sec in the English system). The dimensionless infiltration parameter K^* was defined as

$$K^* = k t_{co}^{(a-3/16)} / [q_{in}^{9/16} (n/C_u)^{3/8}] \tag{7}$$

The solution assumes that the advancing stream has not reached the end of the field at time of cutoff, which is a valid assumption. To find the maximum water depth in a basin, calculate Y and K^* from Eqs. (6) and (7) and then find a value for Y^*_{max} from Fig. 10 that corresponds to the calculated K^* value and appropriate a value. The maximum depth is then

$$Y_{max} = Y Y^*_{max} \tag{8}$$

The procedure for developing level-basin design charts was presented by Clemmens *et al.* (1981). The solution allowed for continuous variation in desired depth of application, the time required to infiltrate the desired application depth, and the Manning roughness coefficient for a wide range of distribution uniformities. A single design chart represented the solution for one value of the infiltration exponent a. Charts for a range of commonly found a values (0.1, 0.3, 0.4, 0.5, 0.6, 0.7, 0.8, and 1.0) were published by Clemmens and Dedrick (1982). The chart for $a = 0.7$ is shown in Fig. 11.

Reference values for X and Q were defined as:

$$X = C_1 Z_n^{7/9} \tau_n^{2/3} / n^{2/3} \tag{9}$$

$$Q = C_2 X Z_n / \tau_n \tag{10}$$

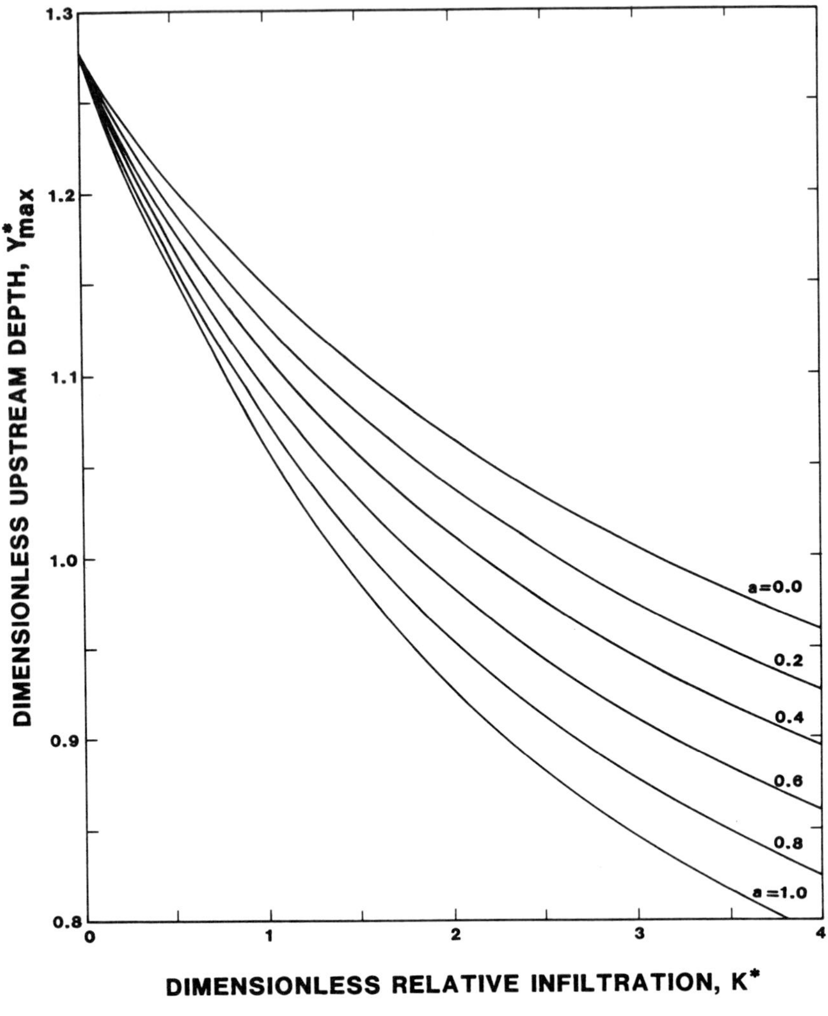

FIG. 10. Dimensionless upstream depth for level basins.

where Z_n is the net or desired depth of application, τ_n is the opportunity time required to infiltrate Z_n, and C_1 and C_2 are units conversion factors. With Z_n in centimeters and τ_n in minutes, $C_1 = 0.426$ and $C_2 = 1/6$ are used to give SI units with X in meters and Q in liters/sec m^{-1}. For English units, $C_1 = 2.89$ and $C_2 = 1/720$ when Z_n is in inches, τ_n in minutes, X in feet, and Q in ft^3/sec ft^{-1}.

At large values of unit flow rate, the curves approach vertical, indicating that increasing the unit flow rate will not allow a proportionate increase in the basin

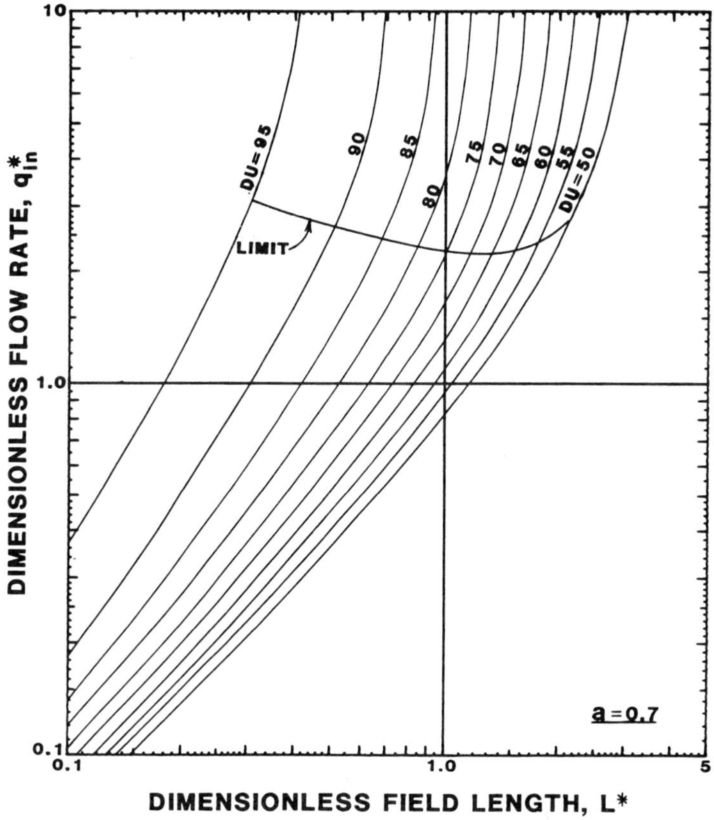

FIG. 11. Dimensionless design curve for $a = 0.7$.

length while maintaining the same uniformity. This is an important fact, indicating that there is a point beyond which it is no longer practical to increase the unit flow rate and basin length. This limit is shown on each chart. The SCS curves like Fig. 9 do not show the vertical trend as markedly and may be the cause of some unsatisfactory design results. The steep portion of the curves in Fig. 11 should be avoided.

The procedure for using Fig. 11 is as follows:

(1) Determine net depth Z_n to apply.

(2) Determine appropriate constants to describe infiltration. The exponent a of the infiltration function and the time τ_n required to infiltrate the net depth Z_n should be determined.

(3) Select appropriate value for roughness.

(4) Select the distribution uniformity (application efficiency) value desired.

(5) Calculate reference values for basin length X and unit flow rate Q using Eqs. (9) and (10).

(6) Determine the maximum basin length by selecting L^* from Fig. 11 at the point where the limit line and distribution uniformity curve intersect. The maximum basin length $L_{max} = XL^*_{max}$. The actual basin length L selected should fit the geometry of the field with L_{max} being used as an approximate limiting condition.

(7) Determine the minimum unit flow rate q_{in} by using the chart again to read q^*_{in} corresponding to the selected basin length L and the desired uniformity. The $q_{in} = Qq^*_{in}$.

(8) Determine the basin width W_b based on the available flow rate Q_{in}. Basin width $W_b = Q_{in}/q_{in}$ and the basin area $A = LW_b$.

The design factors used in the example of Section III,B,3 were $Z_n = 7.6$ cm, $n = 0.15$, an intake family of 0.5, $\tau_n = 217$ min, DU = 80%, and $L = 200$ m. When the 0.5 intake family is fitted to a power function through the points 5.08 and 10.16 cm, the value of a is 0.684 (use $a = 0.7$ in the design) and k is 3.162 cm/hr^a. What is the unit flow rate?

From Eq. (9),

$$X = \frac{C_1 Z_n^{7/9} \tau^{2/3}}{n^{2/3}} = \frac{(0.426)(7.6)^{7/9}(217)^{2/3}}{(0.15)^{2/3}} = 264 \text{ m}$$

and from Eq. (10),

$$Q = \frac{C_2 X Z_n}{\tau_n} = \frac{(264)(7.6)}{(6)(217)} = 1.54 \text{ liters/sec m}^{-1}$$

$$L^* = L/X = 200/264 = 0.76$$

From Fig. 11 with $a = 0.7$ and $L^* = 0.76$, $q^*_{in} = 1.9$ at 80% uniformity and $q_{in} = q^*_{in}\, Q = (1.9)\,(1.54) = 2.9$ liters/sec m^{-1}. This unit flow rate was slightly lower than that found using the SCS design chart of Fig. 9. More important, the design indicates that the basin length should not exceed about 225 m and that an 80% uniformity is unattainable for a 400-m basin regardless of the unit flow rate, i.e., for $L = 400$ m, $L^* = L/X = 400/264 = 1.52$ where the highest degree of uniformity attainable would be approximately 62%.

The maximum water depth for the 200-m basin would be determined as follows: The application time or time of cutoff t_{co} is 1.8 hr (6460 sec). From Eq. (6) with units of m^3/sec m^{-1} for q_{in} and sec for t_{co},

$$\begin{aligned} Y &= (n/C_u)^{3/8} q_{in}^{9/16} t_{co}^{3/16} \\ &= (0.15)^{3/8}\,(0.0029)^{9/16}\,(6460)^{3/16} \\ &= 0.095 \text{ m} \quad \text{or} \quad 9.5 \text{ cm} \end{aligned}$$

and using Eq. (7) with units in meters and seconds, $k = 1.168 \times 10^{-4}$ m/sec^a,

$$K^* = \frac{k t_{co}^{(a-3/16)}}{q_{in}^{9/16}(n/C_u)^{3/8}}$$

$$K^* = \frac{(1.168 \times 10^{-4})(6460)^{(0.684-3/16)}}{(0.0029)^{9/16}(0.15)^{3/8}} = 0.50$$

From Fig. 10 with $K^* = 0.50$ and $a = 0.684$, $Y^*_{max} = 1.17$. The maximum depth, from Eq. (8) is $Y_{max} = YY^*_{max} = (1.17)\ (9.5) = 11.1$ cm (in SCS design $Y_{max} = 10.4$ cm).

The design procedures are illustrated further in the Appendix for two significantly different soils within the 33-ha field of Fig. 2.

C. Evaluation

Once a level-basin system is operational the actual performance should be assessed. The evaluation should be based on actual field measurements under the conditions and practices normally used. The evaluation will indicate if the system was properly designed, if the system is being properly managed, and what management changes might be necessary to attain acceptable performance. The field measurements required will depend on what aspects of the system performance are to be evaluated. Procedures for detailed evaluation schemes have been published by Criddle *et al.* (1956), Merriam (1968), and Merriam and Keller (1978). Procedures for estimating the distribution uniformity of level basins have been published by Clemmens and Dedrick (1981) and the SCS (U.S. Department of Agriculture, 1974).

1. *Detailed Evaluation*

Specific information needed for a detailed evaluation of a level-basin system includes soil moisture deficiency prior to irrigation, flow rate, water application time, rate of advance, recession time, cumulative infiltration, and adequacy of irrigation. A detailed outline of the field procedure was presented by Merriam and Keller (1978) and should be consulted for more information. Erie and Dedrick (1979) presented two examples of using advance–recession–infiltration data to evaluate the water distribution within basins and the distribution uniformity (called *distributions efficiency* in their bulletin) for two differently shaped basins. The procedures should include a comparison of the soil moisture deficit and maximum allowable moisture deficit, to assess whether the irrigation was too early, too late, or correctly timed, and last, whether the depth of water actually applied was as planned. A sketch of the advancing water front and sketches of the water recession on a map of the basin will help evaluate the adequacy of the land-leveling operation and the effect of high and low areas on the actual distribution of water over the basin. A detailed evaluation of soil moisture levels before and after an irrigation can be used to determine how much water was actually applied at different points within the field or if the soil profile was filled to capacity.

Determining the distribution uniformity of an irrigation can be helpful in evaluating the hydraulic design and/or irrigation management procedures. As discussed earlier and shown in Fig. 8, a high distribution uniformity does not necessarily correspond to a high application efficiency. The application efficiency must be evaluated based on knowledge of the available root zone storage. An evaluation of distribution uniformity based on observation of surface flow could be used to determine whether problems are related to hydraulic design, management, soil surface irregularities, or spatial variability of soil properties.

2. *Distribution Uniformity Evaluation*

If distribution uniformity is the only factor to be evaluated, then the procedures developed by Clemmens and Dedrick (1981) can be used. The procedure to use depends upon the infiltration characteristics of the soil.

a. Infiltration and Power Function. For soils whose infiltration can reasonably be described by a power function [Eq. (2)], the distribution uniformity is a function of the ratio of advance time t_t to the time it takes the water to infiltrate at the far end of the basin t_n, the exponents of the infiltration equation [Eq. (2)], and the water advance (power) function, in which

$$x = gt_x^h \tag{11}$$

where x is the advance distance, t_x is the time required for water to advance a distance x, g is a constant, and h is the exponent.

The distribution uniformity can be estimated from Fig. 12. The advance exponent h has a relatively minor effect on the distribution uniformity compared to the infiltration exponent a. A value of $h = 0.5$ was used in Fig. 12. For h values greater than 0.5 the corresponding distribution uniformity would be slightly higher, and it would be slightly lower for h values less than 0.5. If the entire basin is covered in one-fourth to one-half the time needed to fully irrigate, then the distribution uniformity will range from 80% to over 95% (Fig. 12). To assure acceptable distribution, the time ratio should not exceed 1. In an example given by Clemmens and Dedrick (1981) the following factors were either measured or developed from observing the basin when irrigated: advance time $t_t = 80$ min, recession time $t_r = 200$ min, advance exponent $h = 0.76$, and infiltration exponent $a = 0.6$. Then $t_n = t_r - t_t = 200 - 80 = 120$ min and $t_t/t_n = 80/120 = 0.67$. From Fig. 12 the distribution uniformity is seen to be 80%. Since h was greater than 0.5 the distribution uniformity should be slightly higher. In their example the authors estimated the actual distribution uniformity to be 82%.

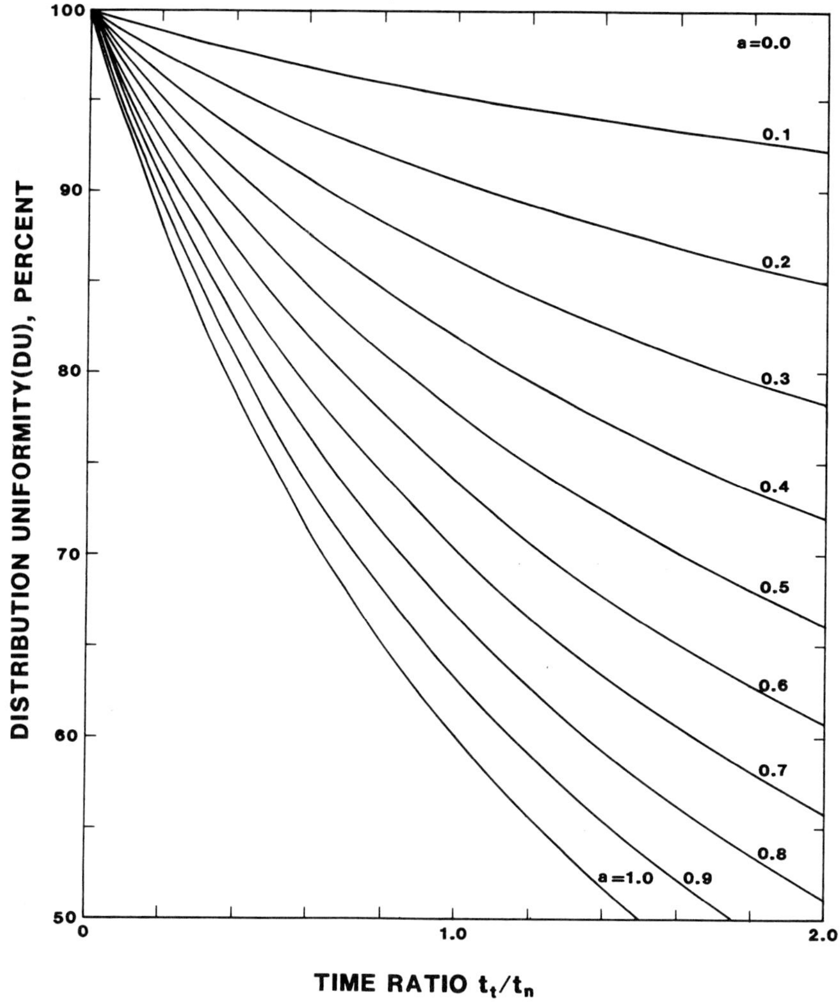

FIG. 12. Distribution uniformity versus time ratios for an advance exponent h of 0.5 and full range of infiltration exponents a.

b. Infiltration and Branch Function. For soils in which the infiltration rate becomes constant, rather than follow the power function until infiltration is complete [Eq. (4)], we can estimate the distribution uniformity as a function of the advance exponent h, the advance time t_t, the final infiltration rate i_f, and the gross depth of water infiltrated Z_g. The distribution uniformity DU can be calculated directly by:

$$\mathrm{DU} = 1 - i_f t_t / Z_g(1 + h) \tag{12}$$

provided that the infiltration time at the far end of the field is greater than the time required to reach the final infiltration rate. Values for h and t_t can be determined by observation of the water advance. The gross depth applied can be found by measuring the volume of water applied $q_{in} \times t_{co}$ and dividing by the field area. The final infiltration rate can be approximated by the rate of fall of the water surface during the latter stages of the irrigation.

In Erie and Dedrick's (1979) advance–recession examples, the infiltration fitted the branch function with a final infiltration rate $i_f = 0.43$ cm/hr, an advance time $t_t = 4.3$ hr, an advance exponent $h = 0.53$, and with a gross depth applied $Z_g = 13.2$ cm. From Eq. (12),

$$\mathrm{DU} = 1 - \frac{i_f t_t}{Z_g(1 + h)} = 1 - \frac{(0.43)(4.3)}{(13.2)(1 + 0.53)} = 0.91$$

The minimum depth infiltrated was about 11.9 cm.

D. Delivery of Water to the Field

Any design must consider the water supply flow rate, distribution system, and methods for delivering water to the field. Various types of turnout techniques are used to turn water into level basins from either open ditches or underground water supplies. Siphon tubes can be used to turn water out from open channels to level basins, provided that the total flow rate is not too large. This technique is used in central Arizona, where flow rates are commonly 50–120 liters/sec. When larger flow rates are used, water is generally distributed from concrete-lined canals to the level basins from either single or multiple outlets. Various types of multiple outlets are currently being used, with slide-type gates being very common. Single outlets are generally equipped with lift gates. If water is distributed to the basins in underground pipes, caps (gates) are used on the end of the pipe rising from the main line to the field surface. The types of turnout to use is a decision generally related to the comparative cost and the owner's personal choice. Some structures, however, are better adapted to row crop irrigation than others and some require less elaborate erosion control techniques. Single or multiple outlets inherently reduce the labor requirements of irrigating. Distribution of the water into the rows or beds can be adequately accomplished with various types of secondary ditches regardless of the number of turnouts (see Section IV,B). The use of siphon tubes has the advantage that they are portable and that the basin size can be changed without disrupting a permanent turnout system. Siphon tubes do lose part of the effectiveness for which they were designed, in that water need not be directed into each individual furrow of a row crop when grown on a level basin.

Lift gates require only one maneuver and minimum labor to direct the entire stream into a basin or to cut it off. These gates can also be designed into a system so that four gates can be located a short distance from each other and still irrigate four separate basins from adjacent corners. Distribution from basin corners can eliminate the supply ditch along the edge of the last basin (at the end of the farm ditch), thus saving considerable cost. Single outlets at the corners can be conveniently automatically controlled. When located in the corner of the basin, the erosion control structures are protected from breakage by machinery.

Scarred areas of potholes may occur where the water is turned into the basin. To prevent this type of erosion, control structures are established in conjunction with the gate so as to create a low discharge velocity or to dissipate the energy of the water before it scours the soil surface. Two structures that have successfully been used to control erosion are illustrated in Figs. 13 and 14. The key parts in these structures are the solid wall in Fig. 13 and the concrete blocks in Fig. 14. The second most important part of each structure is the apron area, which is a strip across the front of the structure that is both level and located below the finished field elevation. In the case of the structure with the solid cutoff wall, the

FIG. 13. Erosion control structure in which the water ponds behind the solid cutoff wall, then cascades over the wall and impinges on the apron. The lift gate is used to turn the water into the basin.

FIG. 14. Erosion control structure in which the water is slowed by the blocks in the foreground. A key part of this structure as well as that shown in Fig. 13 is the apron which is about 1 m wide, level, and below the finished basin elevation. By being below the field surface, the water cannot drop and cause erosion. The lift gates shown are automatically activated by means of air cylinders.

water is spread evenly behind the wall as it is turned out of the gate, and it falls onto the apron, where it goes through a hydraulic jump and then moves upward onto the soil surface. The concrete blocks of the second structure resist the flow of water and cause the water to slow down before moving onto the soil. The width of the structure is related to the flow rate and should be about 1 m per 90 liters/sec of flow. Hence, if the flow rate is 360 liters/sec, then the width should be about 4 m. This guidline is somewhat dependent upon the soil onto which the water is flowing, but it generally gives satisfactory results. The solid cutoff wall structure probably is the superior erosion control structure, but it is somewhat more difficult to construct since the wall must be level and at the proper elevation to work effectively without creating too great a backwater effect into the canal, which would reduce delivery capacity. In contrast, the concrete-block structure is easier to construct, and the placement of the concrete blocks is not critical. The main disadvantage of this energy-dissipating block-type structure is that more maintenance is required, since debris tends to collect on the blocks and restrict the flow. Once flow has been restricted, erosion may occur since the effective width of the structure has been decreased. The structure with the solid wall is essentially self-cleaning.

IV. Management

A. Introduction

Management decisions of when and how much water to apply must be answered for all irrigation methods. A perfectly designed irrigation system is no remedy for poor management decisions. Properly designed, a level basin permits convenient control and distribution of the water in the required amounts. Water turned into a level basin tends to level out and cover all parts of the basin. Management factors to be considered are basin area, flow rate, desired depth of application, and the allowable degree of distribution nonuniformity. Basin area (actual cropped area) can be determined from direct measurement. Flow rate available to a basin must and can be accurately measured. Flow-metering devices are the subject of another article.[1] The quantity of water to apply can be determined from crop requirements or from evapotranspiration projections, water-holding characteristics of the soil, residual moisture levels in the soil, and the crop's effective root zone. Soil- or crop-related differences may exist within a field. In Fig. 2, soil differences (e.g., infiltration rates and water retention capacities) must be considered in both design and management. If different crops were produced concurrently within the field, the frequency and quantity of each irrigation could be affected even further. The water distribution difference from one end of a basin to the other can be estimated from available design information or from actual field evaluation (Section III,C). Once the gross application depth is determined for a level-basin system, the irrigation becomes merely a matter of supplying each basin with water for a certain calculated period of time as long as the flow rate is constant.

In level-basin irrigation, the water applied remains on the surface until it either penetrates into the soil or evaporates. In some areas and with some plants, the manager must be careful not to inundate a crop too long. How long a plant root zone can remain saturated without adversely affecting the plant depends principally on temperature, type of plant, and growth stage. Applying the correct quantity of water can be a paramount importance. Alfalfa, corn, safflower, and many vegetables are particularly sensitive to excessive inundation.

When germinating crops on beds it is undesirable to inundate the seed row. The irrigator should apply only enough water to approach the seed line so as to move moisture and salts past the seed zone. In this way soil crusting and salinity can be minimized.

[1]This volume, article by Replogle and Bos, p. 147.

B. Cultural Practices

As we stated earlier, any crop can be grown with level-basin irrigation. Normally, close-growing crops such as alfalfa, hay, pasture, and small grains are simply flood irrigated, and the level-basin feature provides adequate water distribution. With row or bed crops, however, secondary (temporary) ditches are generally needed at both ends of the furrows. Installing spiles or siphon tubes from a secondary ditch to individual furrows is not necessary for level-basin systems, since all parts of the basin are essentially at the same elevation and water will naturally cover all accessible furrows. Water may move faster along some furrows than others, but upon reaching the end, the small secondary ditch will be filled and water will flow back toward the inlet in furrows where water advances more slowly (Fig. 15). The net result is improved uniformity of application. If large streams of water are diverted through single turnouts into basins with row or bed crops, the onrushing water will overtop and erode the rows or beds in the immediate vicinity of the turnout. To prevent this overtopping, the stream usually must be split or diverted. Erosion and overtopping are minimized but not eliminated when multiple outlets are used.

FIG. 15. Water is being turned into this level basin by multiple outlets at the far end of the basin. Water advances rapidly through some furrows, fills the small secondary ditch in the foreground, and turns and flows back toward the inlet in furrows where the water advanced slowly.

FIG. 16. Secondary ditches are used to divert water from multiple or single turnouts (single turnout in background) into furrows. The secondary ditch must be large enough for the water to be nonerosive past the end of the furrows. A smaller secondary ditch should also be constructed at the far end of the furrows.

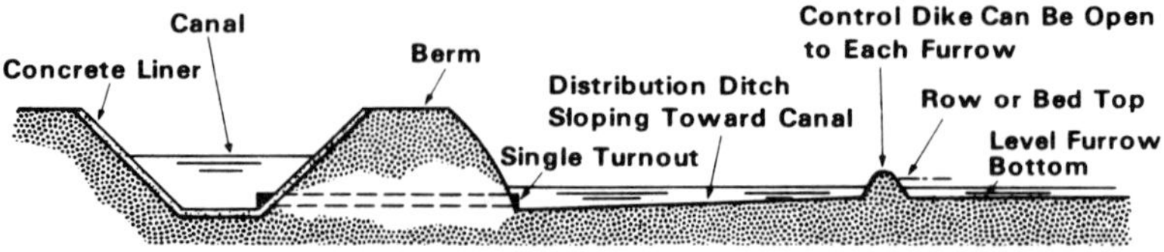

FIG. 17. Cross section of canal and basin showing distribution ditch, usually no less than 3–4 m wide, which assists in maintaining a constant water elevation along inlet end of furrows. Distribution ditch should be at or below the bottom of the furrows and should never slope toward the basin.

To divert the water into the furrows, a secondary distribution ditch should be constructed perpendicular to the direction of irrigation, as shown in Fig. 16. The width of the ditch is determined by the size of stream being turned into the field. It should be wide enough so that the water flowing along the ends of the furrows is not erosive. The width selected must be such that the ditch can be easily constructed with available equipment. The ditches are generally flat bottomed, but if possible they are tilted with the deeper edge away from the field (Fig. 17). The edge of the secondary distribution ditch next to the furrow should be at or below the bottom of the furrow to prevent erosion.

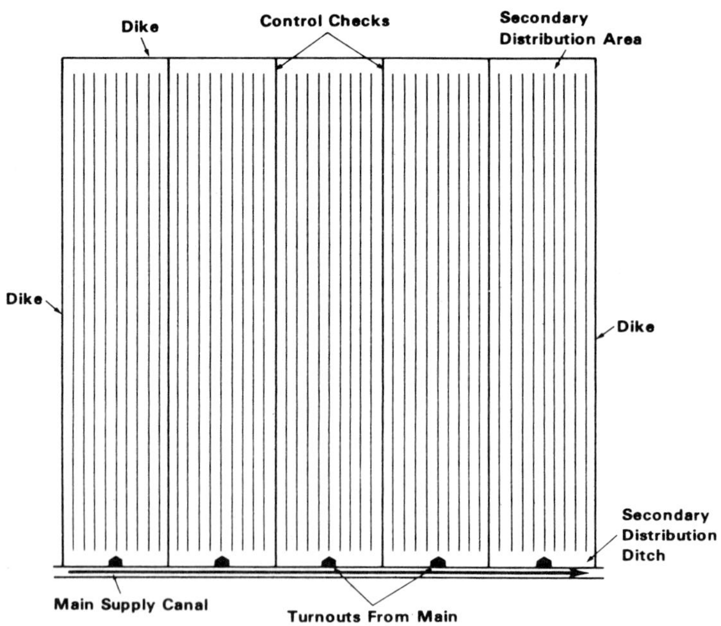

FIG. 18. Basin plan for rows or beds utilizing multiple outlets with temporary control checks to size the basins to fit the flow and crop conditions. In some instances the rows can be used as borders.

Temporary control checks may be used to size basins to fit the flow and crop conditions (Fig. 18). In some instances, a row or bed will satisfactorily confine the water in the designated area, but generally border dikes must be constructed.

Splitting the stream of water from a corner turnout may be necessary. This can be done by diverting part of the flow into a small temporary ditch along the side of the basin leading to a secondary ditch across the far end, while the remainder of the flow is distributed across the turnout end (Fig. 19). When furrows are parallel to the canal and a single turnout is located at midbasin, a temporary ditch leading directly from the turnout to the far end of the basin can be used (Fig. 20). For the same turnout configuration, but with furrows perpendicular to the canal, a temporary dike close to the furrows is usually used to protect the furrows from eroding and to help divert the water into the secondary ditch (Fig. 21). Dividing the stream and getting some water to the far end of the basin quickly is especially important in the case of high-intake-rate soils. If alternate-row irrigation is desired, furrows not to be irrigated can be blocked at both ends of the field.

The water within a level-basin area is controlled by the basin's outer edges, which normally consist of a supply ditch along one side and of rows or border dikes on the other three sides. Borders should be properly maintained. Soil cracking on many clay-type soils, and rodent damage, may necessitate disking and remaking the dikes before each irrigation.

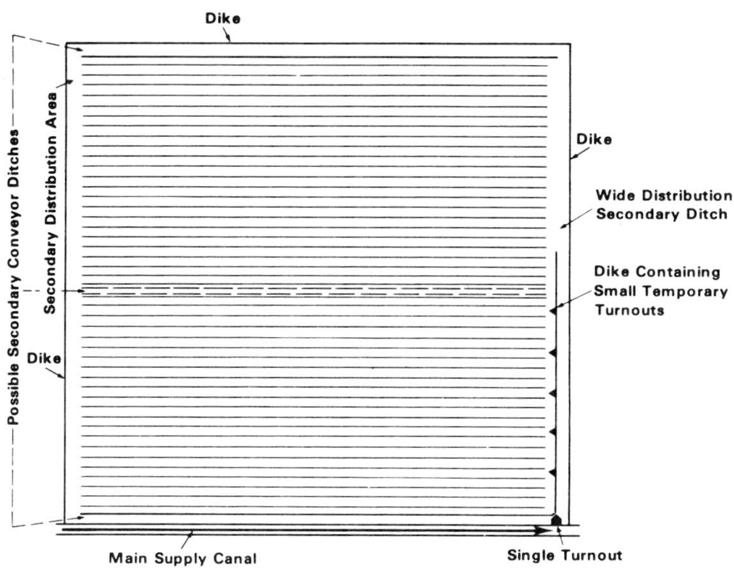

FIG. 19. Basin plan for rows or beds to minimize erosion when single turnout is used in corner of field. Possible locations of secondary conveyor ditches, which may be used, are also indicated.

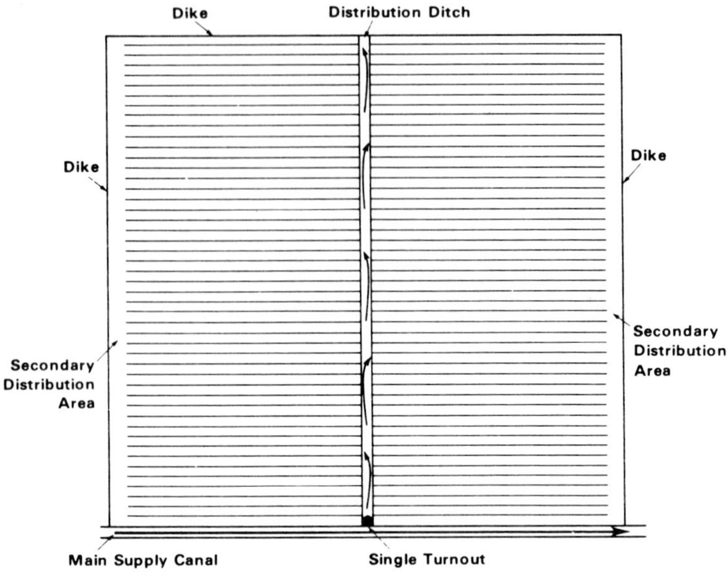

FIG. 20. Basin plan for single, midbasin turnout, with rows parallel to the canal.

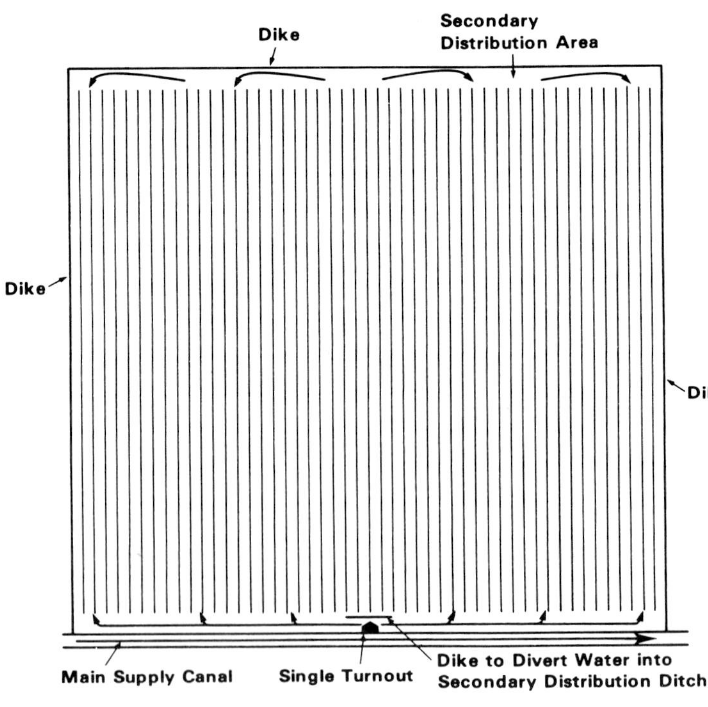

C. Automatic Controls

Automatically controlled field gates can facilitate delivery of a precise quantity of water to each basin, and thus may lead to improved crop production with lower labor requirements. These are the economic advantages of automated controls for level-basin irrigation systems. At present, automatic on-farm irrigation of level basins means that the turnout is operated remotely on a predetermined schedule, with the water diverted sequentially to a series of basins adjacent to a single distribution pipe or canal. Irrigation start-up is not signaled by the crop water status of a particular field, but is determined by some standard manner. An example of a field with automated gates near Yuma, Arizona, is shown schematically in Fig. 22. This 31-ha field is divided into 9 level basins with the water supplied in a concrete-lined canal.

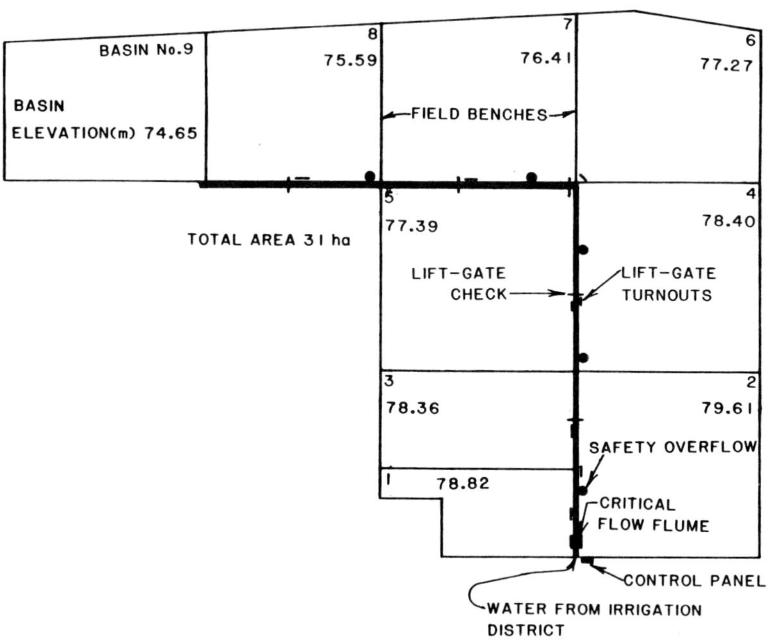

Fig. 22. This 31 ha field is divided into nine separate level basins ranging from 1.3 to 3.7 ha. Water is delivered to the basins via the concrete-lined canal. Gates are controlled by a controller located near the inlet. Irrigation proceeds sequentially from basin to basin after the required amount of water has been applied to each.

Fig. 21. Basin plan for single, midbasin turnout, with rows or beds perpendicular to the canal. Temporary dikes must be constructed in front of turnout to control eroding and row overtopping. Dikes are often protected with sheets of plastic.

FIG. 23. Structure used to open and close turnouts from concrete-lined canal. The movable plate is pressed against the metal ring (bellows inflated) to effect closure. When bellows are deflated the plate moves and flow starts.

To attain the required control, the system must have an actuator to open and/or close the turnout, a timing or volumetric control device to change the irrigation from basin to basin, and some means of signaling the actuator.

Appreciable research and development on gate closures and systems for surface irrigation have been done in the United States since 1963 (e.g., Haise and Kruse, 1969; Dedrick and Erie, 1978; Erie and Dedrick, 1978; Dedrick and Zimbelman, 1981). Three main gate types are currently used. Pipe-gate valves (Fig. 23) are used on pipes leading from an open or closed distribution system

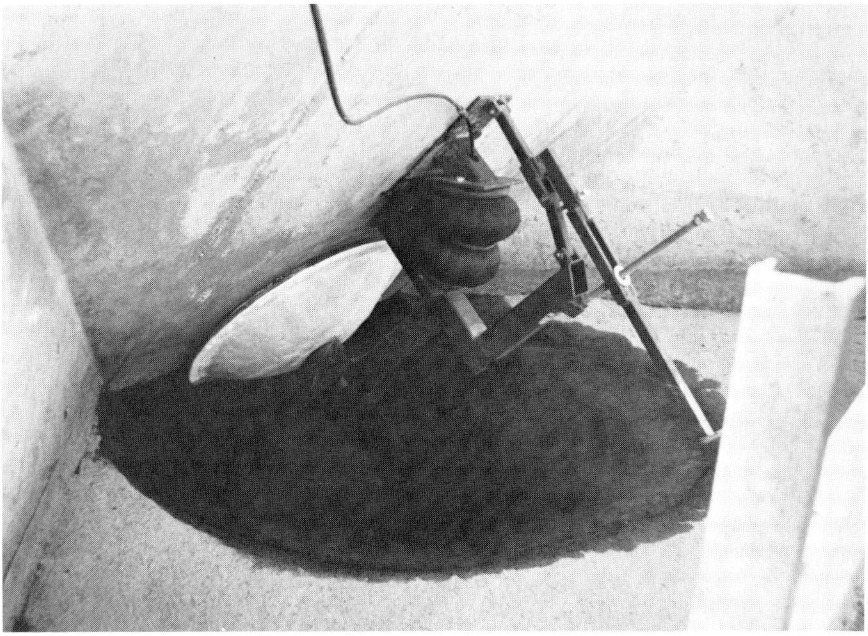

FIG. 24. Irrigation valve that closes a hole in the side of a concrete-lined canal by sealing a dome-shaped lid around the perimeter. Actuation is by an air bellows system.

to the basins. The closures use either inflatable bellows or air pillows to move a rubber-faced metal plate against the end of the turnout to stop the flow of water. Water flows out when the air is released.

Two automatic gate systems have been developed that provide openings directly in the side of concrete-lined canals. One is called an *irrigation flapper valve*[2] (Fig. 24) that closes a hole in the side of the canal by sealing a dome-shaped lid around the outside of the hole. The lid is also actuated by an air bellows system. The valve was developed by a private inventor in collaboration with a farmer owner/operator near Blythe, California. The valve is patented and available commercially.

Lift-type gates also directly open the side of a canal. These commercially available gates have been mechanized by adapting air cylinders to effect the

[2]Manufactured by California Dynamics Corp, Los Angeles, California. Trade names and company names are shown for the benefit of the reader and do not imply endorsement or preferential treatment of the company or products listed.

opening and closing operations (Fig. 14). Compressed air at about 345–410 kPa (50–60 lb/in.^{2}) is diverted through a four-way valve to either the bottom of the cylinder (to open the gate) or the top (to close the gate). Signaling to these systems has been done both electrically and pneumatically.

Other types of gates/actuators have been developed which could be used with level basins. These include center-of-pressure gates, drop gates, apron gates, and drawstring checks (Haise *et al.*, 1981; Humpherys, 1969). Wider adoption of some of these gates may be expected in the future.

Level basins are irrigated by applying a certain quantity of water to the basin without regard to the location of the advancing water front at the time of shut-off. Time represents water quantity as long as the flow rate is constant. Time clocks have been the most common method used to control the systems. Clocks can either be centrally located, with output signals connected to all gates to be controlled, or they can be located at each gate. In nearly all instances, centrally located controllers have been used. They have generally been adapted to surface irrigation systems from turf irrigation applications. The controllers may be electrical-mechanical, electrical, or electronic, with signals which are either electrical or pneumatic.

If the delivery flow rate is not constant throughout an irrigation, the controller should reflect accumulated flow rather than time. Water meters can be used in pipe systems, but when large flows are delivered in open channels, flow sensing must be accomplished with a primary flow-measuring device such as a flume or weir where water depth is associated with flow rate. Duke (1978) evaluated one controller in which a flume equipped with a float-potentiometer system was used to provide a prescribed volume of water. Other types of depth-sensing devices and interface systems with various controllers are presently being evaluated, but their utility is yet to be determined.

Other features of the automated systems may include basin sequence-selection equipment, check-gate actuation signaling, overflow sensing and signaling, canal draindown, and manual override selection switches. Which of these features, if any, will be used depends on the particular farm situation.

As discussed earlier, the proper use of a mechanized system as an irrigation management tool requires that the irrigator know the correct quantity of water to apply to each basin. Based on this information, charts should be constructed to determine the controller settings for various application depths for each basin within the field. Controller settings in Table III are used in conjunction with a controller that adjusts from a time base for some normal flow. In this case the nominal flow was 425 liters/sec, and the tabulated values apply to the field shown in Fig. 16. Such control systems automatically compensate for variable flows. If the flow rate is constant, and at nominal, the numbers in Table III represent minutes, and time is the controlling factor.

TABLE III

CONTROLLER SETTING GUIDE FOR USE WITH AUTOMATED BASINS[a]

Basin (no.)	Basin area (ha)	Controller settings,[b] for gross application depth (cm)			
		5.5	7.5	10	12.5
1	1.3	26	38	51	63
2	3.2	63	94	125	157
3	2.0	40	59	78	98
4	3.3	65	97	130	162
5	3.5	69	103	137	172
6	3.4	67	101	133	167
7	3.7	73	108	145	182
8	3.4	67	101	133	167
9	3.0	59	88	118	148

[a] Illustrated in Fig. 22.
[b] Controller settings are equivalent to minutes for a flow rate of 425 liters/sec.

TABLE A1

DESIGN OF LEVEL BASINS FOR 33-ha FIELD WITH TWO DISTINCTLY DIFFERENT SOILS[a]

	Clay loam	Fine sandy loam
Available water (cm/m)	19	10
Z_n (65% depletion in 1 m) (cm)	12.5	6.5
Intake family	0.3	1
a	0.670	0.710
τ_n (min)	810	86
q (liters/sec)	420	420
n	0.15	0.15
Minimum DU (%)	85	75
X (m) [Eq. (9)]	$0.426(12.5)^{7/9} \times (810)^{2/3}(0.15)^{2/3} = 935$	$0.426(6.5)^{7/9} \times (86)^{2/3}/(0.15)^{2/3} = 126$
Q (liters/sec m^{-1}) [Eq. (10)]	(935)(12.5)/810/6 = 2.4	(126)(6.5)/86/6 = 1.6
L^* at limit (Fig. 11, $a = 0.7$)	0.68	1.0
L (m)	935 × 0.68 = 635	126 × 1.0 = 126
q^*_{in} at limit (Fig. 11, $a = 0.7$)	2.45	2.25
q_{in} (liters/sec m^{-1})	2.4 × 2.45 = 5.9	1.6 × 2.25 = 3.6
W (m)	420/5.9 = 71	420/3.6 = 117
Area/basin (ha)	4.5	1.5

[a] See Fig. 2.

TABLE A2

FINAL DESIGN, NORTH–SOUTH IRRIGATION[a]

	Minimum	Maximum
L (m)	400	520
L^*	400/935 = 0.43	520/935 = 0.56
q^*_{in} for DU = 85% (Fig. 11)	1.05	1.6
q_{in} (liters/sec m^{-1})	2.4 × 1.05 = 2.5	2.4 × 1.6 = 3.8
W (m)[b]	420/2.5 = 168	420/3.8 = 110

[a] See Fig. 2 and Table A1.

[b] Average W is about 140 m; hence use three borders 130-m wide (400/3).

Appendix

From Fig. 2: The area with fine sandy loam soil is about 400 × 400 m which could be divided into nine basins of 130 × 130 m. Final design figures are as shown in Table A1. Placement of the water supply canal could be as shown in Fig. 2. The clay loam area could be irrigated either north–south or west–east, since basin lengths in both directions are within the design limit of 635 m. To irrigate north–south the final design would be as shown in Table A2.

References

Clemmens, A. J. (1979). Verification of the zero-inertia model for border irrigation. *Trans. ASAE* [*N.S.*] **22**(6), 1306–1309.

Clemmens, A. J. (1980). Depths of flow in level basins. *Trans. ASAE* [*N.S.*] **23**(4), 910–912.

Clemmens, A. J. (1981). Evaluation of infiltration measurements for border irrigation. *Agric. Water Manage.* **3,** 251–267.

Clemmens, A. J., and Dedrick, A. R. (1981). Estimating distribution uniformity in level basins. *Trans. ASAE* **24**(5), 1177–1180.

Clemmens, A. J., and Dedrick, A. R. (1982). Limits for practical level-basin design. *J. Irrig. Drain. Div., Am. Soc. CN. Eng.* **108**(IR2).

Clemmens, A. J., and Strelkoff, T. (1979). Dimensionless advance for level-basin irrigation. *J. Irrig. Drain. Div., Am. Soc. Civ. Eng.* **105**(IR3), 259–273.

Clemmens, A. J., Strelkoff, T., and Dedrick, A. R. (1981). Development of solutions for level-basin design. *J. Irrig. Drain. Div., Am. Soc. Civ. Eng.* **107**(IR3), 265–279.

Criddle, W. D., Davis, S., Pair, C. H., and Shockley, D. G. (1956). Methods for evaluating irrigation systems. *U.S., Dep. Agric., Agric. Hand.* **82.**

Dedrick, A. R. (1979). Land leveling—precision attained with laser controlled drag scrappers. *Pap.—Am. Soc. Agric. Eng.* **79-2565.**

Dedrick, A. R., and Erie, L. J. (1978). Automation of on-farm irrigation turnouts utilizing jack-gates. *Trans. ASAE* [*N.S.*] **21**(1), 92–96.

Dedrick, A. R., and Zimbelman, D. D. (1981). Automatic control of irrigation water delivery to and on-farm in open channels. *Symp. Proc. 11th Congr. Int. Comm. Irrig. and Drain.* R.7, 113–128.

Dedrick, A. R., Replogle, J. A., and Erie, L. J. (1978). On-farm level-basin irrigation—save water and energy. *Civ. Eng.* (*N.Y.*) **48**(1), 60–65.

Duke, H. R. (1978). Volumetric control for irrigation automation. *Pap.—Am. Soc. Agric. Eng.* **78-2545.**

Erie, L. J. (1962). Evaluation of infiltration measurements. *Trans. ASAE* [*N.S.*] **5**(1), 11–13.

Erie, L. J., and Dedrick, A. R. (1978). Automation of an open-ditch irrigation conveyance system utilizing tile outlets. *Trans. ASAE* [*N.S.*] **21**(1), 119–123.

Erie, L. J., and Dedrick, A. R. (1979). Level-basin irrigation: A method for conserving water and labor. *USDA Farmers' Bull.* **2261.**

Haise, H. R., and Kruse, E. G. (1969). Automation of surface irrigation systems: The state-of-the-art. *J. Irrig., Drain. Div., Am. Soc. Civ. Eng.* **95**(IR4), 503–516.

Haise, H. R., Kruse, E. G., Payne, M. L., and Duke, H. R. (1981). Automation of surface irrigation: 15 years of USDA research and development at Ft. Collins, Colorado. *U.S., Dep. Agric., Prod. Res. Rep.* **179,** 1–60.

Hinz, W. W., and Halderman, A. D. (1978). "Laser Beam Land Leveling Costs and Benefits," Bull. No. A114. Coop. Ext. Ser., Tucson, Arizona.

Humpherys, A. S. (1969). Mechanical structures for farm irrigation. *J. Irrig. Drain. Div., Am. Soc. Civ. Eng.* **95**(IR4), 463–479.

Israelsen, D. W. (1950). "Irrigation Principles and Practices." Wiley, New York.

Jensen, M. E., and Howe, O. W. (1965). Performance and design of border checks on a sandy soil. *Trans. ASAE* [*N.S.*] **8**(1), 141–145.

Katopodes, N. D., and Strelkoff, T. (1977). Dimensionless solutions of border-irrigation advance. *J. Irrig. Drain. Div., Am. Soc. Civ. Eng.* **103**(IR4), 401–417.

Kruse, E. G. (1978). Describing irrigation efficiency and uniformity. *J. Irrig. Drain. Div., Am. Soc. Civ. Eng.* **104**(IR1), 35–41.

Merriam, J. L. (1968). "Irrigation System Evaluation and Improvement." Blake Printery, San Luis Obispo, California.

Merriam, J. L. (1971). Adjusting cylinder infiltration data for field use. *Pap.—Am. Soc. Agric. Eng.* **71-13.**

Merriam, J. L., and Keller, J. (1978). "Farm Irrigation System Evaluation: A Guide for Management." Utah State University, Logan.

Michael, A. M. (1978). "Irrigation Theory and Practice." Vikas, New Delhi.

Phelan, J. T. (1960). Bench leveling for surface irrigation and erosion control. *Trans. ASAE* [*N.S.*] **3,** 14–17.

Ross, P. E., and Swanson, N. P. (1957). Level irrigation. *J. Soil Water Conserv.* **12**(5), 209–214.

Schwab, G. O., Frevert, R. K., Edminster, T. W., and Barnes, K. K. (1966). "Soil and Water Conservation Engineering," 2nd. ed. Wiley, New York.

Strelkoff, T., and Katopodes, N. D. (1977). Border-irrigation hydraulics with zero-inertia. *J. Irrig. Drain. Div., Am. Soc. Civ. Eng.* **103**(IR3), 325–342.

U.S. Department of Agriculture (1974). "Border Irrigation," National Engineering Handbook, Chapter 4, Sect. 15. Soil Conserv. Serv., USDA, Washington, D.C.

Zimmerman, J. D. (1966). "Irrigation." Wiley, New York.

FLOW MEASUREMENT FLUMES: APPLICATIONS TO IRRIGATION WATER MANAGEMENT

J. A. Replogle

U.S. Department of Agriculture
Agricultural Research Service
U.S. Water Conservation Laboratory
Phoenix, Arizona

M. G. Bos

International Institute for Land Reclamation and Improvement
Wageningen, The Netherlands

ISBN 0-12-24301-6

I. Introduction

A. Irrigation Flow Measurements

Measurements of water flows in irrigation canals have usually been costly, too often of questionable accuracy, and otherwise difficult to apply to field situations. Major problems include the general need to reshape the canals to accommodate a limited assortment of laboratory-calibrated devices, the requirement of considerable water surface drop through the devices, and the general inability to control installation errors and readout errors so as to assure an accurate and convenient reading. Many of these problems have now been reduced or eliminated by special application of critical flow in long-throated flumes.

B. Scope

This article describes the design, theory, and application of long-throated flumes, particularly the hydraulically related broad-crested weirs, to irrigation flow measurements in delivery-system canals and farm ditches, both lined and unlined.

II. Long-Throated Flumes

A. Description and Definition

A free-flowing, critical-flow flume is essentially a streamlined contraction built into an open channel where there must be sufficient water surface fall through the contraction so that critical flow occurs in the throat of the flume. The channel contraction may be formed from the sides only, from the bottom only, or from both the side and the bottom. Of most practical significance is critical flow involving nearly straight-lined, parallel flow both in the approach channel and within the contraction (or throat) itself. When this parallel critical flow exists, classical hydraulic principles can be used to approximate the stage-discharge relationship with minimum observable data. For curvilinear flow the rate of streamline curvature must be known at particular control points. This information is usually difficult to incorporate into practical hydraulic models and is even less practical to obtain in field situations. Consequently, flow-measuring devices based on curvilinear flow are usually laboratory rated, are sensitive to downstream water levels, and must be reproduced in careful detail to obtain reasonably accurate measurements.

In this article we deal with "long-throated flumes" and "broad-crested weirs," which from a hydraulic point of view are similar. The name "weir" is used if the control section is formed by a raised channel floor or sill. The "flume" is formed by side contractions in the channel. If the control section is formed by both side contractions and bottom sill it also is usually called a "flume." Classified under the above family of structures are those weirs or flumes over which the streamlines are nearly parallel to each other in the flume throat, at least over a short distance, and in the upstream approach channel near the station where flow depth is to be sensed (Fig. 1). To obtain this condition at the control section, the throat length L in the direction of flow must be a certain minimum length equal to one to two times the total upstream sill-referenced energy head H_1. In the following sections the limitations on the ratio H_1/L will be specified in more detail.

B. Advantages

The use of this family of structures is recommended for measuring flows in open channels whenever the water surface at the control section must remain free, because they have the following major advantages over other styles of weirs and flumes:

(1) The control section, perpendicular to the direction of flow, can be tailored in such a way that wide ranges of discharge can be measured accurately.
(2) For any prismatic shape of such a control, a rating table can be calculated with an error in the listed discharge of less than 2%.
(3) The head loss over the weir or flume required to have a unique and accurate relationship between measured depth h_1 and discharge Q is the lowest attainable for free-flowing open-channel devices.
(4) This head-loss requirement can be estimated with sufficient accuracy for any structure placed in any given channel.
(5) Under similar hydraulic and other boundary conditions these weirs or flumes are the least expensive structures that measure flow accurately.

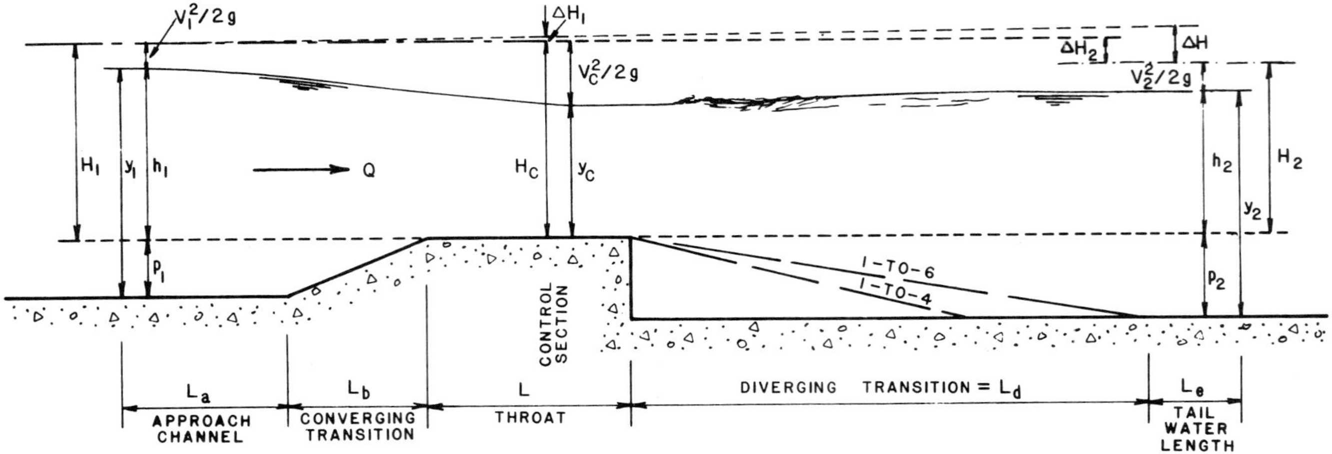

FIG. 1. Illustration of terms used for the dimensions and flow depths in long-throated flumes.

III. Types of Long-Throated Flumes

A. Trapezoidal Flumes

The flumes are generally named according to the shape of the throat, or control section. Thus, a trapezoidal throat combined with a rectangular upstream section would still be referred to as a trapezoidal flume. The limiting cases for trapezoidal flumes are the triangular flumes (when the bottom width is zero) and the rectangular flumes (when the side slopes are zero).

The triangular flumes exhibit the widest ranges between lowest detectable flow and highest detectable flow for a particular flume module. This flow range ratio can exceed 1 : 350. It may, however, require more absolute head loss in a particular channel and be more sensitive to errors in head reading than a rectangular flume, but it offers better sensitivity in the low flow ranges. The wide range is desirable in some applications, such as watershed research, but is seldom needed in irrigation flow metering.

Rectangular flumes have a flow range exceeding 1 : 30. While their sensitivity at high flows is usually adequate, low-flow sensitivity may be a problem. Absolute head loss in a particular channel can be low.

Trapezoidal shapes exhibit intermediate characteristics, combining wide range with acceptable sensitivity and relatively low head loss.

B. Complex Shapes

No particular limitations exist on throat section shape. Parabolic, circular, or any complex shapes that can be described mathematically are usable. Complex shapes can be designed to obtain a particularly wide range of measureable flows. To accommodate small flows, a small triangular or trapezoidal cross section can be used for the bottom few centimeters; then it may be flared outward at much flatter side slopes to accommodate large flows. Sharp breaks in the side walls at these slope changes are permitted, but the side slopes must not flare outward horizontally, lest there be a discontinuity in the rating. Another use for complex shapes is as vertical extenders for the simple shapes, to accommodate an overfull condition. The U shape with a semicircular bottom is particularly adapted to sewer flows at manholes.

Up to two changes in side-wall flare angle—for a total of three sidewall slopes in the same throat—have been mathematically modeled and verified by laboratory rating (Replogle, 1975). Cross-sectional shapes that are not symmetrical with respect to a channel centerline have also been computed and laboratory verified (Replogle, 1978).

Perhaps the most practical applications of tbe capability of complex shapes are to predict ratings for flumes as misconstructed, to retrofit existing flumes to

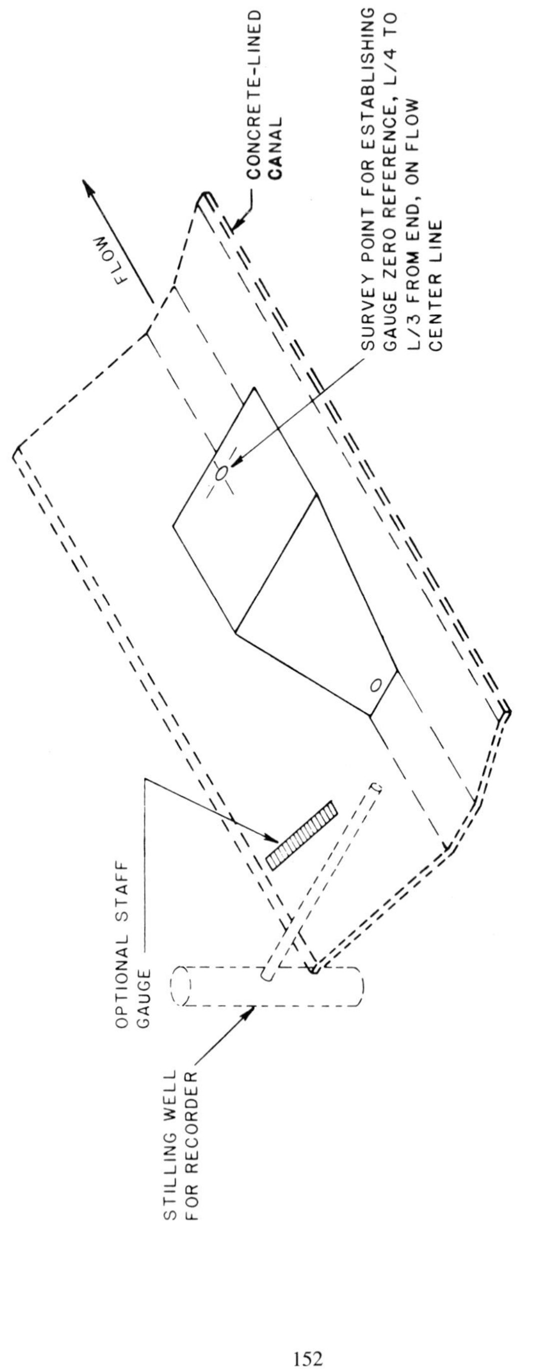
CONCRETE-LINED CANAL
FLOW
SURVEY POINT FOR ESTABLISHING GAUGE ZERO REFERENCE, L/4 TO L/3 FROM END, ON FLOW CENTER LINE
OPTIONAL STAFF GAUGE
STILLING WELL FOR RECORDER

(a)

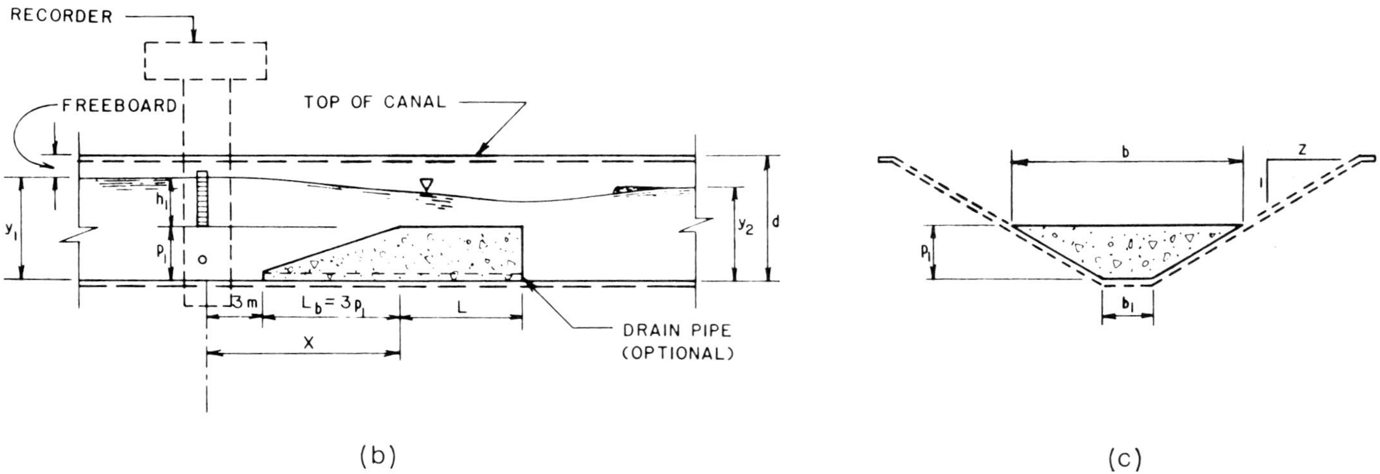

FIG. 2. Modified broad-crested weir is one form of long-throated flume: (b) section along center line; (c) cross section.

handle unexpected flow rates, or to impart additional tolerance to downstream backwater conditions. The procedures for making these applications and computing the ratings for simple and complex flumes using either a discharge coefficient adjustment to a computed ideal flow or direct mathematical modeling are presented in succeeding sections. Long-throated flumes thus offer the designer high flexibility with respect to shape, size, and range of flows to be measured.

C. Broad-Crested Weirs

The modified broad-crested weirs made by a bottom contraction or sill placed in an otherwise unmodified lined canal of nearly any shape are particularly convenient forms of long-throated flumes. The sill is accompanied by an upstream ramp, or transition, which can be either rounded or at a fixed slope of about three units horizontal to one unit vertical (Fig. 2). For irrigation flow measurement, these forms have been singled out as the most promising long-throated flume devices—combining accuracy, ease of construction, and high tolerance to downstream backwater levels. Error control is incorporated in several ways. The weir widths are chosen large enough to easily absorb normal concrete construction errors; therefore existing concrete-lined canals can be used for most of the structure required for long-throated devices. The weir-crest length in the direction of flow and the sloping approach ramp provide flow situations that can be accurately modeled mathematically ($\pm 2\%$) and solved by computer techniques or approximation methods (Section IV). Surveying methods are used in order to accurately determine wall-gauge mounting positions, or stilling-well reference levels, above the elevation of the constructed crest of the weir shown in Fig. 2. This vertical gauge positioning upstream from the weir is the most critical of the field determinations, and special care should be observed. Construction of the modified broad-crested weir is simple. It requires only one carefully constructed surface, namely the weir sill crest. All other surfaces can be approximately dimensioned and rough finished to about $\pm 10\%$ without affecting the calibration beyond about 1%. Because of these simplified construction requirements, the costs of broad-crested weirs are only 10–20% of those of previously described devices, which furthermore tend to be more complicated and offer no advantage in accuracy.

Another advantage of the broad-crested weir is the low drop in water surface through the device needed to make it work properly. This low drop, or head loss, is a direct function of the upstream depth h_1 with respect to the sill crest level. Since broad-crested weirs provide the control-section contraction by a rise in the channel bottom, this flow depth is small compared to flumes that acquire most or all of their control-section contraction by inward movement of the channel side walls. Typical head losses for small canals are on the order of 2–5 cm. This means that these weirs can be retrofitted to almost any canal system without reconstructing the canal. Very large canals may require special foundation considerations, however, because of the massive sill—which is usually constructed of concrete and may exceed a meter in thickness.

Contrary to intuitive expectations, it appears that sediments moving along a relatively narrow trapezoidal canal bottom spread across the ever-widening ramp and pass the crest without causing significant problems in the few dozen installations that have been observed for two to five years (Replogle and Clemmens, 1980).

IV. Theoretical Considerations

A. General Equations for Rectangular Control Sections

Flow in long-throated flumes is governed by the principles of critical flow. Critical flow in open-channel hydraulics is defined as the maximum discharge for minimum energy. Thus, a channel contraction causes just enough increase in upstream channel depth to provide the minimum energy to sustain the flow through the constricted throat. This assumes that the contraction is sufficient so that the downstream backwater does not restrict the flow. Critical flow then occurs at some point within the constricted portion of the flume. Critical flow in the cross section at this point is defined by the flow equation (Chow, 1959)

$$v_c^2/2g = A_c/2B_c \tag{1}$$

where the subscript "c" refers to the critical section. Thus v_c is the velocity at the critical section; it can be replaced by Q/A_c (by continuity considerations), where A_c is the flow cross-sectional area in the reach of critical flow, B_c is the top width of the flow at the same section, and Q is the discharge rate. Incorporating the continuity relationship, Eq. (1) becomes

$$Q = (gA_c^3/B_c)^{1/2} \tag{2}$$

This general discharge equation is valid for all arbitrarily shaped control sections. It is easily applicable if simple mathematical expressions exist for A_c and B_c.

1. *Head-Discharge Equation for Ideal Fluid Flows*

Rectangular Control Section. For critical flow in a rectangular control section, $A_c = by_c$ and $B_c = b$, where b is the channel width and y_c is the flow depth at the critical section. Thus Eq. (1) reduces to

$$v_c^2/2g = y_c/2 \tag{3}$$

The total energy head H, which combines the depth y and velocity head terms $v^2/2g$ at a channel section referenced to the channel bottom, is defined as (Chow, 1959)

$$H = y + (v^2/2g) \tag{4}$$

For critical flow in a rectangular section, Eq. (4) becomes

$$H_c = y_c + \tfrac{1}{2} y_c = \tfrac{3}{2} y_c \tag{5}$$

where subscript "c" refers again to the critical section, or

$$y_c = \tfrac{2}{3} H_c \tag{6}$$

For an idealized flow, the total heads—H_c at the critical section and H_1 for an upstream gauging station—are assumed to be equal with no energy loss between the two locations, the velocity distributions are assumed to be uniform at both sections, and the streamlines at both sections are assumed to be straight and parallel. Substituting Eq. (6) into Eq. (2) and using the rectangular channel expressions for A_c and B_c and $H_1 = H_c$, we obtain

$$Q = g^{0.5} b (\tfrac{2}{3} H_1)^{1.5} \tag{7}$$

2. *Real Fluid Flows*

In reality the above-mentioned assumptions are not entirely correct because there are frictional losses, nonuniform velocity distributions, and deviations from straight parallel flow. Therefore, these deviations from the idealized flow assumptions must be compensated by the introduction of a discharge coefficient C_d. Equation (7) rearranged then reads

$$Q = C_d \tfrac{2}{3} (\tfrac{2}{3} g)^{0.5} b H_1^{1.5} \tag{8}$$

In an open channel it is not convenient to measure the energy head H_1 directly. It is therefore common practice to relate the flow rate to the upstream-sill-referenced water level h_1 (or head) in the following way:

$$Q = C_v C_d \tfrac{2}{3} (\tfrac{2}{3} g)^{0.5} b h_1^{1.5} \tag{9}$$

where C_v is the approach-velocity coefficient which accounts for the velocity head at the gauging location $v_1^2/2g$.

The calibration for any flume cross-sectional shape can be approximated by a generalized form of Eq. (9), expressed as a power function

$$Q = C_v C_d K (h_1)^u \tag{10}$$

where K is a constant that includes flow area and other factors and u is a function of the flume shape (1.5 for rectangular flumes and 2.5 for triangular flumes).

a. Discharge Coefficient C_d. As stated above, the discharge coefficient corrects for such phenomena as energy loss between gauging and control sections, nonuniformity of velocity distribution, and streamline curvature. These factors are closely related to the value of the ratio H_1/L, where L is the length of the contraction section. When the flow is relatively shallow over a particular throat of length L (for example, an H_1/L ratio of 0.1), the frictional effects amount to a relatively large proportion of the value for H_1, approaching perhaps a 5% reduction in idealized discharge, or a C_d value of about 0.95. For deeper relative flow depths, say $H_1/L = 0.5$, the frictional and other effects are a smaller fraction of H_1 and may require a C_d value of perhaps 0.99. Finally, as the flow approaches an H_1/L value of 1, the flow streamlines become so curved through the region of critical flow that there are significant pressure reductions. The discharge then starts to exceed the computed idealized flow in spite of frictional losses and velocity distribution effects, and C_d may slightly exceed 1. Flow conditions with H_1/L exceeding 1 are not reliably predicted. Also, flows with H_1/L exceeding about 0.5 may show increasing sensitivity to downstream backwater effects.

Values for C_d as related to H_1/L are illustrated in Fig. 3. The range of application

$$0.1 \leq H_1/L \leq 1.0 \tag{11}$$

indicated in Fig. 3 is partly due to inadequate confirmation of C_d values outside this range. For $H_1/L \leq 0.1$, minor changes in boundary roughness cause increasingly large variations in the C_d value. For values of $H_1/L > 1$, the streamline

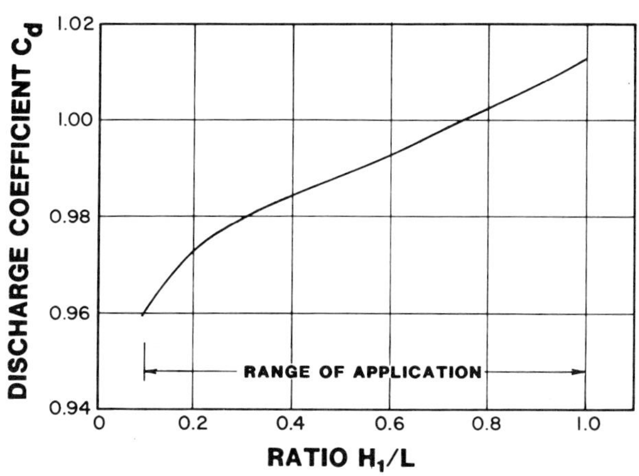

FIG. 3. Values of C_d as a function of the ratio H_1/L (after Bos, 1976).

curvature at the control section is influenced by the downstream transition and water stage. Also, the head loss over the structure—required for a unique relation between h_1 (or H_1) and Q—increases sharply with increasing values of H_1/L, because a high tailwater level decreases the streamline curvature at the control section and thus can produce a variable rating.

b. Approach-Velocity Coefficient C_v. The approach-velocity coefficient corrects for the use of sill-referenced flow depth h_1 instead of total head H_1 in Eq. (9); thus it accounts for neglecting $v_1^2/2g$, where v_1 is the velocity in the approach channel. The exact value of C_v equals the ratio of generalized versions of Eq. (8) over Eq. (9). Thus, C_v is given by

$$C_v = \left(\frac{H_1}{h_1}\right)^u = \left(\frac{h_1 + v_1^2/2g}{h_1}\right)^u = \left(1 + \frac{v_1^2}{2gh_1}\right)^u \tag{12}$$

where u equals the power of h_1 in a head discharge equation similar to Eq. (10) with $u = 1.5$ for a rectangular control section. The power u of h_1 also determines both the range of discharges that can be measured by a structure (Section III,A) and the sensitivity of the structure, which is the change in discharge Q per unit change in h_1 (Section V,B).

The C_v value increases rapidly if v_1 increases. This is related to C_d times the area ratio $C_d A^*/A_1$, where A^* is the imaginary wetted area at the control section if the water depth would equal h_1 there. Thus, for a rectangular control section,

$$A^* = bh_1 \tag{13}$$

and if it is inserted into a trapezoidal approach channel,

$$A_1 = b_1 y_1 + z_1 y_1^2 \tag{14}$$

where y_1 is the total flow depth in the approach channel and z_1 is the channel side-slope ratio, horizontal to vertical.

Values of C_v as a function of C_d times the area ratio, $C_d A^*/A_1$ are shown in Fig. 4 for rectangular and triangular control shapes.

The approach velocity should not be allowed to become too large, lest standing waves formed at the gauging station make accurate head reading difficult. In terms of the ratio of dynamic forces to gravity forces, the Froude number (Chow, 1959) in the approach channel Fr_1 defined as

$$\mathrm{Fr}_1 = v_1/(gA_1/B_1)^{1/2} \tag{15}$$

should not exceed about 0.5, where the subscript "1" refers to the approach channel, A_1 is the approach-channel flow area, B_1 is the flow-top width, v_1 is the flow velocity, and g is the gravitational constant.

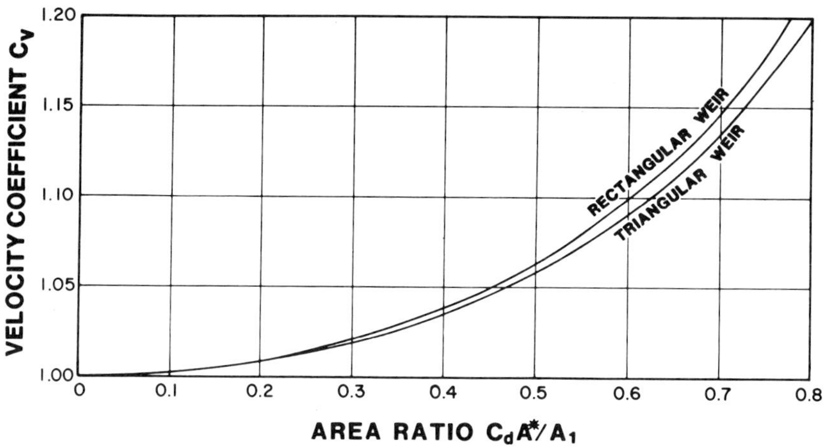

FIG. 4. Approach-velocity coefficient for various control sections.

3. *Head-Discharge Equations for Other Control Shapes*

Following the same procedures which lead to Eq. (7) and making the same assumptions which underlie the C_d and C_v coefficients in Eq. (9), Bos (1976, 1977) derived head-discharge equations for commonly used shapes of the control section. Table I is adapted from Bos (1977) to assist in determining critical depth for trapezoidal control sections. The derived equation is in terms of critical depth y_c, which is obtained from Table I. The discharge equation is

$$Q = C_d[by_c + zy_c^2][2g(H_1 - y_c)]^{1/2} \tag{16}$$

where b is the control section bottom width, z is the channel side slope, and y_c is obtained from Table I for various z values and H_1/b ratios.

B. MODULAR LIMIT

The limiting downstream total head that causes no more than 1% deviation in the real discharge as determined for the particular h_1 is defined as the modular limit for the particular flow device or module (Bos, 1976). The modular limit, sometimes referred to as the limiting submergence ratio, is numerically expressed as the downstream head H_2 (tailwater depth and velocity head) divided by the upstream head H_1 (upstream depth and velocity head), or H_2/H_1. These depths and total heads are referenced to the elevation of the bottom of the flume throat or to the sill top in the case of the broad-crested weir, not to the channel floor.

Modular limits vary widely for various weirs and flumes, usually related to the

TABLE I

VALUES OF THE RATIO y_c/H_1 AS A FUNCTION OF z AND H_1/b FOR TRAPEZOIDAL CONTROL SECTIONS

		Side slopes of channel ratio, horizontal to vertical (z : 1)								
H_1/b	Vertical	1/4 : 1	1/2 : 1	3/4 : 1	1 : 1	3/2 : 1	2 : 1	5/2 : 1	3 : 1	4 : 1
0.00	—	0.667	0.667	0.667	0.667	0.667	0.667	0.667	0.667	0.667
0.01	0.667	0.667	0.667	0.668	0.668	0.669	0.670	0.670	0.671	0.672
0.02	0.667	0.667	0.668	0.669	0.670	0.671	0.672	0.674	0.675	0.678
0.03	0.667	0.668	0.669	0.670	0.671	0.673	0.675	0.677	0.679	0.683
0.04	0.667	0.558	0.670	0.671	0.672	0.675	0.677	0.680	0.683	0.687
0.05	0.667	0.668	0.670	0.672	0.674	0.677	0.680	0.683	0.686	0.692
0.06	0.667	0.669	0.671	0.673	0.675	0.679	0.683	0.686	0.690	0.696
0.07	0.667	0.669	0.672	0.674	0.676	0.681	0.685	0.689	0.693	0.699
0.08	0.667	0.670	0.672	0.675	0.678	0.683	0.687	0.692	0.696	0.703
0.09	0.667	0.670	0.673	0.676	0.679	0.684	0.690	0.695	0.698	0.706
0.10	0.667	0.670	0.674	0.677	0.680	0.686	0.692	0.697	0.701	0.709
0.12	0.667	0.671	0.675	0.679	0.684	0.690	0.696	0.701	0.706	0.715
0.14	0.667	0.672	0.676	0.681	0.686	0.693	0.699	0.705	0.711	0.720
0.16	0.667	0.672	0.678	0.683	0.687	0.696	0.703	0.709	0.715	0.725
0.18	0.667	0.673	0.679	0.684	0.690	0.698	0.706	0.713	0.719	0.729
0.20	0.667	0.674	0.680	0.686	0.692	0.701	0.709	0.717	0.723	0.733
0.22	0.667	0.674	0.681	0.688	0.694	0.704	0.712	0.720	0.726	0.736
0.24	0.667	0.675	0.683	0.689	0.696	0.706	0.715	0.723	0.729	0.739
0.26	0.667	0.676	0.684	0.691	0.698	0.709	0.718	0.725	0.732	0.742
0.28	0.667	0.676	0.685	0.693	0.699	0.711	0.720	0.728	0.734	0.744
0.30	0.667	0.677	0.686	0.694	0.701	0.713	0.723	0.730	0.737	0.747
0.32	0.667	0.678	0.687	0.696	0.703	0.715	0.725	0.733	0.739	0.749
0.34	0.667	0.678	0.689	0.697	0.705	0.717	0.727	0.735	0.741	0.751
0.36	0.667	0.679	0.690	0.699	0.706	0.719	0.729	0.737	0.743	0.752
0.38	0.667	0.680	0.691	0.700	0.708	0.721	0.731	0.738	0.745	0.754
0.40	0.667	0.680	0.692	0.701	0.709	0.723	0.733	0.740	0.747	0.756
0.42	0.667	0.681	0.693	0.703	0.711	0.725	0.734	0.742	0.748	0.757
0.44	0.667	0.681	0.694	0.704	0.712	0.727	0.736	0.744	0.750	0.759
0.46	0.667	0.682	0.695	0.705	0.714	0.728	0.737	0.745	0.751	0.760
0.48	0.667	0.683	0.696	0.706	0.715	0.729	0.739	0.747	0.752	0.761
0.5	0.667	0.683	0.697	0.708	0.717	0.730	0.740	0.748	0.754	0.762
0.6	0.667	0.686	0.701	0.713	0.723	0.737	0.747	0.754	0.759	0.767
0.7	0.667	0.688	0.706	0.718	0.728	0.742	0.752	0.758	0.764	0.771
0.8	0.667	0.692	0.709	0.723	0.732	0.746	0.756	0.762	0.767	0.774
0.9	0.667	0.694	0.713	0.727	0.737	0.750	0.759	0.766	0.770	0.776
1.0	0.667	0.697	0.717	0.730	0.740	0.754	0.762	0.768	0.773	0.778
1.2	0.667	0.701	0.723	0.737	0.747	0.759	0.767	0.772	0.776	0.782
1.4	0.667	0.706	0.729	0.742	0.752	0.764	0.771	0.776	0.779	0.784
1.6	0.667	0.709	0.733	0.747	0.756	0.767	0.774	0.778	0.781	0.786
1.8	0.667	0.713	0.737	0.750	0.759	0.770	0.776	0.781	0.783	0.787
2	0.667	0.717	0.740	0.754	0.762	0.773	0.778	0.782	0.785	0.788
3	0.667	0.730	0.753	0.766	0.773	0.781	0.785	0.787	0.790	0.792
4	0.667	0.740	0.762	0.773	0.778	0.785	0.788	0.790	0.792	0.794
5	0.667	0.748	0.768	0.777	0.782	0.788	0.791	0.792	0.794	0.795
10	0.667	0.768	0.782	0.788	0.791	0.794	0.795	0.796	0.797	0.798
∞	—	0.800	0.800	0.800	0.800	0.800	0.800	0.800	0.800	0.800

degree of parallel flow achieved in the throat section and the friction and eddy losses caused by the structure and the associated downstream expansion section, if one is used. For example, in the case of thin-plated weirs, this ratio is zero or negative, whereas for some long-throated flumes and broad-crested weirs it can approach 0.95.

Long-throated flumes and broad-crested weirs have modular limits that appear to change slowly with changes in cross-sectional flow shape and are influenced primarily by the depth of flow through the throat section. Therefore, the relatively shallow flows over the broad-crested weir result in a low absolute water surface drop, or head loss, in the channel. Thus, almost paradoxically, minimum required drop in canal water surface is achieved by using a raised bottom sill and not by using side-wall contractions.

Bos and Reinink (1981) proposed a method for computing the expected modular limit based on using the discharge coefficient of the flume to estimate the head loss between the upstream gauging station and the point of computed critical flow in the flume throat. Manning's equation is applied for the remaining length, and an expansion coefficient is employed to handle the exit losses. This expansion coefficient is approximated from laboratory tests on flume exit sections. A similar procedure was followed for the broad-crested weirs presented herein, except that the mathematical model (Section IV,D) and the direct computation of head losses based on boundary layer concepts are used instead of the discharge coefficient and Manning's equation. The expansion coefficient values and application techniques of Bos and Reinink (1981) are retained. Computational results are similar for similar flumes.

In new construction, the measuring device can be selected to match the normal depth of the approach channel at design flow and a canal downstep equal to or exceeding the required head loss can be planned to avoid exceeding the modular limit. This will allow normal design freeboard limits to be maintained.

In canal retrofit situations, design freeboard is often infringed upon to obtain the necessary modular limit clearance. Usually this is a safe procedure, since precise knowledge of the flow depth is available because of the measuring device.

C. Numerical Example for Trapezoidal Flume in a Trapezoidal Channel

Given: A broad-crested weir with a sill width $b = 0.20$ m, $p_1 = 0.15$ m, and $L = 0.60$ m is placed in a concrete-lined canal with $b_1 = 0.50$ m and $z = 1.0$.

Required: (1) What will be the flow rate through the flume when the total head $H_1 = 0.240$ m? (2) What sill-referenced flow depth h_1 belongs with this H_1?

Solution: (1) Eq. (16) gives as the head discharge equation

$$Q = C_d[by_c + zy_c^2][2g(H_1 - y_c)]^{1/2} \tag{17}$$

Since $H_1/b = 0.24/0.20 = 1.20$ and $z = 1.0$, Table I yields $y_c/H_1 = 0.747$. Consequently for the required H_1, $y_c = 0.179$ m. Entering Fig. 3 with $H_1/L = 0.24/0.60 = 0.40$ produces $C_d = 0.985$. Substituting all known values into Eq. (17) gives

$$\begin{aligned} Q &= (0.985)[(0.20)(0.179) + (1)(0.179)^2] \\ &\quad \times [19.62(0.240 - 0.179)]^{1/2} \\ &= 0.0732 \text{ m}^3/\text{sec} \end{aligned}$$

(2) The upstream depth $h_1 = H_1 - v_1^2/2g$, which will be measured at the head measurement section if $Q = 0.0732$ m²/sec, can be approximated by assuming that $H_1 \simeq h_1$ and calculating A_1 by using the equation for a trapezoidal area

$$A_1 = z_1(p_1 + h_1)^2 + b_1(p_1 + h_1) \tag{18}$$

Assuming $h_1 = 0.240$ m gives $A_1 = 0.347$ m². By use of the continuity equation $v = Q/A$, we find that $v_1 \simeq 0.0732/0.347 = 0.211$ m/sec. Subtracting $v_1^2/2g$ from H_1 gives $h_1 = 0.238$ m as a first approximation. This first approximation of h_1 can then be substituted into Eq. (18) to obtain a more refined v_1 value of 0.212 m/sec. This value, however, again gives $h_1 = 0.238$ m, so further refinement for h_1 is not needed. Other points on the stage–discharge curve of the flume can be calculated and rating tables can be compiled.

D. Mathematical Modeling of Long-Throated Flumes

The use of C_d is based on average values for many control section shapes. Specific allowances for roughness of construction materials or effects of channel shape are not attempted. A computer model incorporating roughness effects and velocity distribution as a function of flow shape and length of control section was presented by Replogle (1975).

The control-section, sill-referenced energy equations for the flow are developed to incorporate computations for friction losses between the head-measurement section and the critical-flow location in the control section. Also included are the velocity distribution coefficients α_1 and α_c for each of the two locations. These coefficients correct for the mathematical expedient of using the average velocity in the energy expressions instead of the integrated square of the point velocities (King, 1954; Chow, 1959).

Equation (4) is modified to express the total head H_1 at the gauging station, including the velocity distribution coefficient α_1 and the sill-referenced depth h_1,

$$H_1 = h_1 + \alpha_1 v_1^2/2g \tag{19}$$

or

$$H_1 = h_1 + \alpha_1 Q^2/(2gA_1^2) \tag{20}$$

The total energy at the gauging station H_1 is equal to the total energy in the control section H_c except for the head loss ΔH_1 between the two sections, or $H_1 = H_c + \Delta H_1$ but

$$H_c = y_c + \alpha_c(v_c^2/2g) \tag{21}$$

thus

$$H_1 = y_c + \alpha_c(v_c^2/2g) + \Delta H_1 \tag{22}$$

Using the relationship of Eq. (1) and solving for y_c,

$$y_c = H_1 - \alpha_c(A_c/2B_c) - \Delta H_1 \tag{23}$$

Equation (2) in this context becomes

$$Q = (gA_c^3/\alpha_c B_c)^{1/2} \tag{24}$$

Equations (20), (22), and (24) thus relate the upstream depth h_1 to the control-section depth y_c for any arbitrary channel cross-sectional area A_1 and control-section area A_c. Because A_1 and A_c are functions of h_1 and y_c, respectively, the equations must be solved by iteration techniques for Q in terms of h_1 to produce a usable rating table. Expressions for the total head loss ΔH and for evaluating α_c and α_1 follow, and, with their inclusion, complete the framework for a mathematical model that can produce rating curves for long-throated flumes and broad-crested weirs.

1. *Evaluating Head Loss* ΔH_1

Real fluids are affected by channel wall friction and internal energy conversions to heat. The latter is significant in the highly agitated flow at the structure outlet and contributes to the total head loss through the entire structure, but it can be disregarded in the relatively streamlined flow between the head-measuring section and the control section because its magnitude is small compared to the wall friction effects. Also, roughness values and other laboratory-derived basic head-loss data generally include it as part of the wall friction effects.

The friction losses are evaluated separately for the approach-channel reach of length L_a, the converging transition L_b, and the throat L. The friction and eddy losses in the reaches downstream from the control section affect the tolerance of the structure to tailwater levels (Section IV,B).

Several methods have been used to evaluate friction losses through flumes. Ackers and Harrison (1963) tried using variations of the Darcy–Weisbach pipe friction factors, and an approach based on boundary drag on a flat plate. Later, Harrison (1967) used boundary-layer displacement thickness to compensate for the discharge decrease from ideal flow caused by a rough boundary. While all of the procedures are empirical, a boundary-layer drag method has some distinct advantages because it can accommodate changes in friction with flow depth and channel cross-sectional shape that are not well represented by the friction factor approaches. Replogle (1975) expanded upon this boundary-layer work and found a modified boundary drag method to be adaptable to a wide variety of channel and control-section shapes, construction materials, and flow ranges. This approach, with minor modifications, is presented here.

2. *Boundary-Layer Method for Friction Assessments*

For the boundary-layer evaluation of friction losses, the flume reach is assumed to be one side of a thin, flat plate placed in an infinite flow field. The plate is then warped to the shape of the channel section to be evaluated and the fluid drag on the plate results in energy or head losses. Boundary-layer theory indicates that flow in the boundary layer, and hence boundary drag, is not constant but varies along the plate. The boundary layer begins as a laminar flow which develops into a turbulent boundary layer at some distance away from the leading edge of the plate. There is a gradual and somewhat predictable transition from the laminar to the turbulent boundary layer. For computational purposes, it is convenient to assume an abrupt change from laminar to turbulent at a distance L_x from the leading edge of the plate. Also, it is assumed that the approach channel is a zone of fully developed turbulent flow, as is the converging section, but that the control section begins a new boundary layer with its upstream edge being analogous to the leading edge of the plate.

A combined drag coefficient for the laminar and turbulent regions of the control section can be obtained by adding the relative drag coefficients for the laminar and turbulent parts of the boundary layer. The turbulent portion is assumed to act as if the entire boundary layer were turbulent, following the findings of Prandtl, as reported in Schlichting (1960). Thus, the drag coefficient for the nonexistent portion of the turbulent boundary layer $C_{F,x}$ acting over distance L_x (Fig. 5) must be subtracted from the turbulent drag coefficient $C_{F,L}$, computed for the entire plate, and replaced with the computed drag coefficient for the laminar portion $C_{f,x}$. The combined drag coefficient C_F then becomes

$$C_F = C_{F,L} - (L_x/L)C_{F,x} + (L_x/L)C_{f,x} \tag{25}$$

The distance L_x can be expressed in terms of a distance Reynolds number R_x by an empirical relation based on the concept of permissible roughness as described by Schlichting (1960),

$$R_x = 350{,}000 + L/K \tag{26}$$

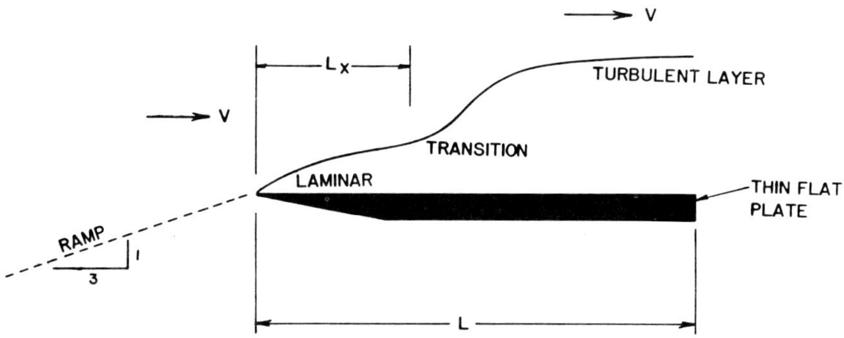

FIG. 5. Transition from laminar to turbulent boundary layer.

where k is the absolute roughness height of the boundary material. This Reynolds number is related to L_x by the definition

$$R_x = v_c L_x / \nu \tag{27}$$

where ν is the kinematic viscosity, and v_c is the average velocity of flow in the section of interest. Here v_c is expressed in terms of the control section. Similarly, the Reynolds number over the entire length L is

$$R_L = v_c L / \nu \tag{28}$$

Specific values for the turbulent drag coefficient are calculated using the relationship presented by Grenville (1958)

$$C_{F,L} = \frac{0.544(C_{F,L})^{1/2}}{5.61(C_{F,L})^{1/2} - 0.638 - \ln\{(R_L C_{F,L})^{-1} + [4.84(C_{F,L})^{1/2} L/k]^{-1}\}} \tag{29}$$

Equation (29) can also be used for $C_{F,x}$ by substituting R_x and L_x for R_L and L, respectively. This equation must be solved by iteration since it is implicit in $C_{F,L}$.

The drag coefficient for the laminar flow contribution can be calculated using the following equation from Schlichting (1960)

$$C_{f,x} = 1.328/R_x^{0.5} \tag{30}$$

If $R_L < R_x$, then the entire boundary layer is laminar and $C_F = C_{f,L}$ and is obtained by inserting R_L into Eq. (30).

For a fully developed turbulent boundary layer, as would be expected in the approach channel, the converging transition section, the diverging transition section, and the tailwater channel, the drag coefficient can be taken as

$$C_F = 0.00235 \tag{31}$$

The transition from drag to head loss ΔH_n is by use of the relation (Replogle, 1975)

$$\Delta H_n = (C_F x/R)(v^2/2g) \tag{32}$$

where R is the hydraulic radius (R = area of cross section/wetted perimeter), and x is the length of each section n considered. Thus, the total head loss H_1 between the gauging station and the control section is

$$\Delta H_1 = \Delta H_a + \Delta H_b = \Delta H_L \tag{33}$$

where the subscripts "a", "b", and "L" refer to the appropriate section, L_a, L_b, and L, respectively (Fig. 1). This result is then used in Eq. (22).

The computation for ΔH_b, which is the friction loss in the converging transition uses the average drag:

$$\Delta H_b = (0.00235/4g)(v_1^2/R_1 + v_b^2)L_b \tag{34}$$

where v_b and R_b are the velocity and hydraulic radius at the entrance to the control, or throat, section of the flume. The depth at the entrance.is approximated by

$$Y_b = y_c + \tfrac{5}{8}(h_1 - y_c) \tag{35}$$

The value for ΔH_L, which is the friction loss in the throat, uses the entire throat length L, rather than the computed location of y_c, which is the location of critical depth for parallel flow. The exact location of critical flow, and thus the actual point of control, is not known. It moves from near the upstream end of the throat for very small H_1/L values to the downstream end of the flume for H_1/L values close to one. To simplify the computation the total length L is used. The differences in friction are not significant enough to warrant exact location of the control section.

3. *Roughness of Construction Materials*

The value of the roughness height k usually used for smooth-troweled concrete is k = 0.00015 m. The absolute roughness height for a number of materials typically used for flume construction is given in Table II.

An analysis of the sensitivity of rating to k was made using the equations above. This showed that a change of several orders of magnitude in k produces less than a 0.5% (and often less than a 0.1%) change in discharge indication, depending on the range of roughness values chosen. Thus, in the usual range of flows for a particular flume, whether one uses materials as smooth as glass or as rough as troweled concrete will have only minor effects on the discharge rate.

4. *Velocity Distribution Coefficients α_1 and α_c*

As mentioned earlier in Section IV,D, a velocity distribution coefficient α corrects for the mathematical expedient of using average velocity in the energy

TABLE II

ABSOLUTE ROUGHNESS HEIGHTS FOR VARIOUS CONSTRUCTION MATERIALS

Material	k (m)
Glass, plastic, fiberglass	0.000001–0.000010
Metal painted or smooth	0.000020–0.000100
Galvanized sheet metal	0.000100–0.001000
Wood sanded to rough sawn	0.000200–0.001000
Concrete, smooth troweled,	0.000100–0.002000
float finished, or broomed	0.000500–0.005000

expressions. Mathematically, α is the ratio between the actual velocity head of the flow based on the summation of the kinetic energy per unit weight per unit time and that calculated using the average velocity (King, 1954; Chow, 1959). For a uniform velocity profile, the results would require no correction, hence $\alpha = 1$. For a laminar (parabolic) profile in a pipe $\alpha = 2.0$, and for laminar flow in a wide rectangular channel it is 1.56. For developed turbulent flows in open channels, a typical value is $\alpha = 1.04$.

For fully developed turbulent flow in wide channels, Chow (1959) presents the following approximation

$$\alpha = 1 + 3\epsilon^2 - 2\epsilon^3 \tag{36}$$

where $\epsilon = (v_m/v) - 1$, with v_m the maximum flow velocity and v the average flow velocity. Also, ϵ can be approximated for a fully developed flow as

$$\epsilon = 0.883\sqrt{f} = 0.883\sqrt{4C_F} = 1.77\sqrt{C_F} \tag{37}$$

(cf. Chow, 1959, pp. 8, 168, 211; Schlichting, 1960, p. 537).

In the present application the flow may not be fully developed, particularly in the control section, and the flow channel may not be sufficiently wide. Equation (36) is modified to include adjustments for these conditions. The adjustment for cross-sectional shape is based on analogies to laminar flow and is expressed in terms of the average, or hydraulic, depth D and the hydraulic radius R. The adjustment for flow length is in terms of the control section length L and the hydraulic radius. The resulting empirical equation (Replogle, 1975) is

$$\alpha_c = 1 + [3\epsilon^2 - 2\epsilon^3][1.5(D/R) - 0.5][0.025(L/R) - 0.05] \tag{38}$$

with limits of $1 \leqslant 1.5\,(D/R) - 0.5 \leqslant 2$ and $0 \leqslant 0.025\,(L/R) - 0.05 \leqslant 1$.

Equation (38) produces velocity distribution coefficients ranging from 1 to about 1.04 for the range of conditions used to calculate the tables presented in this article. Values may exceed 1.04 for some complex shapes.

5. *Computer Model*

The equations described in this section are easily adapted to computer solution. Versions in BASIC were presented by Replogle (1975, 1978). The reader is referred to these references. Versions in FORTRAN were primarily used to compile the tables in Sections VI–VIII, but have not yet been published.

V. Factors Influencing Calibration

Dimensional error of construction, errors in zeroing the readout device, roughness changes with time, seasonal temperature effects on fluid viscosity, and sensitivity of the readout device are the major factors influencing calibration.

A. Dimensional Errors

The accuracy with which a flow can be measured depends on the accuracy of the rating table or equation for the structure and the accuracy with which the upstream sill-referenced head h_1 can be determined.

The error Δh_1, which is the difference between the determined and the true value of h_1, can refer to an unnoticed change in water level, head-error reading on scale or paper, mislocation of gauging station, error in zero setting of the scale or recorder, internal resistance in recorder, etc. Its effect on final discharge will, of course, depend on the control-section shape. The calibration for each shape can be approximated by a power function [Eq. (10)], and therefore the influence of Δh_1 would depend on the u value. Thus, forming an error equation for the effects of misrepresenting h_1 from Eq. (10) by first taking the logarithm of both sides and differentiating, we obtain

$$dQ/Q = u(dh_1/h_1) \tag{39}$$

or

$$\Delta Q/Q = u(\Delta h_1/h_1) \tag{40}$$

where ΔQ is the difference between the determined discharge Q and the true Q. Equation (40) shows that small ΔQ values are obtained with small u values relative to h_1 values. It does not follow, bowever, that highest reliability is necessarily obtained with rectangular control sections ($u = 1.5$), since, for example, a triangular control section ($u = 2.5$) might have a large h_1 value and thus a similar or better sensitivity.

Deriving an error equation for the flumes that involves all of the computational variables used in the computer model is impractical. A straightforward approach is to examine the influence of introducing an incremental change of +1% in each variable, in turn, and noting the corresponding change in indicated discharge rate for the entire calibration range. A problem arises when applying this percentage concept to a variable like side slope. These quantities, for example, are zero for a rectangular flume, so a percentage change is not meaningful. Therefore, the side-slope change chosen was one that caused 1% change in throat flow area A_3. A simlilar problem arises for the bottom width for triangular throated flumes. Again, the change in throat area A_3 was used.

The effects of construction and gauge-registration errors on discharge prediction were examined by using the computer model for two representative shapes of flumes. One was a broad-crested weir 0.3 m high placed in a 0.6-m-wide canal with 1 : 1 side slopes. The other was a trapezoidal flume in the same sized canal with a throat width of 0.3 and 1 : 1 side slopes. The throat section was elevated above the canal bottom by 0.06 m. Each device was designed for maximum flows of about 600 liters/sec. Detailed plots of each variable and the dimensions of each device were presented by Replogle (1977). The variables with most influence were those that affected the determination of the cross-sectional area of the throat A_3, namely b, z_3, and h_1. For both flume styles, a change of +1% in A_3 caused approximately +1% change in indicated discharge at low flows and about +1.3% change for high flows. If the error were totally due to width change, the influence for the weir would be a nearly constant +1% for all discharges, but for the trapezoidal shape it varies from +1% at low flows to +0.6% at high flows. The influence of an error in h_1 was somewhat higher, was nearly constant for all flows, and was about +1.65% for the broad-crested weir and +2.11% for the trapezoidal flume. These latter values correspond to the slope u of the h_1 versus Q log-log plots [Eq. (10)], that have traditionally been developed for weirs and flumes from laboratory-type data.

Variables capable of a 10% change before causing 1% change in discharge were z_1, b_1, L_a, L_b, p_1, and L. Roughness height had an almost undetectable influence in these sizes and flow ranges, as did fluid viscosity. An entire magnitude change in roughness k was required to cause fractional percentage changes in discharge rate. The influence was slightly greater for the relatively longer throated trapezoidal flume.

These results are applicable to estimating the relative effects of siltation in the approach channels for the two styles of flume. For the broad-crested weir, 5 cm of silt would cause between 0% and −1% change in indicated discharge at low and high flows, respectively, whereas for the trapezoidal flume similar siltation would cause −3% and −1% change in indicated discharge. Thus, siltation is more of a problem with the trapezoidal-shaped flumes, particularly at low flows, than it is with the broad-crested weir style.

B. Sensitivity to Zero Setting of the Gauge

As mentioned above, 1% error in gauge setting usually will mean 1.6–2.2% error in discharge. However, a constant zero offset error becomes a higher percentage of h_1 as h_1 becomes small. Thus zero setting becomes a source of major error, particularly at low flows. In field practice side-wall-attached gauges are normally set by surveying techniques to within ±1 mm. This provides more than sufficient accuracy for flow depths greater than 100 mm and still provides acceptable accuracy at flow depths down to about 25 mm.

When a gauge that is to be mounted on the channel side wall has been premarked in direct-reading discharge units, then an error can be introduced if the side-wall slope is misrepresented. To adjust for this in the field, it is good practice to determine if a particular flow rate is most important. If so, the gauge can be mounted so that this important flow rate is accurately located on the wall, regardless of side-wall slope. Thus, this flow rate will be most nearly correct, relegating the larger indication errors to seldom-used regions of the gauge.

C. Measurement of Head

1. *General Considerations*

As discussed above, the accuracy of a flow measurement depends strongly on the true determination of the upstream sill-referenced head. The success of a measuring structure often depends entirely on the effectiveness of the gauge or recorder used.

The sill reference level refers to the effective control section, which is located above the weir crest, or in the flume throat, at a distance of about $L/3$ from the downstream edge of the sill (see Fig. 2). In the direction of the flow the top of the sill (weir crest, or invert of flume throat) must be truly level. If minor undulations in level occur, it is recommended that the average level at the effective control section be used rather than the average level of the entire sill.

2. *Location for Head Measurement*

The gauging or head-measuring station should be located sufficiently far upstream to avoid detectable water surface drawdown, yet close enough for the energy losses to be negligible. This is particularly true if the ratings are derived using the C_d and C_v curves of Section IV,B. In the computer-derived ratings the gauge location drawdown and friction losses are an integral part of the calculation, and therefore the gauge should be located as indicated in the appropriate tables.

For the ratings derived using C_d and C_v, the gauging station should be located at a distance between two and three times the maximum h_1 reading from the beginning of the throat section (sill) or at a distance equal to h_1 from the beginning of the converging transition, whichever is greater (Fig. 1).

3. *Head Measurement Methods*

Basically, the head is sensed either in the channel itself or in a stilling well located to one side of the channel and connected to the channel by means of a small pipe. Many methods can be used to detect the water surface. Some methods exploit the electromagnetic properties of the water and of the water–air interface. Other methods depend on reflecting a sonic wave from the air–water interface. Still other methods detect water depth with a variety of pressure-sensing devices and infer the head from that information.

The most frequently encountered methods, however, are the vertical and side-wall-mounted staff gauges in the canal or in a stilling well, and the float-operated recorder placed in a stilling well.

a. Stilling Wells. Stilling wells serve two purposes: (1) to facilitate the accurate reading of a water level at a gauging station where the water surface is disturbed by wave action, and (2) to house the float of a recorder system or other surface detecting equipment.

The cross-sectional dimensions of the well depend on the method by which the head is to be measured, ranging from a recommended minimum size of 0.1 m for hand-inserted dipsticks to over 0.5 m to accommodate large-diameter floats. The pipe connecting the stilling well to the canal should be large enough to allow the stilling well to respond quickly to water level changes. Usually this pipe diameter is about one-tenth the diameter of the stilling well.

Care should be taken to make the pipe opening flush and smooth where it taps the canal wall. It should also be perpendicular to the direction of flow, and, when possible, perpendicular to the wall plane for about two pipe diameters as well, although on sloping side walls this is not usually practical.

A third precaution that can be observed when a bottom sill is used is to place the stilling-well pipe as high off the canal bottom as practical and still keep it below the water surface at low flow. This helps keep the pipe clear of sediments.

For flumes that have only side contractions, and no bottom sill, sediment plugging can be delayed or avoided by using the sediment trap arrangement shown in Fig. 6. This allows long periods of flow sensing between maintenance operations, depending on the size of the sediment trap and the sediment load in the channel. The stilling-well pipe enters the cavity which is cast into the floor and partly under the side wall. The dimensions are not significant but should be large enough for cleaning equipment. Approximately 30 cm long (in the flow direction), 60 cm wide, and 45 cm deep are suggested minimum dimensions.

The stilling-well pipe enters this cavity and is fitted with a pipe elbow which is turned down so that sediment cannot flow directly into the pipe during initial channel fillings. A small vent bole is drilled near the top of the elbow to help avoid air trapping. The cavity is located so that about half of it is under the side-wall slope, if side-wall slopes are used. It is covered with two pieces of 6-mm-thick aluminum plate, one on the floor and one on the side-wall slope. Sensing holes are drilled into the side-wall plate. Considering the probable increased bedload trapping caused by transverse slots in the plates and the low-

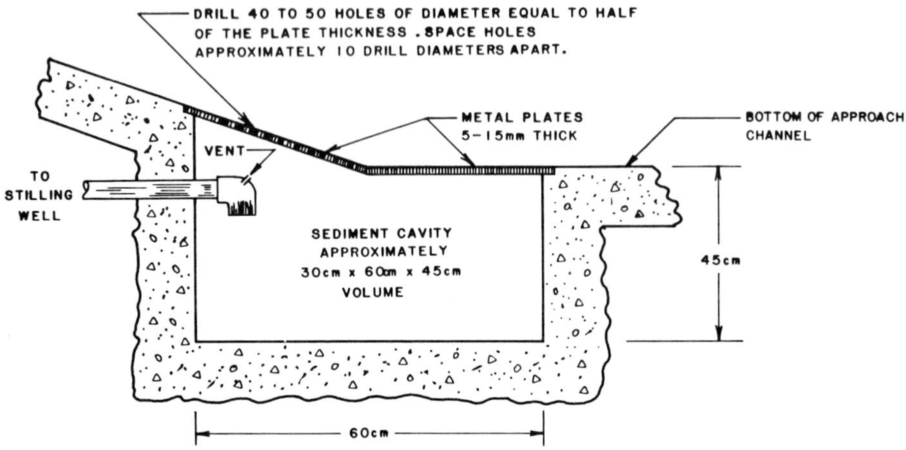

FIG. 6. Sediment trap to delay or avoid sensing-line plugging.

quality pressure detection likely with slots parallel to the flow, a compromise was chosen that consisted of drilling 3-mm-diameter holes into the side-wall plate. No particular hole pattern is recommended, except that enough holes should be drilled to approximate the area of the connecting pipe, and the holes should be spaced 5–10 drill-size diameters apart to reduce mutual interference. This is usually not a significant problem since the flow velocity in most instances for the Froude numbers encountered in these flumes is so low. Installation precautions concerning float size, stilling-well height needed to avoid interference between that and the counterweight, and well sizing are discussed in detail by Bos (1976), and in "Agricultural Handbook No. 224" (U.S. Department of Agriculture, 1979).

b. Staff Gauges. Where no continuous information on the flow rate is needed, as in canals where the fluctuation of flow is gradual, periodic readings on a calibrated staff gauge may provide adequate information. The gauge should be placed in such a manner that the water level can be read from the canal bank and so that its surface can be cleaned by the observer.

For concrete-lined canals, the gauge can be mounted directly on the canal wall as in Fig. 7. In the trapezoidal-shaped canals, appropriate adjustments to the value for h_1 as read on the canal wall must be made to convert the reading to the vertical h_1 value before entering the discharge tables.

Within an irrigation district it is frequently desirable to use a limited number of standardized structures. It is convenient to mark the gauges of these structures directly in discharge units rather than in head units.

Most permanent gauges are enameled steel, cast aluminum, or polyester. Enameled linear scales marked in metric or English units are available from

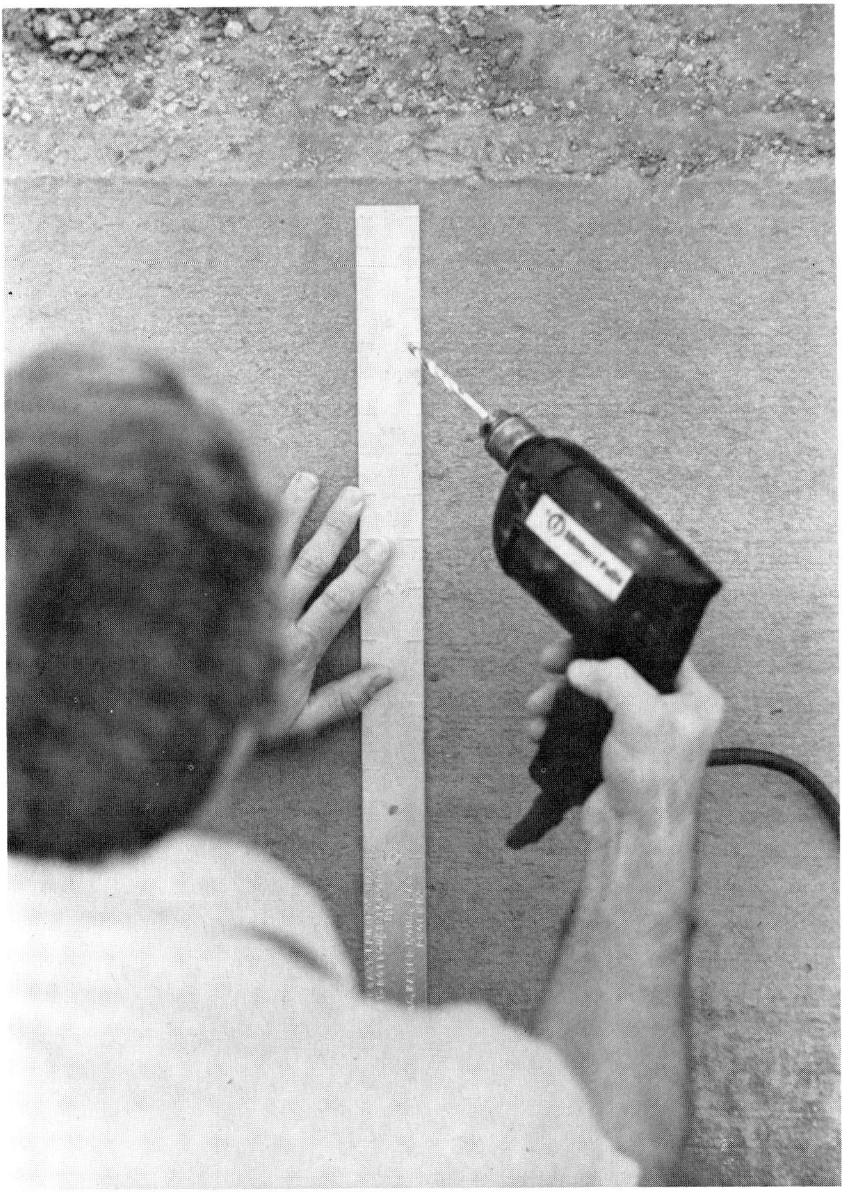

FIG. 7. A shop-stamped gauge is ready for attachment to a canal wall with concrete anchor bolts.

commercial sources. Important flow rates can be noted on these scales by separate markings to avoid the need for tables to be always at hand.

Nonlinear enameled gauges marked directly in discharge units can be custom ordered, but must be purchased in quantities of about 100 or more to keep the unit cost reasonable. Examples are shown in Fig. 8. Individual shop-made gauges marked on aluminum bar stock with chisel and hammer and metal-stamping number die set, can be constructed in about 1 man-hour each. Such a gauge is shown in Fig. 7. Except for the rectangular flumes (Table X) the calibration tables in this article provide for easy calculation of the gauge-mark locations for integer values of discharge. This is also true for the rectangular offerings (Table X) if integer widths are used.

For lined canals, staff gauges can be mounted on the canal side walls. For unlined canals, a vertical support is required. Often the canal side slopes are not constructed to the intended slope. If this occurs, it is recommended that the gauge be mounted so that the correct reading occurs at the correct elevation for a point in the most frequently used flow range. Thus, the greatest reading errors will occur in the flow ranges that are seldom used (Section V,B). If the zero setting is displaced more than about 5 mm, or if accuracy for the full flow range is

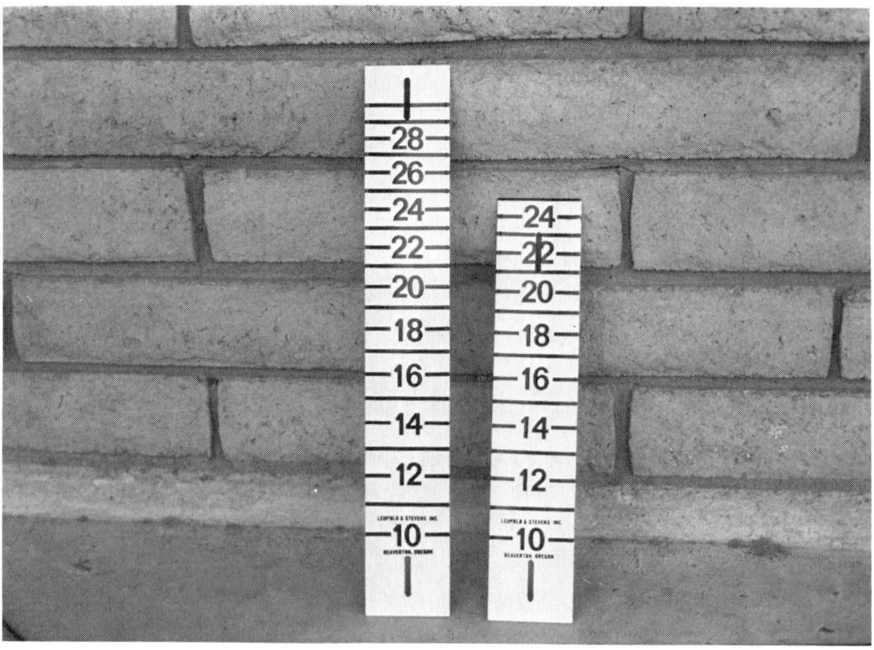

FIG. 8. Baked-enamel special staff gauges that are commercially available for a few sizes of broad-crested weirs and long-throated flumes.

required, then the actual side slope should be separately determined for adjustments to the gauge.

VI. Selecting Flume Style and Site

A. Rangeability

The flow rates in open channels usually vary with time. The range between minimum and maximum discharge ($Q_{\min}$ and $Q_{\max}$) through which the flow is to be measured strongly depends on the use of the channel in which the structure is to be placed. Irrigation canals, for example, have a considerably narrower range of discharges than natural streams and drainage canals. This range of discharges may be expressed as a ratio

$$\gamma = Q_{\max}/Q_{\min} \tag{41}$$

In this section, the broad-crested weirs and flumes described have been developed for a ratio of H_1/L ranging from 0.1 to 1 if the coefficient of discharge approach was used, or from 0.05 to about 0.5 (with permitted extensions of H_1/L to about 0.75) with the direct computation method using the mathematical model. In either case the H_1/L ratio is 1 : 10, which provides the basis for computing the range in γ that can be expected for each broad-crested weir or flume using the tables provided. Alternatively, the power function approximation of the tables in the form

$$Q = C_d K(H_1^u/L) \tag{42}$$

can be used to obtain the useful range. In Eq. (42), K is a dimensional constant. Since u values for rectangular flume devices are about 1.5 and for triangular devices 2.5, the respective discharge rangeabilities γ are 35 and 345. For trapezoidal flumes γ usually will range between 55 and 200.

In irrigation canals, the required ratio rarely exceeds 35, so that almost any shape can be used for the control section (for example, see Fig. 9). In natural drains, however, the range of flows that are to be measured will usually determine the shape of the control section (Fig. 10).

B. Field-Site Considerations

Structures for measuring flow rate should be located in a channel reach where an accurate value of h_1 can be obtained, and where sufficient drop in water surface (head loss) can be created (Fig. 9).

The site survey of the channel to find a suitable location for the structure should include information on several relevant factors that can influence the performance of the future structure. These factors include the following.

FIG. 9. Structures should be located in a channel reach where accurate flow depth can be sensed and where sufficient drop in water surface can be created. Here a straight reach of canal provides a successful site for a broad-crested weir. [Flow shown is 1.13 m^3/sec; canal data, b_1 = 0.61 m, z = 1.25; broad-crested weir data, b = 1.83 m, L = 0.914 m (approximately equal to FC2M in Table III)].

1. *Stream or Canal*

The stream or canal should be straight, with approximately uniform cross section for a length of about three times the maximum h_1 before encountering the depth-measuring location where L_a begins. Channel bends combined with high-velocity flows may require more length, else the measurement may have an additional error of uncertainty of about ±3%. In these cases it is recommended that the canal side corresponding to the outer bend be used for gauging purposes. High-velocity currents and draw-down effects are often associated with the side on the inner bend.

2. *Channel Stability*

The channel reach should have a stable channel bottom. In some channel reaches sedimentation occurs in dry periods and erosion during wet periods. Sedimentation can alter the approach conditions significantly and sometimes will

FIG. 10. Triangular-throated flume selected to measure a wide range of flows from a Florida Forestry Research watershed.

bury the structure. Erosion on the other hand can undercut the foundation of the structure.

3. *Downstream Water Level Controls*

Water level controlling factors for the channel should be determined. Sometimes this level is directly predictable by channel discharge equations, but frequently this level is influenced by downstream confluences with other channels, operation of gates, and reservoirs, etc. The approximate channel water levels are necessary to determine the required vertical location of the long-throated flume or the broad-crested weir to achieve modular flow.

4. *Water Surface Stability*

For the full range of anticipated discharges, the Froude number at the gauging station should be calculated by Eq. (15).

To obtain a reasonably stable water surface without standing waves, which might interfere with accurately determining the value for h_1, the Froude number should not exceed 0.5 at the gauging station and for a distance of (20–30) h_1 upstream.

5. *Subsoil Conditions*

Subsoil conditions should be determined. At the site of the structure, leakage around and beneath the structure must be cut off at reasonable cost, and a stable foundation without significant settling must be obtained.

TABLE III

FLUME CALIBRATIONS FOR WATER DEPTH AT GAUGE LOCATION REFERENCED TO SILL CREST (cm)[a]

	Canal shape A $b_1 = 30.5$, $z = 1:1$				Canal shape B $b_1 = 61.0$, $z = 1:1$		
Flow rate (liters/sec)	FA1M $b = 76.2$ $L = 61.0$ $p_1 = 22.9$	FA2M $b = 91.4$ $L = 45.7$ $p_1 = 30.5$	FA3M $b = 106.6$ $L = 30.5$ $p_1 = 38.1$	Flow rate (liters/sec)	FB1M $b = 121.9$ $L = 91.4$ $p_1 = 30.5$	FB2M $b = 137.2$ $L = 76.2$ $p_1 = 38.1$	FB3M $b = 152.4$ $L = 61.0$ $p_1 = 45.7$
10	3.89	3.50	3.17	25	5.32	4.94	4.61
15	5.06	4.53	4.11	30	5.97	5.53	5.17
20	6.06	5.42	4.93	35	6.58	6.11	5.70
25	6.93	6.24	5.68	40	7.15	6.65	6.20
30	7.75	6.99	6.38	45	7.70	7.16	6.68
35	8.52	7.70	7.03	50	8.22	7.65	7.15
40	9.25	8.36	7.64	60	9.21	8.59	8.02
45	9.94	8.99	8.23	70	10.14	9.46	8.85
50	10.59	9.59	8.79	80	11.01	10.28	9.63
60	11.81	10.73	9.85	90	11.83	11.06	10.37
70	12.94	11.78	10.83	100	12.62	11.81	11.08
80	14.0	12.77	11.76	125	14.5	13.6	12.75
90	15.0	13.7	12.65	150	16.1	15.2	14.3
100	15.9	14.6	13.5	175	17.7	16.7	15.7
125	18.1	16.7	15.4	200	19.2	18.1	17.0
150	20.1	18.6	17.2	250	21.8	20.6	19.5
175	21.9	20.3	18.9	300	24.3	23.0	21.8
200	23.6	21.9	20.4	350	26.5	25.2	23.9
225	25.2	23.4	21.9	400	28.6	27.2	25.9
250	26.7	24.9	23.3	450	30.6	29.2	27.7
275	28.1	26.2	24.6	500	32.5	31.0	29.5
300	29.4	27.5	25.8	600	36.0	34.4	32.9
325	30.7	28.8	27.0	700	39.2	37.5	35.9
350	32.0	30.0	28.2	800	42.1	40.4	38.7
375	33.2	31.2	29.3	900	44.9	43.1	41.4
400	34.3	32.3	30.4	1000	47.5	45.7	43.9
450	36.5	34.4	32.5	1100	50.0	48.1	46.3
500	38.6	36.4	34.4	1200	52.4	50.5	48.6

TABLE III—*Continued*

	Canal shape C $b_1 = 61.0$, $z = 1.25:1$				Canal shape C $b_1 = 61.0$, $z = 1.25:1$		
Flow rate (liters/sec)	FC1M $b = 156.2$ $L = 106.6$ $p_1 = 38.1$	FC2M $b = 175.3$ $L = 91.4$ $p_1 = 45.7$	FC3M $b = 194.7$ $L = 76.2$ $p_1 = 53.3$	Flow rate (liters/sec)	FC1M $b = 156.2$ $L = 106.6$ $p_1 = 38.1$	FC2M $b = 175.3$ $L = 91.4$ $p_1 = 45.7$	FC3M $b = 194.7$ $L = 76.2$ $p_1 = 53.3$
50	7.10	6.61	6.17	600	31.7	30.1	28.7
60	7.96	7.41	6.93	700	34.5	32.9	31.4
70	8.76	8.16	7.64	800	37.2	35.5	33.9
80	9.52	8.88	8.31	900	39.7	38.0	36.3
90	10.23	9.56	8.95	1000	42.1	40.3	38.6
100	10.92	10.21	9.57	1200	46.5	44.6	42.8
150	14.0	13.1	12.34	1400	50.5	48.5	46.6
200	16.7	15.7	14.8	1600	54.2	52.2	50.2
250	19.0	17.9	16.9	1800	57.7	55.6	53.6
300	21.2	20.0	18.9	2000	60.9	58.8	56.7
350	23.2	22.0	20.8	2200	64.1	61.9	59.7
400	25.1	23.8	22.5	2400	67.0	64.8	62.6
450	26.9	25.5	24.2	2600	69.8	67.5	65.4
500	28.5	27.1	25.8				
550	30.1	28.6	27.3				

[a] Add 15 cm to L if depths and discharges below dashed line (larger flow rates) are used.

6. *Sedimentation*

To avoid aggravating a possible sedimentation situation, it is desirable that sufficient head be available to allow a drop in the channel for maintaining the modular limit relationships. Then the placement of the flume in the channel will not significantly change the basic channel flow velocities and sediment will more likely pass through the flume.

VII. Trapezoidal Lined Canals

A. Selecting the Flume Dimensions

The installation procedures and calibrations for a series of broad-crested weirs designed for on-farm concrete-lined canals were presented by Clemmens and Replogle (1980) and Replogle and Clemmens (1981). Another group was selected for larger canals (Replogle and Clemmens, 1980).

1. *Broad-Crested Weirs for On-Farm Applications*

The farm-sized units were selected for 30.5- and 61-cm canal-bottom widths b_1 corresponding to 1- and 2-ft bottom widths in English units. Two side slopes z_1 were selected, 1 unit of horizontal displacement to 1 vertical, and 1.25 to 1.

TABLE IV

FLUME CALIBRATIONS FOR WATER DEPTH AT GAUGE LOCATION REFERENCED TO WEIR CREST (ft)[a]

	Canal shape A $b_1 = 1.0$, $z = 1:1$				Canal shape B $b_1 = 2.0$, $z = 1:1$		
Flow rate (ft³/sec)	FA1 $b = 2.50$ $L = 2.00$ $p_1 = 0.75$	FA2 $b = 3.00$ $L = 1.50$ $p_1 = 1.00$	FA3 $b = 3.50$ $L = 1.00$ $p_1 = 1.25$	Flow rate (ft³/sec)	FB1 $b = 4.00$ $L = 3.00$ $p_1 = 1.00$	FB2 $b = 4.50$ $L = 2.50$ $p_1 = 1.25$	FB3 $b = 5.00$ $L = 2.00$ $p_1 = 1.50$
0.5	0.160	0.143	0.130	1	0.189	0.175	0.164
0.6	0.179	0.161	0.146	2	0.292	0.271	0.254
0.7	0.197	0.177	0.161	3	0.375	0.350	0.328
0.8	0.214	0.192	0.175	4	0.447	0.418	0.393
0.9	0.230	0.207	0.189	5	0.511	0.480	0.452
1.0	0.246	0.221	0.202	6	0.570	0.536	0.506
1.2	0.275	0.248	0.226	7	0.625	0.589	0.556
1.4	0.302	0.273	0.249	8	0.676	0.638	0.604
1.6	0.327	0.296	0.271	9	0.725	0.685	0.649
1.8	0.351	0.318	0.292	10	0.771	0.729	0.691
2.0	0.374	0.340	0.312	11	0.814	0.771	0.732
2.5	0.427	0.389	0.358	12	0.856	0.812	0.771
3.0	0.476	0.434	0.400	13	0.896	0.851	0.809
3.5	0.520	0.477	0.440	14	0.935	0.889	0.845
4.0	0.562	0.516	0.477	15	0.973	0.925	0.881
4.5	0.601	0.553	0.512	16	1.009	0.960	0.915
5.0	0.638	0.588	0.546	17	1.044	0.994	0.948
6.0	0.707	0.654	0.608	18	1.078	1.027	0.980
7.0	0.771	0.715	0.666	20	1.143	1.091	1.042
8.0	0.829	0.772	0.720	22	1.205	1.152	1.101
9.0	0.885	0.825	0.773	24	1.26	1.21	1.158
10.0	0.937	0.875	0.821	26	1.32	1.27	1.21
11.0	0.986	0.923	0.867	28	1.38	1.32	1.26
12.0	1.033	0.968	0.911	30	1.43	1.37	1.31
13.0	1.078	1.014	0.954	32	1.48	1.42	1.37
14.0	1.121	1.056	0.994	34	1.53	1.47	1.41
15.0	1.162	1.096	1.034	36	1.58	1.52	1.46
16.0	1.20	1.135	1.071	38	1.62	1.56	1.50
17.0	1.24	1.173	1.108	40	1.67	1.60	1.55

TABLE IV—*Continued*

	Canal shape C $b_1 = 2.0$, $z = 1.25:1$				Canal shape C $b_1 = 2.0$, $z = 1.25:1$		
Flow rate (ft^3/sec)	FC1 $b = 5.125$ $L = 3.50$ $p_1 = 1.25$	FC2 $b = 5.75$ $L = 3.00$ $p_1 = 1.50$	FC3 $b = 6.375$ $L = 2.50$ $p_1 = 1.75$	Flow rate (ft^3/sec)	FC1 $b = 5.125$ $L = 3.50$ $p_1 = 1.25$	FC2 $b = 5.75$ $L = 3.00$ $p_1 = 1.50$	FC3 $b = 6.375$ $L = 2.50$ $p_1 = 1.75$
2	0.252	0.234	0.217	32	1.31	1.25	1.196
4	0.387	0.362	0.340	34	1.35	1.29	1.24
6	0.495	0.465	0.437	36	1.40	1.34	1.28
8	0.589	0.554	0.523	38	1.44	1.38	1.32
10	0.673	0.634	0.599	40	1.48	1.42	1.36
12	0.749	0.708	0.670	45	1.58	1.51	1.45
14	0.819	0.776	0.735	50	1.67	1.60	1.54
16	0.885	0.839	0.796	55	1.75	1.69	1.62
18	0.947	0.899	0.854	60	1.84	1.77	1.70
20	1.006	0.956	0.909	65	1.92	1.85	1.78
22	1.061	1.010	0.962	70	1.99	1.92	1.85
24	1.115	1.062	1.012	75	2.06	1.99	1.92
26	1.166	1.112	1.061	80	2.13	2.06	1.99
28	1.215	1.159	1.107	85	2.20	2.13	2.06
30	1.26	1.206	1.153				

[a] Add 6 in. to L if depths and discharges below dashed line (larger flow rates) are used.

Tables III and IV list the discharge–depth relationships for the farm-sized units in SI and English measuring systems, respectively. Figure 2 illustrates the terms used in the tables. Note that the discharge–depth relations are in columns reversed from that often shown for flumes and weirs. This format allows easy determination of gauge length at integer values of discharge without interpolation. This facilitates marking shop-constructed readout gauges (Clemmens and Replogle, 1980). The depths in the table need to be multiplied by the side-slope ratio (1.414 for 1 : 1 side slopes and 1.601 for 1.25 : 1 side slopes) if a wall-mounted gauge is to be used for direct readout.

Direct-reading side-wall gauges are encouraged even if recorders are installed. Figure 11 shows a broad-crested weir with a side-wall gauge in a canal of the type listed in Tables III and IV as "shape C." This allows instant flow-rate monitoring without the necessity of referring to a table, and prevents using the wrong table, or misreading in a proper table, once the gauge is correctly made and installed. Also, the instantaneous flow-rate reading is available to the farmer without waiting to intercept canal-operation personnel for information.

In the small sizes of broad-crested weirs selected for on-farm applications, the designations and selections were based on convenient increments of sill height,

FIG. 11. A delivery-system canal with a modified broad-crested weir.

since sill height is an important design parameter to determine canal overflow safety. It is not significantly important for maintaining an accurate (±2%) calibration because it causes only a minor change in canal velocity of approach to the weir. The SI values for sill heights p_1 in Table III are direct conversions from the English system of units, where the sill heights are 9, 12, 15, 18, and 21 in. (Table IV).

2. *Broad-Crested Weirs for Large Canals*

The most important broad-crested weir dimensions are the crest width across the canal and the canal side slopes, as discussed in Section V,A. The relative insensitivity to sill height is used to reduce the number of discharge ratings required to cover a wide range of large canal sizes. By selecting a width of weir and accepting the resulting sill heights for different canals, we can adapt a flume-rating table to two or more canal widths (both SI and English unit sizes) without significant error. For all large canals selected, a side slope of 1.5 horizontal to 1 vertical was used. Table V lists the selected bottom widths. Criteria from the U.S. Bureau of Reclamation and from the International Commission on Irrigation and Drainage (1979) were considered in determining these selections.

TABLE V
CANAL SIZES CHOSEN AS BASIS FOR FLUME DESIGN USING 1.5:1 SIDE SLOPES

Metric (m)			English (ft)		
Bottom width, b_1	Constructed depth, d	b_1/d	Bottom width, b_1	Constructed depth, d	b_1/d
0.6	1.0	0.60	2	3.5	0.57
0.8	1.2	0.67	3	4.5	0.67
1.0	1.4	0.71	4	5.5	0.73
1.5	1.8	0.83	5	6.0	0.83

In these larger sizes, broad-crested weirs are selected with modular widths in both the English and SI system of units, and are not necessarily equivalents, although either set would suffice for both systems of canal-width designs. Table VI gives the weir sizes (widths across the canal), each of which can accommodate canals with different bottom widths b_1. The designation M1.5, for example, indicates a metric, or SI-dimensioned, broad-crested weir with a 1.5-m crest width. Similarly, E5 indicates an English-unit-dimensioned weir with a 5-ft crest width b (Fig. 2).

Table VI also gives the sill height p_1 and the throat length L. Note that Table VI indicates that the throat length L varies for the same weir width designation when used in different canal configurations. For example, for weir M2.0 the throat lengths L are 1.0, 1.4, and 1.4 m for the 0.6-, 0.8-, and 1.0-m bottom widths respectively (Table VI). These selected bottom widths reflect the maximum flow depths expected in different canal sizes and the relationship of maximum flow depth and throat length as discussed in Section VI,A.

Tables VII and VIII give the calibrations for the broad-crested weirs listed in Table VI. The value x is the distance from the beginning of the throat section to the gauge location. This produces a more consistent adherence to the calibration tables for the various canal–weir combinations than designating a distance from the upstream end of the sloping ramps, which vary in length. [For the smaller flumes of Tables III and IV, a fixed distance from the end of the ramp of 0.3 m (1 ft) was used.]

As with the smaller broad-crested weirs, Tables VII and VIII are presented with discharge rate as the independent variable, the reverse of usual flume and weir tables. This procedure provides more table values in the lower flow ranges with evenly spaced flow rates than would be available with evenly spaced depth readings. Again, it allows simple calculation of values for marking side-wall gauges in discharge units by multiplying the listed depth for any selected flow rate by the side-slope ratio (1.803 for 1.5:1 canals).

TABLE VI

BROAD-CRESTED WEIRS THAT ACCOMMODATE LINED CANALS WITH 1.5 : 1 SIDE SLOPES

SI units[a]							
b_1	Weir	p_1	L	b_1	Weir	p_1	L
0.6	M1.50	0.300	1.1	1	M2.00	0.333	1.4
	M1.75	0.383	1.1		M2.25	0.417	1.6
	M2.00	0.467	1.0		M2.50	0.500	1.6
	M2.25	0.550	0.8		M3.00	0.667	1.4
	M2.50	0.633	0.7		M3.50	0.833	1.1
0.8	M1.75	0.317	1.2	1.5	M3.0	0.500	2.0
	M2.00	0.400	1.4		M3.5	0.667	2.2
	M2.25	0.483	1.3		M4.0	0.833	2.0
	M2.50	0.567	1.2		M4.5	1.000	1.6
	M3.00	0.733	0.9		—	—	—
English units[b]							
b_1	Weir	p_1	L	b_1	Weir	p_1	L
2	E5	1.00	3.5	4	E8	1.33	6.0
	E6	1.33	3.5		E9	1.67	6.5
	E7	1.67	3.0		E10	2.00	6.5
	E8	2.00	2.5		E12	2.67	5.5
	E9	2.33	2.0		E14	3.33	4.5
3	E6	1.00	4.5	5	E10	1.67	7.0
	E7	1.33	5.0		E12	2.33	7.0
	E8	1.67	5.0		E14	3.00	6.0
	E9	2.00	4.5		E16	3.67	5.0
	E10	2.33	4.0		E16	3.67	5.0

[a] All units in meters.
[b] All units in feet.

B. MODULAR LIMITS FOR BROAD-CRESTED WEIRS

The modular limits, or limiting submergence ratios, are expressed numerically as the downstream total head, divided by the upstream total head, at the gauging location (see Section IV,B). Both heads are referenced to the elevation of the weir sill crest and are the respective sums of the flow depth referenced to the sill elevation and the velocity head.

As mentioned in Section IV,B, modular limits vary widely and are usually related to the degree of parallel flow achieved in the throat section and the friction and eddy losses caused by the structure and the associated downstream expansion. They can approach 0.95 with special attention to the downstream expansion. However, the absolute water surface elevation that can be gained by this special attention is usually so small that it is seldom worth the effort or expense, particularly in the applications for the broad-crested weirs.

For design purposes a value for the modular limit of 0.85 is used with the small weirs of Tables III and IV. For the larger sizes of Tables VI–VIII a computed

TABLE VII

DISCHARGE OF MODIFIED BROAD-CRESTED WEIRS FOR USE IN LINED CANALS WITH 1.5:1 SIDE SLOPES (SI UNITS)

M1.5 b = 1.5 m x = 1.0 m z = 1.5		M1.75 b = 1.75 m x = 1.3m z = 1.5		M2 b = 2.0 m x = 1.6 m z = 1.5		M2.25 b = 2.25 m x = 1.9 m z = 1.5	
Q (m^3/sec)	h_1 (m)	Q (m^3/sec)	h_1 (m)	Q (m^3/sec)	h_1 (m)	Q (m^3/sec)	h_1 (m)
0.05	0.072	0.05	0.066	0.1	0.094	0.1	0.088
0.10	0.110	0.10	0.102	0.2	0.145	0.2	0.136
0.15	0.141	0.15	0.130	0.3	0.185	0.3	0.175
0.20	0.167	0.20	0.155	0.4	0.219	0.4	0.208
0.25	0.191	0.25	0.177	0.5	0.250	0.5	0.237
0.30	0.212	0.30	0.197	0.6	0.278	0.6	0.264
0.35	0.231	0.35	0.216	0.7	0.304	0.8	0.312
0.40	0.250	0.40	0.233	0.8	0.328	1.0	0.355
0.45	0.267	0.45	0.250	0.9	0.350	1.2	0.393
0.50	0.283	0.50	0.265	1.0	0.372	1.4	0.429
0.55	0.298	0.60	0.294	1.1	0.392	1.6	0.462
0.60	0.313	0.70	0.321	1.2	0.411	1.8	0.493
0.65	0.327	0.80	0.346	1.3	0.430	2.0	0.522
0.70	0.341	0.90	0.369	1.4	0.448	2.2	0.550
0.75	0.354	1.00	0.391	1.5	0.465	2.4	0.576
0.80	0.366	1.10	0.412	1.6	0.481	2.6	0.602
0.90	0.390	1.20	0.432	1.8	0.513	2.8	0.626
1.00	0.413	1.30	0.451	2.0	0.543	3.0	0.649
1.10	0.434	1.40	0.469	2.2	0.571	3.2	0.672
1.20	0.455	1.50	0.487	2.4	0.598	3.4	0.693
1.30	0.474	1.60	0.503	2.6	0.623	3.6	0.715
1.40	0.493	1.70	0.520	2.8	0.648	3.8	0.735
1.50	0.511	1.80	0.536	3.0	0.672	4.0	0.755
1.60	0.528	2.00	0.566	3.2	0.695	4.2	0.774
1.70	0.545	2.20	0.595	3.4	0.717	4.4	0.793
1.80	0.561	2.40	0.622	3.6	0.738	4.6	0.811
ΔH = 0.030 m or $0.1H_1$		ΔH = 0.035 m or $0.1H_1$		ΔH = 0.040 m or $0.1H_1$		ΔH = 0.047 m or $0.1H_1$	

continued

TABLE VII—*Continued*

M2.5 $b = 2.5$ m $x = 2.2$ m $z = 1.5$		M3 $b = 3.0$ m $x = 2.5$ m $z = 1.5$		M3.5 $b = 3.5$ m $x = 2.9$ m $z = 1.5$		M4 $b = 4.0$ m $x = 2.9$ m $z = 1.5$		M4.5 $b = 4.5$ m $x = 3.5$ m $z = 1.5$	
Q (m^3/sec)	h_1 (m)	Q (m^3/sec)	h_1 (m)	Q (m^3/sec)	h_1 (m)	Q (m^3/sec)	h_1 (m)	Q (m^3/sec)	h_1 (m)
0.2	0.129	0.2	0.117	0.5	0.189	0.5	0.174	0.5	0.162
0.4	0.197	0.4	0.179	1.0	0.288	1.0	0.267	1.0	0.250
0.6	0.251	0.6	0.229	1.5	0.367	1.5	0.342	1.5	0.321
0.8	0.298	0.8	0.273	2.0	0.435	2.0	0.407	2.0	0.382
1.0	0.339	1.0	0.311	2.5	0.495	2.5	0.464	2.5	0.437
1.2	0.376	1.2	0.347	3.0	0.550	3.0	0.516	3.0	0.488
1.4	0.411	1.4	0.379	3.5	0.600	3.5	0.564	3.5	0.534
1.6	0.443	1.6	0.410	4.0	0.647	4.0	0.610	4.0	0.578
1.8	0.474	1.8	0.439	4.5	0.691	4.5	0.652	4.5	0.619
2.0	0.502	2.0	0.466	5.0	0.732	5.0	0.692	5.0	0.658
2.2	0.529	2.2	0.492	5.5	0.772	5.5	0.730	5.5	0.695
2.4	0.555	2.4	0.517	6.0	0.809	6.0	0.767	6.0	0.730
2.6	0.580	2.6	0.541	6.5	0.845	6.5	0.802	6.5	0.764
2.8	0.604	3.0	0.586	7.0	0.880	7.0	0.835	7.0	0.797
3.0	0.627	3.5	0.638	7.5	0.913	7.5	0.867		
3.2	0.649	4.0	0.687	8.0	0.945	8.0	0.899		
3.4	0.671	4.5	0.732	8.5	0.976	8.5	0.929		
3.6	0.692	5.0	0.775	9.0	1.006	9.0	0.958		
3.8	0.712	5.5	0.815	9.5	1.035	9.5	0.987		
4.0	0.731	6.0	0.854	10.0	1.064	10.0	1.014		
4.4	0.769	6.5	0.891	10.5	1.092				
4.8	0.805	7.0	0.927	11.0	1.118				
5.2	0.839	7.5	0.961	11.5	1.145				
5.6	0.872	8.0	0.994	12.0	1.170				
6.0	0.904	8.5	1.025	12.5	1.195				
6.4	0.934	9.0	1.056	13.0	1.220				
$\Delta H = 0.053$ m or $0.1H_1$		$\Delta H = 0.063$ m or $0.1H_1$		$\Delta H = 0.072$ m or $0.1H_1$		$\Delta H = 0.080$ m or $0.1H_1$		$\Delta H = 0.090$ m or $0.1H_1$	

value of head loss ΔH is listed at the bottom of the columns that will be satisfactory for the narrow range of canal sizes it can accommodate but with the restriction that the computed modular limit will not be greater than $0.90H_1$. This means that the minimum head loss through the weir is 10% of H_1. Thus the minimum design head loss is either $0.1H_1$ or the listed value for ΔH, whichever is greater. These values for ΔH are derived from a computational procedure suggested by Bos and Reinink (1981) (Section IV,B). Technically, the modular limit is based on the drop in total head through the weir (H_1/H_2), but the velocity head components are small, and are within the range of uncertainties in the calculated modular limits, so that h_1 may be used instead. Thus,

$$\text{Modular limit} = \frac{h_1 + v_1/2g}{h_2 + v_2/2g} = \frac{H_1}{H_2} \simeq \frac{h_1}{h_2} \qquad (44)$$

TABLE VIII

DISCHARGE OF MODIFIED BROAD-CRESTED WEIRS FOR USE IN LINED CANALS WITH 1.5:1 SIDE SLOPES (ENGLISH UNITS)

E5 $b = 5.0$ ft $x = 3.5$ ft $z = 1.5$		E6 $b = 6.0$ ft $x = 4.5$ ft $z = 1.5$		E7 $b = 7.0$ ft $x = 5.5$ ft $z = 1.5$		E8 $b = 8.0$ ft $x = 7.0$ ft $z = 1.5$	
Q (ft^3/sec)	h_1 (ft)	Q (ft^3/sec)	h_1 (ft)	Q (ft^3/sec)	h_1 (ft)	Q (ft^3/sec)	h_1 (ft)
2	0.253	2	0.231	5	0.373	5	0.347
4	0.387	4	0.353	10	0.569	10	0.532
6	0.494	6	0.452	15	0.725	15	0.679
8	0.585	8	0.537	20	0.859	20	0.806
10	0.667	10	0.613	25	0.977	25	0.920
12	0.741	12	0.683	30	1.085	30	1.023
14	0.809	14	0.747	35	1.184	35	1.119
16	0.872	16	0.807	40	1.277	40	1.208
18	0.932	18	0.864	45	1.364	45	1.292
20	0.988	20	0.918	50	1.446	50	1.371
22	1.042	25	1.041	55	1.523	55	1.446
24	1.093	30	1.153	60	1.598	60	1.519
26	1.141	35	1.256	65	1.669	70	1.654
28	1.188	40	1.351	70	1.737	80	1.780
30	1.234	45	1.440	75	1.803	90	1.898
32	1.277	50	1.525	80	1.867	100	2.009
34	1.320	55	1.605	85	1.928	110	2.115
36	1.360	60	1.681	90	1.987	120	2.215
40	1.439	65	1.754	95	2.045	130	2.312
45	1.531	70	1.823	100	2.101	140	2.404
50	1.618	75	1.891	105	2.156	150	2.493
55	1.700	80	1.955	110	2.209	160	2.578
60	1.778	85	2.018	120	2.312	170	2.661
65	1.853	90	2.079	130	2.410	180	2.741
70	1.924	95	2.137	140	2.504	190	2.819
75	1.993	100	2.194	150	2.594	200	2.894
HL = 0.11 ft or $0.1H_1$		HL = 0.13 ft or $0.1H_1$		HL = 0.16 ft or $0.1H_1$		HL = 0.18 ft or $0.1H_1$	

continued

TABLE VIII—*Continued*

E9 $b = 9.0$ ft $x = 8.0$ ft $z = 1.5$		E10 $b = 10.0$ ft $x = 8.0$ ft $z = 1.5$		E12 $b = 12.0$ ft $x = 9.0$ ft $z = 1.5$		E14 $b = 14.0$ ft $x = 11.5$ ft $z = 1.5$		E16 $b = 16.0$ ft $x = 12.5$ ft $z = 1.5$	
Q (ft^3/sec)	h_1 (ft)	Q (ft^3/sec)	h_1 (ft)	Q (ft^3/sec)	h_1 (ft)	Q (ft^3/sec)	h_1 (ft)	Q (ft^3/sec)	h_1 (ft)
5	0.327	10	0.471	10	0.424	10	0.385	10	0.353
10	0.499	20	0.720	20	0.651	20	0.594	20	0.547
15	0.640	30	0.919	30	0.835	30	0.765	30	0.706
20	0.761	40	1.090	40	0.994	40	0.914	40	0.845
25	0.870	50	1.242	50	1.137	50	1.047	50	0.971
30	0.969	60	1.380	60	1.267	60	1.170	60	1.087
35	1.061	70	1.508	70	1.388	70	1.284	70	1.195
40	1.148	80	1.627	80	1.500	80	1.392	80	1.296
45	1.229	90	1.739	90	1.607	90	1.493	90	1.392
50	1.306	100	1.845	100	1.708	100	1.589	100	1.484
60	1.449	110	1.946	120	1.896	120	1.769	120	1.655
70	1.581	120	2.042	140	2.069	140	1.935	140	1.814
80	1.704	130	2.134	160	2.230	160	2.090	160	1.963
90	1.820	140	2.223	180	2.382	180	2.236	180	2.104
100	1.929	150	2.308	200	2.525	200	2.374	200	2.237
110	2.032	160	2.390	220	2.661	220	2.506	220	2.363
120	2.131	170	2.470	240	2.790	240	2.632		
130	2.225	180	2.547	260	2.914	260	2.752		
140	2.316	190	2.622	280	3.033	280	2.868		
160	2.487	200	2.695	300	3.148	300	2.979		
180	2.648	220	2.835	320	3.258				
200	2.799	240	2.969	340	3.365				
220	2.942	260	3.097	360	3.469				
240	3.078	280	3.219	380	3.570				
260	3.208	300	3.337	400	3.667				
280	3.332	320	3.450	420	3.763				
$HL = 0.20$ ft or $0.1H_1$		$HL = 0.22$ ft or $0.1H_1$		$HL = 0.26$ ft or $0.1H_1$		$HL = 0.30$ ft or $0.1H_1$		$HL = 0.34$ ft or $0.1H_1$	

It should be pointed out that the absolute change in water surface elevations for the modular limit of broad-crested weirs is very low because of the high sills and shallow flows, and is usually from one-half to one-fourth the elevation difference required for a flume with side-wall contractions and deep flow. This makes the broad-crested weir especially attractive for irrigation applications in retrofit situations where existing freeboard is usually limiting.

C. Freeboard Recommendations

An obvious restriction on sill height is that it not cause so much of an obstruction that the upstream canal is overtopped. The Soil Conservation Service has long recommended that the freeboard for stable, subcritical flow in small trapezoidal canals be about 20% of the energy head at normal depth to allow for overdischarges, wave action, and sediment accumulations. For approach velocities usually found in these small canals, this is nearly equivalent to using 20% of the flow depth upstream from the sill crest. Because of the stable nature of the flow upstream from broad-crested weirs, and the reliable depth–discharge relationship, the freeboard requirement can be safely reduced from 20% of Y_1 to 20% of h_1, which is the sill-referenced flow depth. This recommendation is somewhat flexible because freeboards less than these can and have been tolerated in specific cases where limited possibility for overdischarges was assured and no problems were anticipated from wave action and sediment deposition.

D. Numerical Examples

1. *Small Canal (English Units)*

A 2.67-ft-deep canal has been previously constructed with 2-ft bottom and 1 : 1 side slopes. It is to convey 20 ft^3/sec and is to retrofitted with a broad-crested weir. The canal flow at 20 ft^3/sec is determined to be 2.0 ft deep before the weir is constructed. Select a broad-crested weir that will maintain a freeboard equal to or greater than 20% of h_1 for the maximum flow.

Table IV offers three precomputed weir sizes for this shape of canal: FB1, FB2, and FB3. We first compute the respective upstream canal flow depths that each weir will cause for 20 ft^3/sec and compare it to the required minimum increase in depth, which should equal a 15% increase in h_1 to allow for a modular limit of 0.85. The upstream depth is calculated as the sum of the table values of h_1 for 20 ft^3/sec and the sill height p_1, both from Table IV. These flow depths are given in Table IX. Thus, FB1 will not provide enough increase in upstream depth to assure free discharge. Both FB2 and FB3 meet this requirement.

The available freeboard for the two workable sill sizes is the difference between the upstream flow depth and the constructed canal top at 2.67, or 0.33, and 0.13 ft, respectively, for FB2 and FB3. Thus FB2 provides more than the recommended 20% of h_1, or 0.22 ft as shown in the tabulation, and FB3 is somewhat short of this recommendation.

The selected broad-crested weir is then FB2, which best meets the requirements on both modular limit and freeboard. If freeboard limits could not have been met, then the upstream canal would need to have its sides extended upward a few centimeters with a concrete edging over sufficient distance to protect the

TABLE IX

FLOW DEPTHS

Specification (ft)	Weir FB1	FB2	FB3
Still height	1.000	1.250	1.500
h_1 for 20 ft^3/sec	1.143	1.091	1.042
Upstream canal depth	2.143	2.341	2.542
Original depth	2.000	2.000	2.000
Increase in flow depth	0.143	0.341	0.542
Required increase ($0.15h_1$)	0.171	0.164	0.156
Recommended freeboard ($0.2h_1$)	0.22	0.22	0.21

canal embankments. This distance varies with the canal slope but is usually only 20–50 m. Concrete blocks have been satisfactorily used to provide a 20-cm vertical extension to trapezoidal canals.

2. *Large Canal (SI Units)*

Larger canals, which are not usually used inside farm units, usually require access to engineering assistance; therefore, the following example will be presented in an engineering format.

Given: An existing concrete-lined canal is constructed with depth $d = 1.4$ m, side slope $z = 1.5$, and bottom $b_1 = 1$ m. It carries a maximum discharge estimated to be 3 m^3/sec at an observed flow depth $y = 1.1$ m. Manning's equation is assumed to control the discharge-stage relationships for the flows. However, no direct canal data other than cross-sectional shape are available.

Required: Select a broad-crested weir using Tables VI and VII that will measure this maximum flow as well as measure lower discharge rates down to at least 0.3 m^3/sec (a 10:1 flow range).

Solution: From Table VI five weirs are offered that will fit into a canal with $b_1 = 1.0$ m

Try: Weir M2.50, sill width = 2.5 m:

$$\begin{aligned} p_1 &= 0.500 \text{ m (from Table VI)} \\ h_1 &= 0.627 \text{ m (from Table VII, } Q = 3 \text{ m}^3\text{/sec)} \\ y_{1_a} &= p_1 + h_1 = \underline{1.127 \text{ m}} \\ \text{But the required } y_1 &= 1.1 \text{ m (given)} + \Delta H \\ \Delta H &= 0.1\,(Y_1) = 0.0627 \text{ m } (>0.053 \text{ m, Table VII}) \\ \text{Required } y_{1_r} &= 1.1 + 0.063 = \underline{1.163 \text{ m}} \end{aligned}$$

Since $y_{1_a} < y_{1_r}$, the weir will not function (modular limit exceeded); try a higher sill (wide flume).

Next, try: Weir M3.50, sill width = 3.5 m:

$$\begin{aligned} p_1 &= 0.833 \text{ m} \\ h_1 &= 0.550 \text{ m} \\ y_{1_a} &= 1.383 \text{ m} \\ \Delta H &= 0.1 y_1 = 0.550 \ (<0.072) \\ \text{Required } y_{1_r} &= 1.1 + \Delta H = 1.1 + 0.072 = \underline{1.172 \text{ m}} \end{aligned}$$

Since $y_{1_a} > y_{1_r}$, this flume might be adequate. However, it leaves nearly zero freeboard, about 17 mm (1.4–1.383 m) which is too little to meet the recommendation that freeboard be about 20% of h_1, which would be 110 mm

Next, try: Weir M3, sill width = 3.0 m:

$$\begin{aligned} p_1 &= 0.667 \text{ m} \\ h_1 &= 0.586 \text{ m} \\ y_{1_a} &= 1.252 \text{ m} \\ \Delta H &= 0.1 y_1 = 0.0586 \ (<0.063) \\ \text{Required } y_{1_r} &= 1.1 + \Delta H = 1.1 + 0.063 = \underline{1.163 \text{ m}} \\ y_{1_a} &> y_{1_r} \end{aligned}$$

Thus, this flume will be satisfactory. It leaves about 0.15-m freeboard (1.4–1.252 m), which is about 25% of h_1 and should be adequate, considering the reliable surface stability and the fact that overflow will not occur until h_1 exceeds 0.73 m ($y_1 = 1.4 - p_1 = 0.73$ m), which represents a discharge of about 4.5 m^3/sec (from Table VII, weir M3).

Since normal depths are assumed for the canal (Mannings equation is valid), lower flows will create more and more differences in water surface elevations through the weir and thus will be safe from submergence. The required range in flows can be readily achieved (Table VII).

Other engineering values of interest are the upstream velocity $v_1 = Q/A =$ 0.833 m/sec, and the Froude number = 0.32 [Eq. (15)]. Froude numbers below about 0.5 indicate a reasonably stable water surface suitable for side-wall gauges or stilling-well installations, and free from standing waves that would affect pressure readings. Velocities above about 0.3 m/sec aid in controlling the snail, which in tropical countries hosts the parasite of schistosomiasis.

3. *Canal with Slight Drop*

As mentioned previously, sediment movement can be expected to be less interrupted if the measuring structure is installed in such a way as to assure that normal depth in the upstream reach is maintained for the most usual flow rate. In this example, we select a structure that will approximate this.

Given: The canal and flows of the previous example.

Required: Select a weir that will maintain normal depth upstream of the weir and design a canal drop downstream from the weir to take care of the head loss.

Solution: The problem reduces to selecting a weir that passes 3 m^3/sec at an upstream canal depth that matches or slightly exceeds the normal depth of 1.10 m. Consulting Table VI, it is found that five choices of weirs are available, ranging in width from 2.0 to 3.5 m for this canal size. Now, consulting the appropriate calibrations in Table VII, consider the five combined sums of $h_1 + p_1$, obtaining the p_1 values from Table VI. These are, respectively,

$$0.677 + 0.333 = 1.005 \text{ m}, \quad \text{for a weir 2 m wide}$$
$$0.699 + 0.417 = 1.070 \text{ m}, \quad \text{for a weir 2.25 m wide}$$
$$0.627 + 0.500 = 1.127 \text{ m}, \quad \text{for a weir 2.50 m wide}$$

The remaining weirs are quickly seen to offer sums that are too great. Therefore, the nearest selection that slightly exceeds the normal depth of 1.10 m is that for the 2.5-m width (weir 2.5 m, Table VII).

Since the weir must operate at maximum flow without exceeding its modular limit, the drop in the canal at the structure must match or exceed the ΔH value for the maximum flow of 3 m^3/sec. From Table VII, Weir M2.5, this is $0.1H_1 \simeq 0.1h_1 = 0.063$ m.

Thus the canal bottom should drop about 0.06 m at the downstream edge of the sill. This may be accomplished as an abrupt drop, or, if concrete slipforming equipment is used, as a sloping section of canal that achieves the desired drop in 10 or 20 m, beginning at the downstream end of the weir.

Again calculate the upstream velocity for the maximum flow rate, which is 0.99 m/sec, and the Froude number, which is 0.38. This should provide stable flow conditions for easy water surface level detection.

Sill heights that require some drawdown below normal depth may be used, but additional allowance in the canal drop must be provided. This example indicates that normal depth in many canals might require weir sills lower than those that are offered in the precomputed tables. The lowest sills offered usually result in Froude numbers approaching 0.5 at the maximum discharge in that table, because the Froude number for trapezoidal channels increases with depth. In this example, the lowest sill results in a Froude number that is still less than the recommended maximum of about 0.5. If desired, a lower sill can be computed by the methods of Section IV to more clearly satisfy the normal depth flow conditions of the example.

A word of caution is in order. If the computed normal depth indicates sills lower than those offered in the precomputed tables, then special care must be exercised in examining the stability of the flow surface (Froude number, ≤ 0.5). If the velocity for normal depth is such that the Froude number is greater than 0.5, then a sill higher than that for normal depth is needed. In this case the head loss can be accommodated with a combination of canal drop and raised upstream water surface.

4. *Discussion of Examples*

A remark concerning the calculation of normal depth using Manning's equation may be in order, to show how it can be used for extrapolating to flows other than the existing known flows.

In the example, neither channel slope nor friction factor were used directly. If the friction factor is assumed to remain constant for a reasonable range of flows, and the channel slope for normal discharges parallels the true energy slope needed for Manning's equation, then these factors are identical for both flow conditions, and are not needed when determining the normal discharge-depth relations at some other flow rate.

This process with Manning's equation can be used to convert a known depth value at one discharge to the necessary information for the required flow. The process is essentially the trial and error solution of algebraic equations, which can be tedious, and may be expedited by use of the hydraulics tables and nomograph solutions to Manning's equation that appear in many hydraulics design manuals.

In all of these examples, the hydraulic roughness has been assumed constant with flow stage and time. Neither assumption is strictly true, but the change with stage in the ranges encountered in lined irrigation canals is much smaller than the seasonal and maintenance changes that occur. Thus probable maintenance procedures, biological growth, chemical precipitants, and sediment accumulations can all affect the necessary elevation of the sill, and they must be considered to avoid exceeding the modular limit. Usually a value of n should be selected for use in Manning's equation that reflects the worst expected seasonal and maintenance conditions.

E. Construction of Broad-Crested Weirs

1. *Cast-in-Place Concrete Versions*

Once a suitable broad-crested weir has been selected, construction is relatively easy and straightforward. The weir has two sections, the control section made by the sill of length L and height p_1 (Fig. 2) and a sloping ramp of length $3p_1$ that forms the transition from the channel shape to that of the sill. The sill top should be level with no large irregularities. The constructed width of the weir b should be as close to the value in the appropriate tables as the design accuracy of the measurement, because a 1% error in b will produce about a 1% error in discharge (see Section V).

While the sill height p_1 is important for selecting the weir, because it controls the water surface elevation and thus the modular limit and freeboard relationships, its precise vertical dimension is not critical to the weir calibration relation.

Also, the length L can be adjusted $\pm 10\%$ without noticeable effect. The ramp length of $3p_1$ is also approximate and its purpose is to convey the water smoothly to the weir crest. Thus, it may be straight or gently curved, but the straight kind is usually more readily constructed. These liberal tolerances should facilitate construction, but should not be an excuse for slipshod construction.

A drain tube should be installed as shown in Fig. 2 for control of mosquitoes, particularly if the canal is intended for intermittent use, as would be usual for farm systems. Flow through the drain tube is negligible if its diameter is about 30 mm or less and the canal flow is near capacity for the weir. If the tubes are larger, they should be at least partly blocked during measured deliveries. This can be done with valves, bricks, or rags. The latter can be quickly pushed out of the tube with a stick or rod when drain-down of the canal is desired. Complete blockage is not required. Sediments, which tend to accumulate at the base of the ramp, may plug the drain pipe and should be cleaned out periodically. Sediments will not significantly accumulate in the canal upstream from the weir if they did not normally accumulate in the canal before the weir was installed. Canals that normally accumulate significant depths of sediments will continue to accumulate similar depths upstream from the weir.

Cast-*in-situ* construction is simple. Make two forms from forming material, such as plywood, both matching the cross section of the sill to be installed (Fig. 2). Cut a notch in the center bottom of each piece to accommodate the drain tube. The drain tube may be located anywhere along the bottom, and not necessarily at the center line. The bottom corners of the resulting trapezoidal shape will usually have to be trimmed to fit the canal, because slipformed canals frequently have slightly rounded corners. Alternatively, these corner areas may be used for the drain tubes. Next make three to five temporary wood spacers of length L to hold the two identical forms apart until concrete can be poured into them. With the spacers temporarily nailed in place, drill small holes through the forms and use wire to tie the forms together. Usually with the smaller sills, on the order of 1–2 m wide, two wire ties near the bottom and three near the sill top are sufficient. After the tied and spaced forms are in place in the canal, insert the length of drain tube, which should be at least as long as the sill plus the ramp length. Alternately, threaded steel bars can be used to replace both the spacers and wire ties.

If the canal has construction joints, the sill should be placed so that the gauge location is at least 0.5 m downstream from the joint. If that is difficult, the gauge should be located slightly upstream of the joint, and the sill positioned at distance x downstream as listed in Tables VII and VIII and shown in Fig. 2 for large canals, and at 0.3 m from the end of the ramp, as listed for the small canals in Tables III and IV. If possible, construction joints should be avoided in the area of the sill, although they are acceptable in the ramp section. Care should be exercised to assure that if a construction joint is located between weir sill and gauge, no vertical movement occurs, or else the zero settings may not remain reliable.

The forms should be checked with a carpenter's level to assure that the top edges are level in the cross-canal direction and with respect to each other (Fig. 12). Minor shimming with thin strips of wood or other material is usually necessary. The forms should be filled with a concrete mix that is suitable for local waterborne chemicals and prevalent weather conditions. These factors are more important than monolithic strength, since stresses on the weir are usually low. As the forms are filled with fresh concrete, the spacers are carefully removed while the wire ties are left in the concrete. Wooden spacers, particularly, should be removed because they tend to swell and may break the concrete. The concrete will hold the forms apart against the stress of the wire ties. The tops of the plywood forms are used as guides for screeding and troweling the sill to a level surface.

After the concrete in the sill sets sufficiently (usually after a few hours), the upstream form can be carefully removed and the ramp hand-troweled into place. Both the ramp and the sill may be broom finished. The upper edge of the ramp should join the sill edge so that there is no abrupt rise or drop that might

FIG. 12. Concrete forms should be level laterally and longitudinally.

cause unpredictable flow separation. Precision edge matching of the two sections or slight rounding of the joint corner is desirable. This is easily done with the concrete trowel. The ramp should be fairly uniform with no large humps or depressions. Its surface should approximate a slope of 3 horizontal to 1 vertical as seen in sectional view. For sediment passage, it is best that the ramp be slightly sunken or concave rather than rounded upward, or convex. The ramp need not taper to zero thickness but can be ended abruptly when it becomes about 50 mm thick.

Sometimes, it is not convenient to make the forms as described above, and to wait for the sill to harden, before constructing the ramp. If a single concrete pouring operation is desired, then the upstream form can be made from a length of steel angle-iron, flat-steel stock, or pipe of sufficient rigidity that it will remain straight to within the construction tolerance during the screeding and troweling operations (Fig. 13). This upstream form can be a complete trapezoidally shaped frame or a single bar cut to span the width of the particular weir. Spacers and wire ties are not very effective for this construction method, but can be used to maintain top-edge spacing. Soil or other material should be piled against the outside of the downstream form to hold it in place against the pressures of the fresh concrete. This will usually have to be done in stages as the concrete is poured, since the soil and concrete each serve to hold the form against the other. The concrete, of course, will spill through the metal upstream frame, or under the single bar, as appropriate, and tend to automatically form the ramp. Most usual concrete mixes are sufficiently resistant to flow that the 3 : 1 ramp is easily made and the metal bar provides the guide for the sill top finishing. The metal bar is left

FIG. 13. A steel-structured angle or a bar can be used to define the upstream sill top so that the sill and the ramp can be poured simultaneously.

permanently in the concrete. Thus, both the sill and the ramp can be constructed at the same time without requiring the builders to wait at, or return to, the site.

2. *Precast Concrete Versions*

For the smaller canals, precast versions of the permanent weirs can be made. Reinforcing wire and two 15-mm-diameter reinforcing rods are used in thin shell copies of the weirs. The concrete thickness is usually about 5 cm with concrete crossbeams cast into the sill piece. A groove is formed on the upstream edge of the sill piece to support the precast ramp section.

For the smallest size in the 1-ft bottom-width canal (designated canal shape A) the concrete mass can be kept below 40 kg and can be hand carried. Sizes for canal shapes B and C usually require machinery or 3- and 4-man crews for moving the pieces. The canal side walls and bottom support three edges of the ramp section, whereas the fourth is supported by the sill piece, thus reducing its structural requirements. These pieces have been successfully made, and handled without breakage, in the size for FB2, specified in Table IV. The sill pieces contain one reinforcing bar each, centered about 25 mm from the bottom of the beam. The floor part and the ramp need contain no reinforcing. Concrete design calculations show that on the longest beams (Weir FC3 and FC3M, with b = 6.375 ft or 1.95 m) a point mass on each beam exceeding 160 kg (350 lb mass) would be required to exceed design stress. For a uniformly distributed load the mass can be 320 kg (700 lb). Maximum loading at maximum design flow (a

FIG. 14. Adjustable concrete forms for precast sills and ramps with reinforcing wire and bars in place.

stream 0.65 m deep and 1.95 m wide) will not exceed these design limits. For safety in handling, simple wire reinforcing in the flat portions of the sill piece and in the ramp piece is recommended.

Figure 14 shows adjustable forms used to make several sizes of the above with reinforcing bars and wire mesh in place. Figure 15 shows a precast broad-crested weir in place.

3. *Sheet-Metal Fabrications*

Sheet metal can also be used. Fig. 16 shows an installation made using galvanized sheet steel. The support cross members can usually be made from the same sheet material bent at right angles, particularly for the smaller sizes, although here (Fig. 16) commercial angles were used. The ends of the cross members were bent to form tabs for screw anchoring the sill to the channel. The ramp was then secured to the upstream edge of the sill with sheet-metal screws.

4. *Portable Broad-Crested Weirs for Lined Canals*

Modified broad-crested weirs can be constructed in portable versions for use in concrete-lined canals. Only the sill crest and ramp pieces are needed. The canal provides the remaining structure, as it did for the more permanent broad-crested weirs described above. For convenient use, however, the need for precision surveying to maintain proper zero referencing between the sill crest and the upstream gauge should be eliminated. Thus, in addition to being lightweight, a portable weir should provide for a practical method to maintain proper zero registration.

FIG. 15. Precast concrete sill and ramp placed in canal ready for grouting with mortar.

A method for accomplishing the two goals for portability and convenient flow-depth determination was reported by Replogle and Clemmens (1979) and is shown in Figs. 17 and 18. The weir sill (1) causes a flow obstruction in the canal that in turn induces a rise in the upstream water surface (3). This water surface elevation above the sill is accurately sensed through a ring of holes (4) in a piece of pipe (5) laid in the canal at a point upstream that is a selected distance from the flume. In this example, the sensing holes were located 0.305 m upstream from the rest of the structure. If the sensing holes are located at some other distance, for example on a larger flume, the computer calibration should be based on the actual distance selected. The sensing pipe is constructed with a rounded upstream end (6) and the pressure-sensing holes placed 10–20 pipe diameters from that point to ensure accurate pressure detection (Rouse, 1961), uninfluenced by the point of the pipe itself. This pressure is transferred by a siphon tube (14) to a small version of a stilling well (7) placed, not in the ditch bank as usual, but above the flume throat, or weir crest, near the crest centerline and about ¼–⅓ the length of the throat, or crest, from the flow outfall (8). This represents the approximate location of critical depth for parallel flow, which varies slightly with the discharge. Above this same location, a point gauge (9) is mounted rigidly to the flume frame so that it can be quickly lowered and permanently adjusted to read zero when the point touches the crest. The gauge is then raised and the small stilling-well cup moved under the point, thus accurately measuring the upstream water surface above the flume-throat reference point. A

FIG. 16. Galvanized sheet metal used to make sills.

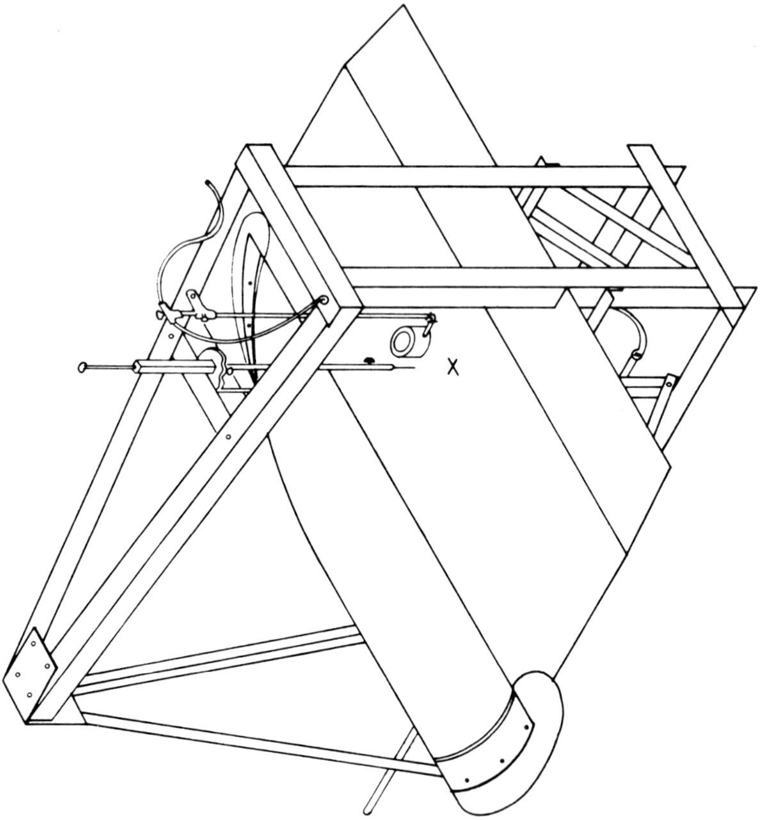
X

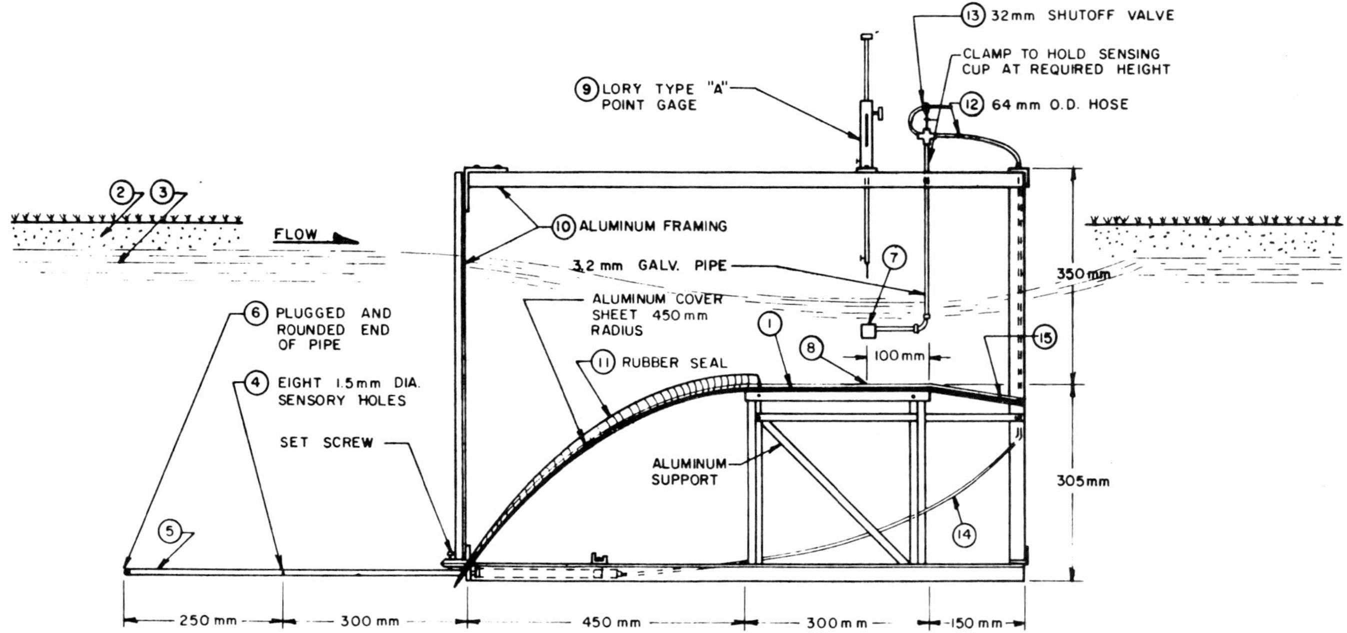

FIG. 17. Schematic of portable flume operation. (1) Flume sill; (2) concrete canal; (3) upstream water surface; (4) sensing holes; (5) retractable sensing pipe; (6) closed, pointed end of pipe; (7) stilling-well cup; (8) sill floor reference point for zeroing point gauge; (9) point gauge; (10) upper flume frame; (11) rubber seal strips; (12) siphon priming tube; (13) shutoff valve for siphon priming tube; (14) siphon tube; (15) diverging section—truncated ramp (optional).

FIG. 18. Portable flume in a lined canal.

siphon priming tube (12) with a valve (13) is used to remove air from the siphon tube (14).

Leakage past the flumes is controlled by rubber strips (11) attached to the front of the converging section (ramp) and another strip attached across the bottom. It is helpful if, during construction, this sheeting is warped slightly so that it springs strongly upstream, especially across the bottom of the flume. Water pressure then causes a good seal against the canal wall and floor. Too little spring allows the rubber to turn under the flume during placement, thus negating its effects.

Design details are described by Replogle and Clemmens (1979). The calibration closely approximates that for canal shape A (Tables III and IV for size FA2 and FA2M, respectively) to within ±1% of the slightly more accurate calibration given in the original paper.

VIII. Structures for Unlined Canals

A. Adaptations to Use Lined-Canal Structures

One approach for inserting structures in unlined canals is to simply line a short length of canal and use one of the lined-canal structures from Tables III, IV, VII, and VIII. If this is done, the lined section should provide an approach length to the gauging location that is a minimum of about two to three canal depths long, and provide adequate downstream erosion protection.

Usually, unlined canals have different width-to-depth ratios than lined canals, being wider by a factor of 2 or more, and flow at lower velocities for the same discharge rates. Thus, unlined channels usually require backfilling before constructing a lined section. The backfill must be adequately compacted. An upstream head wall and downstream toe wall are recommended to prevent piping under and around the inserted canal section and structure. An installation of this kind is shown in Fig. 19, but without the recommended flow cutoff walls. Some have been undercut during the initial uses.

B. Rectangular Structural Shapes

Because of the variety of earthen canal shapes that can be encountered, the choice of appropriate preselected structures is very large. One approach to nearly unlimited offerings are rectangular flume openings in the form of rectangular broad-crested weirs (Fig. 20).

The rectangular broad-crested weirs are offered in terms of discharge per unit of sill-crest width. The problems of the side wall effects is narrow flumes are

FIG. 19. Earthen channel with a short section of canal lining fitted with a precast concrete sill and ramp (Parker, Arizona).

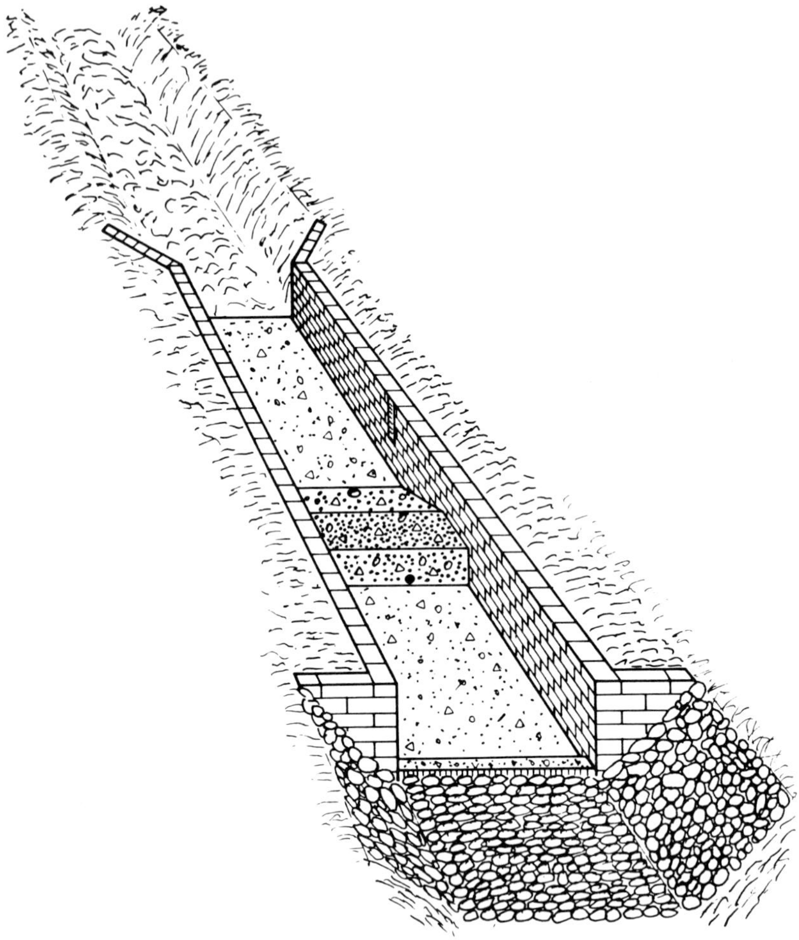

FIG. 20. Rectangular broad-crested weir in an earthen channel.

handled by offering slightly different calibrations for limited ranges of total width (Table X). As the crest width becomes very wide, the side wall effects can no longer be detected. Thus Table X offers one to three sill heights p_1 for specific ranges of rectangular width b. The discharge rate q in liters per second, is for a unit width of 1 m and is to be adjusted to the precise weir width selected.

Special hydraulic considerations are involved in placing a rectangular control in a trapezoidal or parabolic channel. Checking the modular limit at low flows is usually necessary because the water levels through the rectangular control drop faster than they can be expected to drop in a trapezoidal channel. This may require a sill that is higher than the high-flow requirements demand, to avoid low-flow problems. This point is discussed in more detail by Replogle (1968).

The specific use of Table X will be illustrated in the numerical examples which follow.

Numerical Example 1

Given: An earthen canal constructed to about a 0.75-m depth has an approximately trapezoidal shape with a bottom width of about 0.3 m and a top width of about 1.5 m at a flow depth y_1 of 0.45 m. Separately determined discharge estimates place the maximum flow at 200 liters/sec. A rectangular broad-crested weir 0.5 m wide is to be installed. Recommend a suitable sill height p_1.

Solution: Convert the existing discharge to discharge per meter of width for use in Table XI: $q = 400$ liters/sec m^{-1}. Enter Table XI for $b = 0.5$ to 1.0 m and for the given value for q of 400 liters/sec m^{-1}. Two possibilities are offered.

Possibility (a) $p_1 = 0.2$ m
$h_1 = 0.357$ m for maximum flow
$\Delta H = 0.047$ m or $0.1h_1$; use 0.047 m

The resulting flow depth is $p_1 + h_1 = 0.557$ m and exceeds the required flow depth, which is $y_1 + \Delta H = 0.497$ m.

Possibility (b) $p_1 = 0.3$ m
$h_1 = 0.365$ m for maximum flow
$\Delta H = 0.058$ m

The resulting flow depth is $p_1 + h_1 = 0.665$ m and exceeds the required flow depth, which is $y_1 + \Delta H = 0.508$ m. Neither choice causes freeboard problems, so normally the lower sill, $p_1 = 0.2$ m, would be chosen unless high tailwater existed to interfere with lower flows. Note that rectangular control sections placed in trapezoidal channels should be checked for low-flow tailwater problems, because the flow levels through the rectangular control decrease faster than those in a trapezoidal channel.

Numerical Example 2

Given: A flow in a poorly defined earthen canal is 1 m^3/sec. The normal depth in the canal is 0.8 m measured from the lowest point in the canal cross section. The maximum usable canal depth is about 1 m. Select a rectangular broad-crested weir that can be fitted to this situation.

Solution: From the information given, $h_1 + p_1$ must be less than 1.0 m but more than the required channel depth of $0.8 + \Delta H$, or $0.8 + \Delta H \leqslant h_1 + p_1 \leqslant 1.0$. For best sensitivity the width should be relatively narrow. Try a weir with $p_1 = 0.5$ m and a probable total width greater than 2 m [Table X]. In the columns under $b \geqslant 2.0$ m and $p_1 = 0.50$ m find an h_1 value that satisfies $0.8 + 0.107 \leqslant h_1 + 0.5 \leqslant 1.0$. Thus, h_1 should be between 0.407 and 0.5 m.

TABLE X

UNIT WIDTH DISCHARGES IN LITERS PER SECOND PER METER OF WIDTH FOR RECTANGULAR BROAD-CRESTED WEIRS[a]

$b = 0.1$–0.2 m ($L = 0.2$ m, $L_a = 0.05$ m)		$b = 0.2$–0.3 m ($L = 0.4$ m, $L_a = 0.1$ m)		$b = 0.3$–0.5 m ($L = 0.5$ m, $L_a = 0.2$ m)			$b = 0.5$–1.0 m ($L = 0.8$ m, $L_a = 0.3$m)			
					h_1 (m)			h_1 (m)		
q (liters/sec m^{-1})	h_1 (m) ($p_1 = 0.05$)	q (liters/sec m^{-1})	h_1 (m) ($p_1 = 0.1$ m)	q (liters/sec, m^{-1})	$p_1 = 0.1$ m	$p_1 = 0.2$ m	q (liters/sec m^{-1})	$p_1 =$ 0.2 m	$p_1 =$ 0.3 m	$p_1 =$ 0.4 m
2	0.012	6	0.024	10	0.034	0.034	20	0.053	0.053	0.053
3	0.015	8	0.029	20	0.052	0.053	40	0.083	0.083	0.083
4	0.018	10	0.034	30	0.067	0.068	60	0.108	0.109	0.109
5	0.021	12	0.038	40	0.080	0.082	80	0.129	0.131	0.132
6	0.024	14	0.041	50	0.093	0.095	100	0.149	0.151	0.152
7	0.026	16	0.045	60	0.104	0.107	120	0.168	0.170	0.171
8	0.029	18	0.049	70	0.115	0.118	140	0.185	0.188	0.189
9	0.031	20	0.052	80	0.125	0.129	160	0.201	0.204	0.206
10	0.033	25	0.060	90	0.135	0.139	180	0.217	0.220	0.222
12	0.037	30	0.067	100	0.144	0.149	200	0.231	0.236	0.238
14	0.041	35	0.074	110	0.153	0.158	220	0.246	0.250	0.253
16	0.044	40	0.080	120	0.161	0.167	240	0.260	0.265	0.267
18	0.048	45	0.087	130	0.170	0.176	260	0.273	0.278	0.281
20	0.051	50	0.093	140	0.178	0.184	280	0.286	0.292	0.295
22	0.054	55	0.098	150	0.185	0.192	300	0.299	0.305	—
24	0.057	60	0.104	160	0.192	—	320	0.311	0.317	
26	0.060	65	0.110	170	0.200		340	0.323	0.330	

28	0.062	70	0.115	180	0.208		360	0.335	0.342	
30	0.065	75	0.120	190	0.215		380	0.346	0.354	
35	0.072	80	0.125	200	0.222		400	0.357	0.365	
40	0.078	90	0.135	220	0.235		420	0.368	0.377	
45	0.084	100	0.144	240	0.248		440	0.379	0.388	
50	0.089	110	0.153	260	0.261		460	0.390	0.399	
55	0.095	120	0.161	280	0.273		480	0.400	—	
		130	0.170	300	0.285		500	0.410		
		140	0.178	320	0.296		550	0.435		
		150	0.185				600	0.459		
		160	0.193				650	0.483		
$\Delta H =$	0.011 m		0.022 m		0.025 m	0.033 m	$\Delta H = 0.1h_1$ or[b]	0.047 m	0.058 m	0.059 m

$b = 1.0$–2.0 m ($L = 1.0$ m, $L_a = 0.3$ m)				$b \geqslant 2.0$ m ($L = 1.5$ m, $L_a = 0.3$ m)			
	h_1 (m)				h_1 (m)		
q (liters/ sec m^{-1})	$p_1 =$ 0.25 m	$p_1 =$ 0.50 m	$p_1 =$ 0.75 m	q (liters/ sec m^{-1})	$p_1 =$ 0.50 m	$p_1 =$ 0.75 m	$p_1 =$ 1.00 m
20	0.054	0.054	0.054	50	0.098	0.098	0.099
40	0.083	0.084	0.084	100	0.154	0.155	0.155
60	0.109	0.110	0.110	150	0.200	0.201	0.202
80	0.131	0.132	0.133	200	0.240	0.242	0.243
100	0.151	0.153	0.154	250	0.277	0.280	0.281
120	0.169	0.172	0.173	300	0.311	0.315	0.316
140	0.187	0.190	0.191	350	0.344	0.347	0.350
160	0.203	0.207	0.209	400	0.374	0.379	0.381
180	0.219	0.224	0.225	450	0.403	0.409	0.411

continued

TABLE X—*Continued*

b = 1.0–2.0 m (L = 1.0 m, L_a = 0.3 m)				$b \geqslant 2.0$ m (L = 1.5 m, L_a = 0.3 m)			
	h_1 (m)				h_1 (m)		
q (liters/sec m^{-1})	p_1 = 0.25 m	p_1 = 0.50 m	p_1 = 0.75 m	q (liters/sec m^{-1})	p_1 = 0.50 m	p_1 = 0.75 m	p_1 = 1.00 m
200	0.234	0.239	0.241	500	0.431	0.437	0.440
250	0.269	0.276	—	550	0.458	0.465	0.468
300	0.302	0.311		600	0.484	0.491	0.495
350	0.333	0.343		650	0.509	0.517	—
400	0.362	0.373		700	0.533	0.542	
450	0.389	0.402		750	0.557	0.567	
500	0.416	0.430		800	0.580	0.590	
550	0.441	0.457		900	01625	0.636	
600	0.465	0.483		1000	0.667	0.680	
650	0.489	—		1100	0.708	0.723	
700	0.512			1200	0.748	—	
750	0.535			1300	0.786		
800	0.556			1400	0.823		
850	0.578			1500	0.860		
900	0.598			1600	0.895		
950	0.619			1700	0.929		
1000	0.639			1800	0.963		
1100	0.677			1900	0.996		
1200	0.714						
$\Delta H = 0.1h_1$ or[b]	0.061 m	0.083 m	0.068 m		0.107 m	0.124 m	0.119 m

[a] Ramp lengths (not tabulated) equal $3p_1$.
[b] Use $0.1h_1$ or listed value, whichever is greater.

From Table X, unit width discharges are 500 and 600 liters/sec for h_1 values of 0.431 and 0.484 m, respectively. Thus the total width for 1 m^3 at 500 and 600 liters/sec m^{-1} width would be 2 and 1.67 m, respectively.

A compromise is to use:

$$q = 550 \text{ liters/sec m}^{-1}$$
$$b = (1000/550) = 1.818 \text{ m}$$
$$h_1 = 0.457 \text{ m}$$
(from Table X, for $b_1 = 1\text{–}2$ m and $p_1 = 0.50$ m)

Note that the table column changed with the revised width change. Also, the value for ΔH changed to 0.083 m (use of the greater of ΔH listed or $0.1h_1 = 0.0457$ m) and would now allow h_1 to range between 0.383 and 0.5 m, which in turn seems to permit consideration of 400 liters/sec m^{-1}. However, this causes a return to the original columns to accommodate the required width of 2.5 m and thus to the original ΔH.

Other combinations that will work include:

$p_1 = 0.25$ m, $\quad q = 950$ and 1000 liters/sec m^{-1}
$h_1 = 0.619$ and 0.639 m, $\quad b = 1.053$ and 1.00 m

Higher sills of 0.75 m require impractical widths, thus causing unacceptably low sensitivity. Note again that nonmodular flow conditions should be checked in the lowflow ranges.

Like the slip-formed canal versions, the basic concept is to convert a section of earthen canal into a section of rectangular lined canal. Because of the usual major change in cross section, more elaborate head walls and exit configurations are recommended. Figure 20 shows such a structure as it might be constructed in masonry. The outlet end shows wing walls at right angles to the flow. This is not always the most suitable configuration from the hydraulics and erosion standpoint, but offers structural support to the side walls. Alternative construction in reinforced concrete might permit the side walls to extend into a basin lined with properly graded riprap without any wing walls on the downstream end. A toe wall under the outlet end is recommended and an end sill of 2–5-cm height (and in some cases higher) is useful to help break up floor jets and control hydraulic jumps. The entire length of the structure should provide parallel flow for a minimum distance equal to about two to three canal flow depths upstream of the gauge location, which is at distance L_a from the end of the ramp of length $3p_1$. The length of protected channel depends on the type of protective work that is anticipated and could vary from an abrupt truncation of the side walls and floor to an extension of side walls and floor for a distance equal to two to four times the canal flow depth. More details for the outlet end of the structure can be found in Bos (1976).

As with all long-throated flume structures, the final bottom width and side-wall slopes in the throat section (sill section) and the degree to which the sill is level are the most critical dimensions, ranking second in importance only to errors in gauge zero referencing.

C. Small Portable Flumes for Furrow Flow Measurements

Frequently in research situations and sometimes in farm operations, the discharge rate in individual furrows, or the runoff from a field requires monitoring. Small portable broad-crested weirs somewhat resembling the versions for lined canals can be effectively used.

A variety of sizes is offered as a family of five portable flumes designed for use in furrows or small earthen channels. The four smaller versions are shown in Fig. 21. The shape of the flume is simple. As a result it can be readily constructed from galvanized sheet steel. Again, major attention should be paid to the width of the crest and the accuracy of the side-wall slopes in the region of the crest, because 1% error in the flow cross-sectional area here causes about 1% error in indicated discharge.

Fig. 21. A family of portable flumes (modified broad-crested weirs) for temporary flow measurements in furrows and small earthen channels.

TABLE XI

PROPERTIES OF FIVE FLUMES

Still width b (exact size)		Weir sill length		Flow range		
				liters/sec	ft³/sec	
mm	in.	mm	in.	min max	min	max
75	2.95	112	4.40	0.13–5.0	0.005	0.18
100	3.95	150	5.90	0.30–9.5	0.011	0.34
150	5.90	225	8.85	0.75–26	0.027	0.92
250	9.85	375	14.75	2.3–95	0.081	3.4
450	17.70	675	26.60	12.2–400	0.43	14

Because of the requirement for portability, the use of a translocated stilling well is highly desirable. A clear plastic tube 1–2 cm in diameter is mounted near the outlet overfall of the weir crest. A small pipe from the gauging location upstream leads water into the bottom of the clear plastic tube. In this way, the upstream water depth is translocated to a point sufficiently close to the sill-crest reference point that an accurate, direct depth measurement can be made even if the flume is slightly out of level in any direction. Experienced users have found that if it appears to be level it is level enough. New users can check their detection of levelness and the sensitivity of the flume by placing it in a steady flow using a carpenter's level, then introducing a variety of tilts until flow error can be detected. Sufficient time between changes must elapse to compensate for slight ponding changes upstream. Since these flumes are not sensitive to slight tilting or settling during use, they can be reset to position without detectable errors or raised to avoid a changing downstream water level.

Table XI lists the basic throat width b and flow ranges. Figure 22 shows the construction configuration in terms of the sill width b. As shown in Fig. 22, the pressure tap is approximately level with the sill crest. In practice, it should be as far as practical from the channel bottom to avoid clogging, yet low enough to remain under water at low flow rates. Its vertical location does not affect the water surface sensing function it is to perform, as long as it is under the flow surface. Trapped air in the pipeline is most easily dislodged if the pipeline has no high spots between its two ends. The connecting pipe may be made of copper or plastic tubing. The pressure tap at the upstream end should be made perpendicular to the flume wall in order to reduce velocity component errors in the sensed head reading (Rouse, 1961). The downstream end of the pipe goes through a rubber stopper which is inserted into the bottom end of the clear plastic tube. This tube is secured to supporting angles so that the top of the stopper coincides with the weir sill crest.

TABLE XII

DISCHARGE RATINGS OF SMALL PORTABLE FLUMES FOR MEASURING FLOW IN IRRIGATION FURROWS

b = 75 mm		b = 100 mm		b = 150 mm		b = 250 mm		b = 450 mm	
h_1 (m)	Q (liters/sec)	h_1 (m)	Q (liters/sec)	h_1 (m)	Q (liters/sec)	h_1 (m)	Q (liters/sec)	h_1 (m)	Q (liters/sec)
0.010	0.1287	0.014	0.2929	0.020	0.7572	0.030	2.317	0.060	12.20
0.012	0.1771	0.016	0.3692	0.022	0.8912	0.032	2.582	0.062	12.88
0.014	0.2322	0.018	0.4532	0.024	1.035	0.034	2.860	0.064	13.58
0.016	0.2941	0.020	0.5450	0.026	1.188	0.036	3.150	0.066	14.29
0.018	0.3628	0.022	0.6448	0.028	1.350	0.038	3.453	0.068	15.02
0.020	0.4385	0.024	0.7524	0.030	1.522	0.040	3.768	0.070	15.77
0.022	0.5213	0.026	0.8682	0.032	1.704	0.042	4.095	0.072	16.54
0.024	0.6113	0.028	0.9921	0.034	1.896	0.044	4.434	0.074	17.32
0.026	0.7086	0.030	1.124	0.036	2.097	0.046	4.785	0.076	18.11
0.028	0.8135	0.032	1.265	0.038	2.308	0.048	5.150	0.078	18.93
0.030	0.9259	0.034	1.414	0.040	2.529	0.050	5.526	0.080	19.76
0.032	1.046	0.036	1.572	0.042	2.761	0.052	5.915	0.082	20.60
0.034	1.174	0.038	1.738	0.044	3.002	0.054	6.317	0.084	21.47
0.036	1.310	0.040	1.912	0.046	3.254	0.056	6.731	0.086	22.34
0.038	1.455	0.042	2.098	0.048	3.516	0.058	7.158	0.088	23.24
0.040	1.607	0.044	2.291	0.050	3.788	0.060	7.597	0.090	24.15
0.042	1.769	0.046	2.493	0.052	4.071	0.062	8.050	0.092	25.08
0.044	1.938	0.048	2.705	0.054	4.365	0.064	8.515	0.094	26.03
0.046	2.117	0.050	2.927	0.056	4.669	0.066	8.993	0.096	26.99
0.048	2.304	0.052	3.158	0.058	4.984	0.068	9.485	0.098	27.97
0.050	2.501	0.054	3.399	0.060	5.311	0.070	9.989	0.100	28.97
0.052	2.706	0.056	3.649	0.062	5.648	0.072	10.51	0.105	31.53
0.054	2.921	0.058	3.910	0.064	5.996	0.074	11.04	0.110	34.21
0.056	3.145	0.060	4.180	0.066	6.356	0.076	11.58	0.115	36.98
0.058	3.379	0.062	4.461	0.068	6.726	0.078	12.14	0.120	39.87
0.060	3.623	0.064	4.752	0.070	7.109	0.080	12.71	0.125	42.87
0.062	3.876	0.066	5.054	0.072	7.502	0.082	13.29	0.130	45.97
0.064	4.139	0.068	5.366	0.074	7.908	0.084	13.89	0.135	49.19
0.066	4.413	0.070	5.689	0.076	8.325	0.086	14.50	0.140	52.51
0.068	4.696	0.072	6.022	0.078	8.754	0.088	15.13	0.145	55.95
0.070	4.990	0.074	6.367	0.080	9.195	0.090	15.77	0.150	59.50
		0.076	6.722	0.082	9.648	0.092	16.42	0.155	63.16
		0.078	7.089	0.084	10.11	0.094	17.09	0.160	66.94
		0.080	7.467	0.086	10.59	0.096	17.77	0.165	70.84
		0.082	7.856	0.088	11.08	0.098	18.46	0.170	74.85
		0.084	8.257	0.090	11.58	0.100	19.17	0.175	78.98
		0.086	8.669	0.092	12.10	0.105	21.01	0.180	83.22
		0.088	9.094	0.094	12.62	0.110	22.93	0.185	87.59
		0.090	9.529	0.096	13.16	0.115	24.95	0.190	92.07
				0.098	13.72	0.120	27.06	0.195	96.68
				0.100	14.28	0.125	29.27	0.200	101.4
				0.105	15.75	0.130	31.56	0.210	111.2

TABLE XII—*Continued*

b = 75 mm		b = 100 mm		b = 150 mm		b = 250 mm		b = 450 mm	
h_1 (m)	Q (liters/sec)	h_1 (m)	Q (liters/sec)	h_1 (m)	Q (liters/sec)	h_1 (m)	Q (liters/sec)	h_1 (m)	Q (liters/sec)
				0.110	17.31	0.135	33.96	0.220	121.6
				0.115	18.95	0.140	36.45	0.230	132.4
				0.120	20.67	0.145	39.05	0.240	143.7
				0.125	22.48	0.150	41.74	0.250	155.6
				0.130	24.38	0.155	44.53	0.260	167.9
				0.135	26.37	0.160	47.43	0.270	180.9
						0.165	50.43	0.280	194.3
						0.170	53.54	0.290	208.3
						0.175	56.75	0.300	222.9
						0.180	60.06	0.310	238.0
						0.185	63.49	0.320	253.7
						0.190	67.03	0.330	269.9
						0.195	70.67	0.340	286.7
						0.200	74.43	0.350	304.2
						0.205	78.28	0.360	322.2
						0.210	82.26	0.370	340.8
						0.215	86.35	0.380	360.0
						0.220	90.55	0.390	379.9
						0.225	94.88	0.400	400.3

Several methods can be used to measure the depth in the stilling well. One is to place a marked dipstick in the pipe so that the bottom rests on the stopper that is at the sill elevation. Because of the volume of the dipstick, a waiting period is required after insertion into the tube in order to obtain a correct flow rate.

Another method is to reference the elevation of the top of the tube to the sill crest, measure down to the water surface and calculate the net flow depth. This does not require positioning of the rubber stopper at sill-crest elevation, and is accurate and especially effective on the smaller two or three flumes. The larger two versions can be fitted with direct-reading marked inserts that are left permanently mounted in the tube or alongside the tube; however, a mirror may be required to conveniently read the level. Figure 23 shows the smallest of the group installed in operation.

Computer-generated rating tables for the selected widths of Table XI are presented in Table XII using the computer modeling procedures of Section IV,D. The rating error in Table XIII is less then ±2%. To avoid a large additional error in the "measured" flow rate, the table must be entered with the correct value of h_1. A reading error of say 2 mm in h_1 will cause a relatively small error if h_1 is

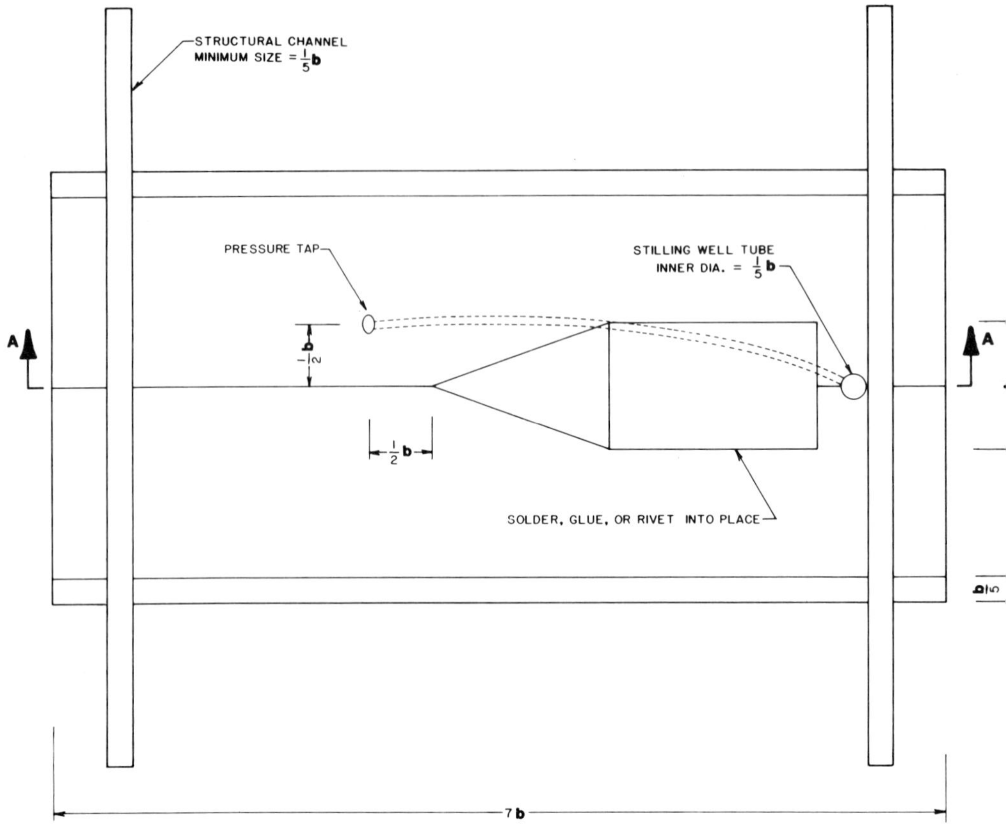

TOP VIEW

FIG. 22. Construction details for the portable flumes for furrow flow measurements.

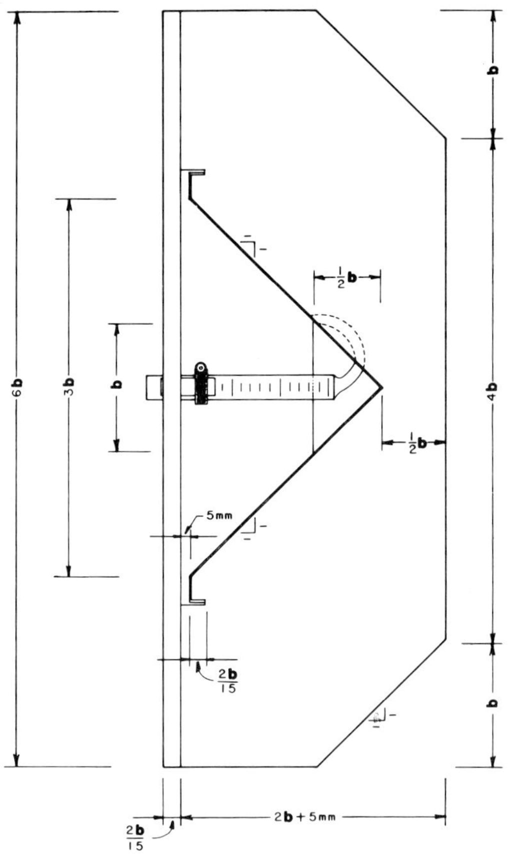

END VIEW

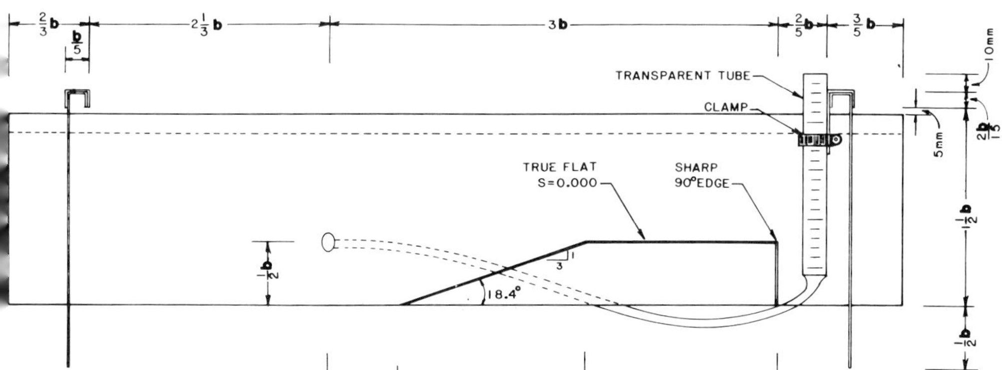

SECTION A-A

FIG. 23. Small portable flume installed in a furrow.

large, but will be very significant if h_1 is small. Thus, to avoid a large error in the measured flow rate, it is best to select a flume that will have a large h_1 reading in the most frequently used flow range.

ACKNOWLEDGMENTS

Special recognition is due Albert J. Clemmens, Research Hydraulic Engineer at the U.S. Water Conservation Laboratory, who sorted out many of the precomputed flume sizes offered in this article, particularly the rectangular offerings for the unlined canals and the selections for the trapezoidal large and small broad-crested weirs.

References

Ackers, P. A., and Harrison, A. J. M. (1963). "Critical-depth Flumes for Flow Measurements in Open Channels," Hydraul. Res. Pap. No. 5. Department of Science and Industrial Research, Hydraulics Research Station, Wallingford, Berkshire, England.

Bos, M. G., ed. (1976). "Discharge Measurement Structures," Publ. No. 20. International Institute for Land Reclamation and Improvement/ILRI, Wageningen, The Netherlands.

Bos, M. G. (1977). The use of long-throated flumes to measure flows in irrigation and drainage canals. *Agric. Water Manage.* **1,** No. 2, 111-126.

Bos, M. G., and Reinink, Y. (1981). Required head loss over long-throated flumes. *J. Irrig. Drain. Div., Am. Soc. Civ. Eng.* **107** (IR1), 87-102.

Clemmens, A. J., and Replogle, J. A. (1980). Constructing simple measuring flumes for irrigation canals. *Farmers' Bull.* **2268,** 1-13.

Chow, V. T. (1959). "Open Channel Hydraulics." McGraw-Hill, New York.

Grenville, P. S. (1958). "The Frictional Resistance and Turbulent Boundary Layer of Rough Surface," Rep. No. 1024. David Taylor Model Basin, Washington, D. C.

Harrison, A. J. M. (1967). Boundary layer displacement thickness on flat plates. *J. Hydraul. Div., Am. Soc. Civ. Eng.* **93** (HY4), 79-91; closure: **95** (HY3), 1048-1051 (1969).

International Commission on Irrigation and Drainage (1979). Recommendations for design criteria and specifications for man-made lined canals. *Int. Comm. Irrig. Drain. (ICID) Bull.* **28** (2), 43-45.

King, H. W. (1954). "Handbook of Hydraulics," 4th ed. McGraw-Hill, New York.

Replogle, J. A. (1968). Discussion of: Rectangular cutthroat flow-measuring flumes (by Skogerboe, G. V. and M. L. Hyatt). *J. Irrig. Drain. Div., Am. Soc. Civ. Eng.* **94** (IR3), 359-362.

Replogle, J. A. (1975). Critical flow flumes with complex cross section. *Proc.—Spec. Conf. Irrig. Drain. Age Competition Resour., Am. Soc. Civ. Eng.* pp. 366-388.

Replogle, J. A. (1977). Compensating for construction errors in critical-flow flumes and broad-crested weirs. *NBS Spec. Publ. (U.S.)* **484** (1), 201-218.

Replogle, J. A. (1978). Flumes and broad-crested weirs — mathematical modeling and laboratory ratings. *In* "Flow Measurement of Fluids" (H. H. Dijstelbergen and E. A. Spencer, eds.), pp. 321-328. North-Holland Pub., Amsterdam.

Replogle, J. A., and Clemmens, A. J. (1979). Broad-crested weirs for portable flow metering. *Trans. ASAE [N.S.]* **22** (6), 1324-1328.

Replogle, J. A., and Clemmens, A. J. (1980). Modified broad-crested weirs for lined canals. *Proc.—Spec. Conf. Irrig. Drain., Today's Challenges, Am. Soc. Civ. Eng.* pp. 463-479.

Replogle, J. A., and Clemmens, A. J. (1981). Measuring flumes of simplified constructions. *Trans. ASAE [N.S.]* **24** (2), 362-366, 374.

Rouse, H. (1961). "Laboratory Instructions in the Mechanics of Fluids," Stud. Eng., Bull. 41. Iowa Institute of Hydraulic Research, State University of Iowa, Iowa City.

Schlichting, H. (1960). "Boundary Layer Theory." McGraw-Hill, New York.

U.S. Department of Agriculture (1979). "Field Manual for Research in Agricultural Hydrology," Agric. Handb. No. 224 (rev.). USDA, Washington, D.C.

PRINCIPLES, PRACTICES, AND POTENTIALITIES OF TRICKLE (DRIP) IRRIGATION

D. A. Bucks and F. S. Nakayama

U.S. Department of Agriculture
Agricultural Research Service
U.S. Water Conservation Laboratory
Phoenix, Arizona

A. W. Warrick

Department of Soils, Water and Engineering
University of Arizona
Tucson, Arizona

ADVANCES IN IRRIGATION, VOLUME 1

ISBN 0-12-24301-6

I. Introduction

Irrigation advancements within the last decade have been astounding. Trickle or drip irrigation is one of the latest innovations for applying water, and it represents a definite advancement in irrigation technology. It can be defined as the precise, slow application of water in the form of discrete drops, continuous drops, tiny streams, or miniature sprays through mechanical devices called emitters or applicators located at selected points along water delivery lines. Fertilizer or other chemical amendments can be efficiently applied to individual or several plants using trickle irrigation. Types of emitters include those used on the surface, subsurface, bubbler, spray, mechanical-move, and pulse irrigation systems. Although "trickle" and "drip" are often used interchangeably, the term *trickle* will be used throughout this article.

Trickle irrigation has few characteristics in common with flood or sprinkler irrigation. Water advances in the soil around a trickle emitter only after the amount of water applied exceeds the infiltration rate at a point. Then it advances until the infiltration rate of the ponded area equals the emitter flow rate. Typically, a wetted diameter of less than 1 m, depending upon the soil properties and emitter application rate, will become saturated on the soil surface. Trickle and sprinkler irrigation systems are similar in that both use a pipe network to deliver water to emission points in the field; however, a trickle system can be more precise in replacing soil water to individual plants compared to flood or sprinkler irrigation.

II. Present Developments and Expansion of Trickle Irrigation

The first reported trickle irrigation experiments began in Germany in 1860, where subsurface clay pipes were used in combination with irrigation and drainage systems (Davis, 1974). In the United States, around 1913 House (Davis, 1974) tried to irrigate with perforated subsurface pipes, but he indicated that the method was too expensive. Irrigation of plants through narrow openings in pipes can also be traced back to greenhouse operations in the United Kingdom in the late 1940s (Davis, 1974).

Current trickle irrigation technology dates back to the work of Blass (1964). Based on the observation that a large tree near a leaking faucet exhibited more vigorous growth than other trees in the area, he developed the first patented trickle irrigation system. From Israel the trickle irrigation concept spread to Australia, North America, and South Africa by the late 1960s, and eventually throughout the world. The availability of low-cost plastic pipe for water delivery lines helped to speed the use of trickle irrigation systems. By the 1970s, equipment development and manufacturing tended to overshadow basic research studies directed toward an understanding of the optimum design, management, and maintenance of this new method.

At present, the United States has the greatest land area under trickle irrigation. Coverage has increased from about 4000 ha in 1972 (Bessesen, 1972) to 175,000 ha in 1980 (McNeill, 1980). California, Florida, Hawaii, and Texas account for 91% of this area (Fig. 1). Although the trickle-irrigated land represents less than 1% of the total irrigated area in the United States, it is of tremendous economic importance. The total trickle irrigation in the United States could go over one million hectares by 1990, based on earlier predictions (Frasier, 1977). Trickle irrigation has also grown rapidly in Australia, Israel, Mexico, and South Africa. Gustafson (1976) indicated the worldwide trickle area to be about 110,000 ha in 1975 and estimated a 1980 figure of 350,000 ha. Many industrial leaders feel that these surveys have significantly underestimated the actual area being trickle irrigated.

Trickle irrigation, like other irrigation methods, will not fit every agricultural crop, specific site, or objective. Presently, trickle irrigation has the greatest potential where (1) water is expensive or scarce; (2) soils are sandy, rocky, or difficult to level; and (3) high-value crops are produced. The principal agricul-

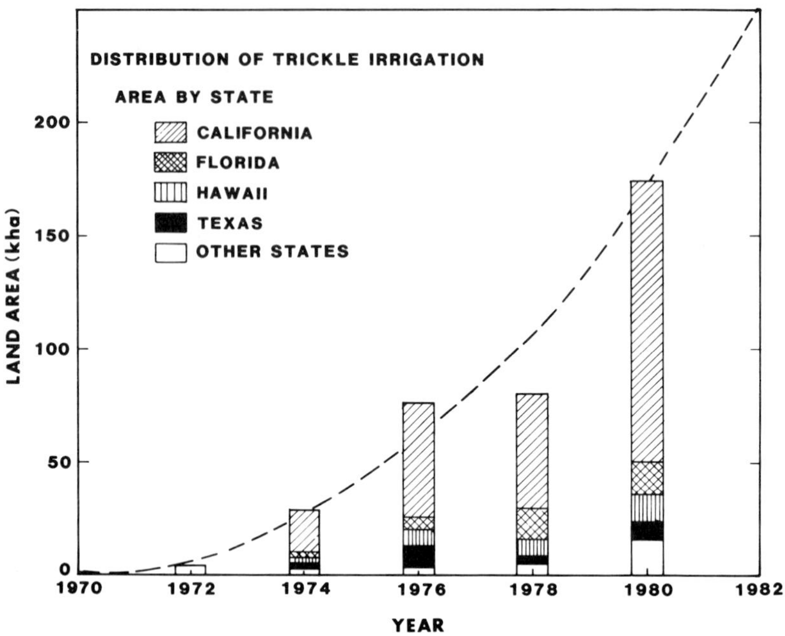

FIG. 1. Trickle irrigation land area and distribution in the United States (based on *Irrigation Journal* surveys).

tural crops under trickle irrigation are avocado, citrus, stone fruits, grape, strawberry, sugarcane, and tomato. This method of irrigation continues to be important in the greenhouse production of tomato, cucumber, and flowers. Trickle irrigation is also used for landscaping of parks, highways, commercial developments, and residences. As water, labor, and land preparation costs increase, more trickle irrigation systems will be substituted for conventional irrigation methods.

III. Potential Advantages of Trickle Irrigation

Many reports have listed and summarized potential advantages of trickle irrigation compared to other methods (Cole, 1971; Davis, 1975; Bresler, 1977; Hillel, 1980; Howell *et al.*, 1980). Each irrigation method has possible advantages and limitations with respect to technical, economical, and crop production factors. We will attempt to discuss some of the important benefits of trickle irrigation.

A. Increased Beneficial Use of Available Water

There is general agreement that irrigation water requirements can be less with trickle irrigation than with traditional irrigation methods. The savings, of course, depend on the crop, soil, environmental conditions, and the attainable on-farm irrigation efficiency. Primary reasons given for the water savings include irrigation of a smaller portion of the soil volume, decreased surface evaporation, reduced irrigation runoff from the field (the dry soil between rows could also store more precipitation), and controlled deep percolation losses below the crop root zone (Aljibury, 1974; Davis, 1975; Shoji, 1977).

Direct evaporation from the soil surface (Dan, 1974) and water uptake by weeds (Lemon, 1956) could be reduced by not wetting the entire soil surface between rows or trees, especially on young tree or row crops. Sprinkler irrigation is subject to water loss by wind drift, increased evaporation, or poor application uniformity, especially with strong winds (Seginer, 1969). Trickle irrigation can prevent runoff even for contour farming practices on steep hills (Marsh *et al.*, 1975), and particularly for low-permeability or crusted soils (Kemper and Noonan, 1970). The development of surface crust and destruction of surface soil structure (Lemos and Lutz, 1957) can be avoided, whereas water infiltration into the soil (Rose, 1961) can be improved by using low-application-rate trickle systems. Deep percolation losses can be controlled (Rawlins, 1973), especially on sandy soils (Roth, 1974).

B. Enhanced Plant Growth and Yield

The soil water content in a portion of the plant root zone remains fairly constant because irrigation water can be supplied slowly and frequently at a predetermined rate. Generally, the total soil water potential increases (the soil water suction decreases) with elimination of the wide fluctuations in the soil water content, which typically result from traditional flood and sprinkler irrigation methods. Bresler (1977) describes how a rather high, soil water potential can be advantageous where the yield response curve is convex and the effects of fluctuating soil water contents behave in a manner similar to the theory of Zaslavsky and Mokady (1967) and Zaslavsky (1972). Other discussions (Hillel, 1972; Childs and Hanks, 1975; Rawlins and Raats, 1975) also imply that the best irrigation policy is to apply water as frequently as possible. This is assuming that no problems of soil aeration (Dasberg and Steinhardt, 1974), plant disease (Hanson and Patterson, 1974), or restricted plant rooting (Willoughby and Cockroft, 1974) will occur with high-frequency irrigation.

Howell *et al.* (1981) reviewed over 50 research reports on crop response to trickle irrigation published over the last decade. Where trickle irrigation was compared with either no irrigation or other water application methods, yields were equal or better in all cases; and the amounts of water applied were equal or less with only two exceptions. Regardless of the difficulties in making adequate comparisons among the different irrigation methods and management practices, the potentiality exists for improved plant growth and yield by using an irrigation method such as trickle that is capable of frequent delivery of water to the soil. Optimization of the soil water regime for maximum production or water-use efficiency depends upon proper irrigation scheduling (frequency and quantity of irrigation) and water placement with the trickle irrigation method.

C. Reduced Salinity Hazard to Plants

Considerable evidence exists showing that waters of higher salinity can be used with trickle irrigation than with other methods without greatly reducing crop yields. Minimizing the salinity hazard to plants irrigated by trickle irrigation can be attributed to: (1) dilution of the soil solution's salt concentration, as a consequence of the high-frequency irrigation used to maintain high soil water contents in the root zone (Bernstein and François, 1973; Rhoades *et al.*, 1974); (2) elimination of leaf damage caused by foliar salt absorption with sprinkler irrigation (Bernstein and François, 1975; Gornat *et al.*, 1973); and (3) movement of salts beyond the active plant root zone (Patterson and Wierenga, 1974; Tscheschke *et al.*, 1974).

D. Improved Fertilizer and Other Chemical Applications

Trickle irrigation offers considerable flexibility in fertilization (Lindsey and New, 1974; Isobe, 1974). Frequent or nearly continuous application of plant nutrients along with the irrigation water is feasible and appears to be beneficial for crop production. Several researchers (Bester *et al.*, 1974; Marsh *et al.*, 1975; Shani, 1974) have proposed various reasons for the increased efficiency of fertilization: (1) decreased quantity of applied fertilizer, because fertilizer is applied only to the root zone; (2) improved timing of fertilization, because the more frequent applications make it possible to match plant requirements at various growth stages; and (3) improved distribution of fertilizer with minimum leaching beyond the root zone or runoff. Besides fertilizer, other water amendments such as herbicides, insecticides, fungicides, and carbon dioxide can be supplied to improve crop production (Chesness *et al.*, 1976; Lange *et al.*, 1974; Nakayama and Bucks, 1981b; Overman, 1974). As a research tool and for practical applications, trickle irrigation can supply many more chemicals directly to the plant than have previously been tried.

E. Limited Weed Growth

Weed infestation may be reduced under trickle irrigation because only a fraction of the soil surface is wetted. Roth *et al.* (1974) noticed remarkable savings in weed control for citrus using trickle irrigation. Since water was applied mostly under the shaded area of the trees (rainfall was minimal), less cultivation was required than with other irrigation methods. Also, filtration of irrigation water for trickle irrigation may result in delivery of fewer weed seeds to the field in comparison with other methods. Numerous investigators (Davis, 1974; Marsh *et al.*, 1975) have mentioned the overall reduction in weed growth; however, many trickle irrigation users have experienced increased weed growth and control problems within the small wetted portion of the soil surface, principally when the crop is young.

Selective herbicides have been applied through trickle systems, and manufacturers are cooperating in developing water soluble products (Goldberg *et al.*, 1976). Lange *et al.* (1974) found that conventional herbicide applications in orchards and vineyards under trickle irrigation did not provide persistent weed control in areas of high soil water content near the emitters. Roth *et al.* (1974) controlled weeds in these areas by spraying contact herbicides around the emitters with ground equipment.

F. Reduced Operation Labor

Trickle irrigation systems can be easily automated where labor is limited or expensive. Simple automation equipment (Davis and Nelson, 1970) has consisted of electrical, mechanical, or battery-operated time clocks that activate pumps and solenoid valves at selected times during the day. Paldi (1974) discussed a more sophisticated system that included soil moisture sensing devices, computer interfacing, and remote controllers.

In addition to labor savings from automation, greater efficiency may result from the fact that other cultural operations can be conducted while the plants are being irrigated. Localized wet and dry patterns allow this to occur. Labor and operational costs can be reduced by the simultaneous application of water, fertilizer, herbicide, insecticide, or other additives through the trickle system. While some of the operational costs may be lower, other requirements such as field inspection and maintenance of the system may be higher. For example, Schaper (1977) reported that annual maintenance requirements for a large-scale grower in California ranged from \$60 to \$200 per hectare per year.

G. Decreased Energy Requirements

Trickle irrigation has a potential for reducing pumping energy costs since operating pressures are considerably lower compared to other types of pressurized systems. However, real energy conservation under trickle irrigation should come from reducing the amount of water pumped (Davis, 1975; Shoji, 1977) because of a higher on-farm irrigation efficiency. Batty *et al.* (1975) compared installation energy (including the energy to produce plastic tubing) and annual pumping costs for nine irrigation systems on a specific site. Gravity or flood irrigation appeared to conserve more energy than trickle irrigation when the same amount of water was applied. A definite trend seems to exist toward lower operating pressures with trickle systems provided that the economics of small-diameter pipes can be maintained in the system design.

H. Improved Cultural Practices

Continuous cultural operations such as spraying, weeding, thinning, and harvesting of tree and row crops are possible without interrupting the trickle irrigation cycle for any prolonged period of time. With trickle-irrigated row crops, for example, furrows remain relatively dry and provide firm footing for farm workers (Marsh *et al.*, 1975). Wind does not affect water distribution when using trickle irrigation compared to sprinkler irrigation (Davis, 1975). Trickle irrigation systems have been used to supply water to crops on slopes as steep as 50–60% because surface water runoff is often minimal (Shoji, 1977). The adaption of cultural practices to match improved trickle irrigation concepts will be discussed in more detail later in this article.

IV. Potential Disadvantages of Trickle Irrigation

Despite observed successes, several problems have been encountered in the mechanics of applying water with trickle equipment for some soils, water qualities, and environmental conditions. Some of the more important possible disadvantages of trickle irrigation compared with other irrigation methods include (1) emitter clogging, (2) rodent or other animal damage, (3) salt accumulation near plants, (4) inadequate soil water movement and plant root development, and (5) economical–technical limitations.

A. Persistent Maintenance Requirements

Emitter clogging is considered the most serious problem in trickle irrigation unless preventive measures are taken (Bucks *et al.*, 1977; Pelleg *et al.*, 1974). Clogging will adversely affect the rate and uniformity of water application, increase maintenance costs (as it becomes necessary to check, replace, or reclaim clogged emitters), and result in crop damage and decreased yield if not detected and corrected early. Researchers and equipment manufacturers have chosen two approaches to solve the clogging problem. The first is to develop emitter devices which may require less maintenance (Wilson, 1972, 1977; Solomon, 1977). A second approach is to focus attention on improving the quality of water before it reaches the emitters (Bucks *et al.*, 1979; Ford, 1976; Ford and Tucker, 1974;

McElhoe and Hilton, 1974; Nakayama *et al.*, 1978). All agree that preventive maintenance (including filtration, chemical water treatment, flushing trickle lines, and field inspection) is probably the most effective solution to emitter clogging.

Other maintenance problems, not as serious as emitter clogging, include pipeline leaks and accessory equipment failures. Rodents, coyotes, rabbits, dogs, etc., can chew and damage trickle lines; and ants or other insects have occasionally enlarged openings in trickle tubing. Trickle lines can be cut or dug up accidentally when weeding, replacing plant material, or when repairing or replacing other pipelines or utilities located in nearby areas. Filters, chemical injectors, time clocks, pressure regulators, water meters, and pumps are also subject to malfunctioning.

B. Salt Accumulation near Plants

Where high-salinity water is used in arid regions, salt tends to accumulate at the soil surface and toward the periphery of the wetted soil volume (Goldberg and Shmueli, 1970; Tscheschke *et al.*, 1974; Yaron *et al.*, 1973). Rain may wash harmful amounts of accumulated salts into the plant root zone and cause osmotic shock to the plants (Bernstein and François, 1973). Nelson and Davis (1974) reported effective leaching of accumulated salts by continuing to trickle irrigate the trees during the rainy season in southern California. An accumulation of salts could be a problem if seeds are germinated in zones of high salt concentrations from the previous row crop. Extensive leaching may be necessary before the next crop is planted (Patterson and Wierenga, 1974). In some situations, it may be necessary to apply a large preplant irrigation using sprinkler or flood methods.

C. Restricted Soil Water Distribution and Plant Root Development

The smaller wetted soil volume under trickle irrigation—compared to that of other irrigation methods—can be both advantageous and disadvantageous. Restricting the available water can provide for efficient use of water, enhanced plant growth, reduced salinity hazards, and improved use of chemical water amendments (see Section III,A–D). On the other hand, a certain minimum wetted volume appears to be necessary for optimum plant response (Dasberg and Steinhardt, 1974; Bresler, 1977). Many factors involving the soil (texture, infiltration rate, and heterogeneous characteristics), the plant (uptake or evapotranspiration rates), the management (amount and frequency of water applications), and the system design (number of emitters per plant, placement, and discharge rate of trickle emitters) affect soil water distribution and plant root development.

D. Economic–Technical Limitations

Because equipment requirements are numerous with trickle irrigation, initial investment and annual costs may be high. Estimated initial costs are generally in the same range as solid-set sprinklers (Marsh and Gustafson, 1971) for a modern system. The actual costs will vary considerably depending on the crop, grade of pipelines, filtration equipment, fertilization equipment, and the degree of automation desired. Generally, trickle irrigation is more suited to widely spaced crops. Although portable systems are being developed for row crops, they are generally not applicable to cereal grains or grasses.

Technical improvements in the designs of emitters, fittings, filters, and system controls have been necessary; the development of procedures for preventing or correcting emitter clogging and equipment failure have been difficult; and the development of proper methods for injection of fertilizers and other chemicals has sometimes been a problem. Overall, a higher level of design, management, and maintenance is required with trickle than other irrigation methods.

V. Advances in System Components and Innovations

Many significant advancements have occurred in the design of components and trickle irrigation systems. The basic components of a trickle irrigation system include a pump, filters, distribution lines, emitters, and other control and monitoring equipment (Fig. 2).

A. Emitters

Many different emitters have been devised and manufactured within the last decade. Some of the more distinctive designs are the short-path, long-path, short-orifice, vortex, pressure-compensating, self-flushing, perforated single- and double-chamber tubing, as well as porous-tubing emitters (Solomon, 1977). These designs can be classified into two types, point source and line source. Point-source systems discharge water from individual or multiple outlets that are spaced at least 1 m apart. Line-source systems have perforations, holes, or porous walls in the irrigation tubing that discharge water at close spacings, or even continuously along a lateral line (Keller and Karmeli, 1975). Typically, point-source emitters are used for tree crops, vines, ornamentals, and shurbs, whereas line-source emitters are used for small fruits, vegetables, or other closely spaced row crops. Better materials and manufacturing have improved extrusion and molding of point-source emitters. Multilayer bonding and laser techniques have enhanced the reliability of line-source emitters. A miscellaneous group of emitters (bubblers, low-head sprayers, spitters, foggers, etc.) can also be included, since they resemble both a point-source emitter and a small sprinkler.

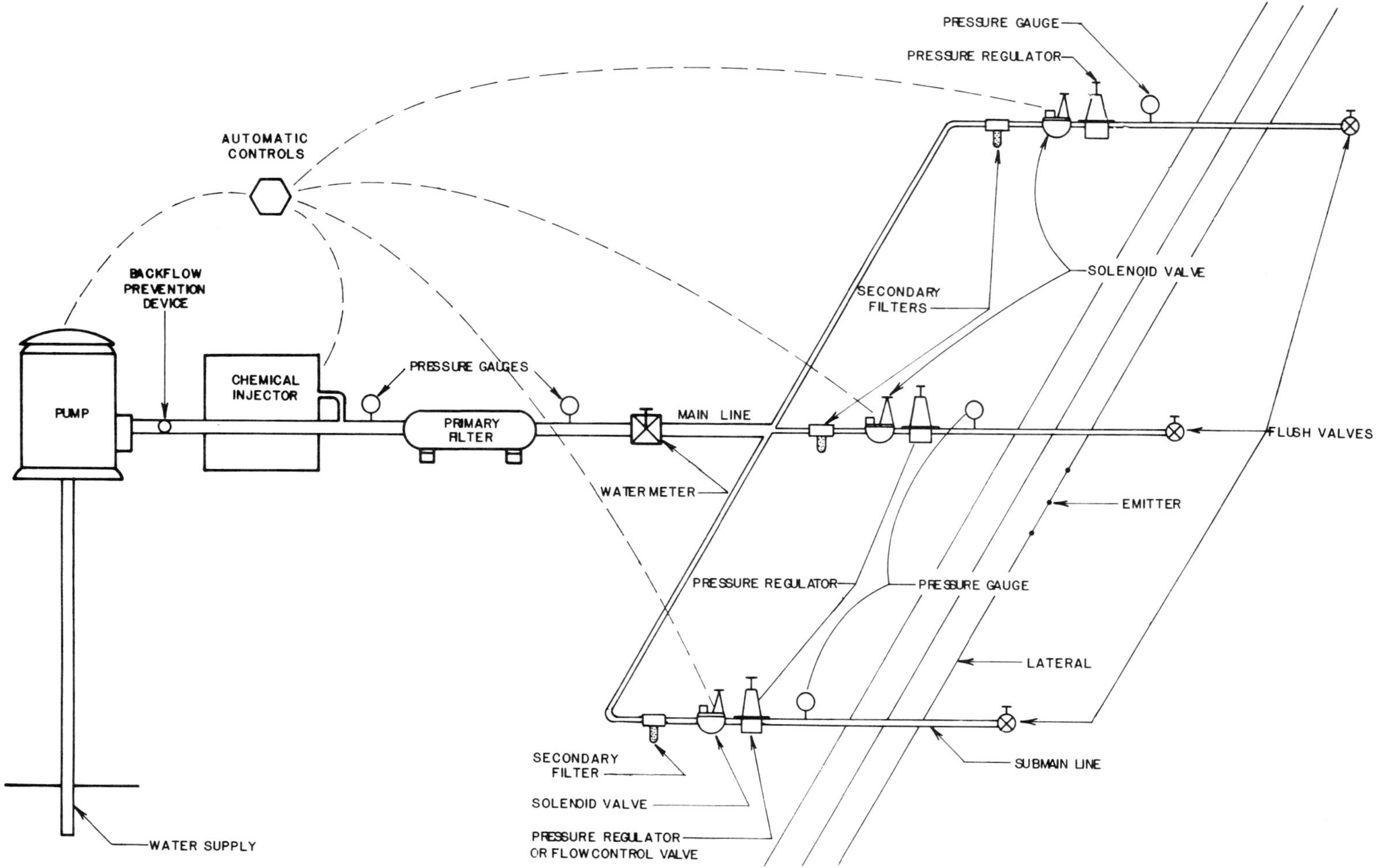
PRESSURE GAUGE
PRESSURE REGULATOR
AUTOMATIC CONTROLS
BACKFLOW PREVENTION DEVICE
SOLENOID VALVE
SECONDARY FILTERS
CHEMICAL INJECTOR
PRESSURE GAUGES
PUMP
PRIMARY FILTER
MAIN LINE
FLUSH VALVES
WATERMETER
EMITTER
PRESSURE REGULATOR
PRESSURE GAUGE
LATERAL
SECONDARY FILTER
SUBMAIN LINE
SOLENOID VALVE
PRESSURE REGULATOR OR FLOWCONTROL VALVE
WATER SUPPLY

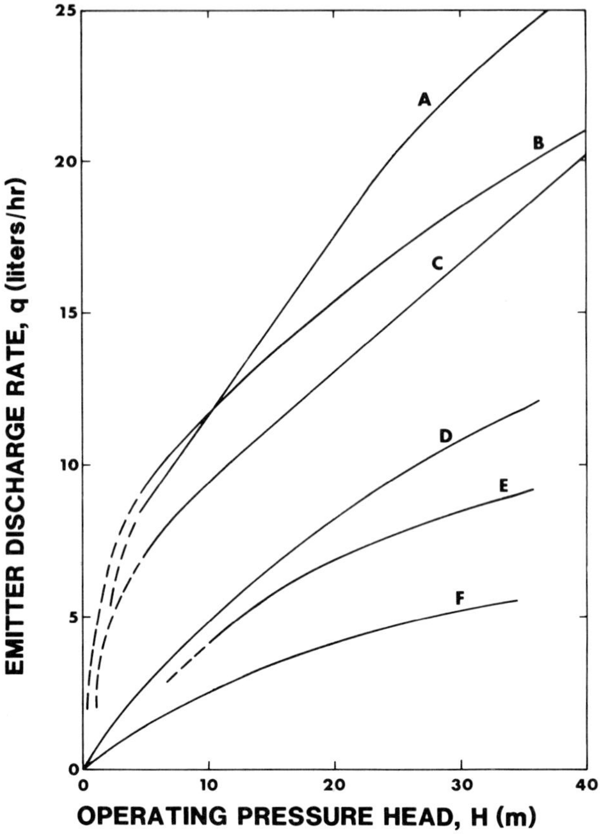

FIG. 3. Average discharge rates q for various emitter designs at different operating pressure heads H (after Keller and Karmeli, 1975):

Curve	Emitter type	Discharge rate (liters/hr)[a]
A	Long path	10
B	Single orifice	10
C	Long path	8
D	Long path	4
E	Multiple outlet	3
F	Long path	2

[a] All rated emitter discharge rates were based on a standard operating pressure of about 10 m.

Hydraulically, most emitter flow regimes can be characterized by the Reynolds number Re defined as

$$\mathrm{Re} = \upsilon d/\nu \tag{1}$$

where υ is the emitter flow velocity (m/sec), d is the emitter diameter (m), and ν is the kinematic viscosity (m^2/sec). Four flow regimes are defined as: (1) laminar, $\mathrm{Re} < 2000$; (2) unstable, $2000 \leqslant \mathrm{Re} \leqslant 4000$; (3) partially turbulent, $4000 \leqslant \mathrm{Re} \leqslant 10{,}000$; and (4) fully turbulent, $10{,}000 \leqslant \mathrm{Re}$.

Keller and Karmeli (1974, 1975) and Howell and Hiler (1974) have suggested that the emitter flow function can be given as

$$q = cH^e \tag{2}$$

where q is the emitter discharge rate (liters/hr), c is the emitter discharge coefficient that depends on the choice of units, H is the emitter pressure head (m), and e is the emitter discharge exponent that characterizes the flow regime. For fully turbulent flow, $e = 0.5$; for partially turbulent flow, $0.5 < e < 0.7$; for the unstable flow regime, $0.7 < e < 1.0$; and for laminar flow, $e = 1.0$. Short-orifice emitters are always fully turbulent, whereas long-path or other emitters may vary in flow characteristics.

Many manufacturers provide standard curves that show average emitter discharge rates for different operating pressure heads under laboratory conditions. Figure 3 presents discharge–pressure curves for several emitter designs (Keller and Karmeli, 1975).

1. *Surface Trickle Irrigation Systems*

Trickle irrigation systems with the lateral lines laid on the soil surface remain the most popular application method, particularly on crops with widely spaced plants. Surface trickle can also be used for row crops with a pattern on the soil surface as shown in Fig. 4. Advantages of surface trickle systems include the ease of installing, inspecting, changing, and cleaning emitters, plus the possibility of checking soil surface wetting patterns and measuring individual emitter discharge rates. However, surface trickle lines can interfere with cultivation, harvesting, and other cultural operations. Generally, discharge rates are less than 12 liters/hr for single-outlet point-source emitters, and less than 12 liters/hr m^{-1} for line-source emitters. Examples of installation methods for surface trickle systems on tree, vine, and row crops are shown in Figs. 5–10.

2. *Subsurface Trickle Irrigation Systems*

Lately, trickle irrigation systems with lateral lines buried below the soil surface have gained wide acceptance. Earlier problems with clogging have been

FIG. 4. Typical wetting pattern on the soil surface from a line-source, surface trickle irrigation system.

reduced, and these systems are now being used primarily on small fruit and vegetable crops in the western United States (Fig. 11) and on sugarcane in Hawaii. Advantages of subsurface trickle irrigation systems include freedom from the necessity of anchoring of the tubing at the beginning and removing it at the end of the growing season, little interference with cultivation or other cultural practices, and a longer economic life. Bucks *et al.* (1981a) used a subsurface trickle system to produce several consecutive row crops without major conventional cultural operations or replacement of the tubing. Experience has shown that emitter outlets should be pointed upward and that maintenance requirements are similar to surface trickle systems. Discharge rates for subsurface systems are in the same range as surface trickle. This method of water application should not be confused with irrigation through or by a groundwater table, called *subirrigation*.

Numerous research studies have been directed toward optimizing subsurface depths and spacings between lateral lines (Gilley and Allred, 1974; Thomas *et al.*, 1974; Van Bavel *et al.*, 1973; Warrick *et al.*, 1979). Optimum placement depends on the soil, crop, and discharge rate of the system. Most investigators

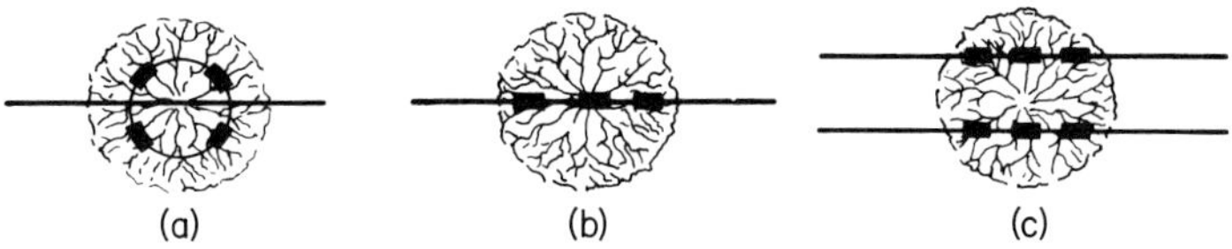

FIG. 5. Point-source emitters for tree crops: (a) loop or pig-tail with four emitters per tree; (b) single lateral line with three emitters per tree; (c) dual lateral line with six emitters per tree.

FIG. 6. Point-source emitters for vine crops: (a) single emitter per vine on soil surface or trellis; (b) two emitters per vine on soil surface or trellis.

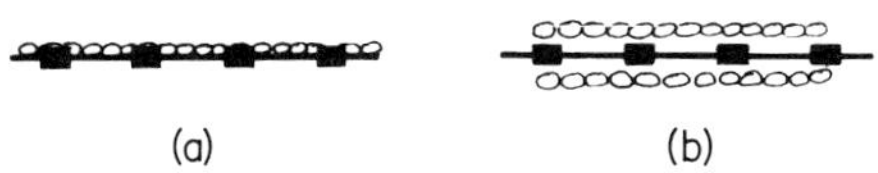

FIG. 7. Point-source emitters for row crops: (a) single lateral line per row; (b) single lateral line per two rows.

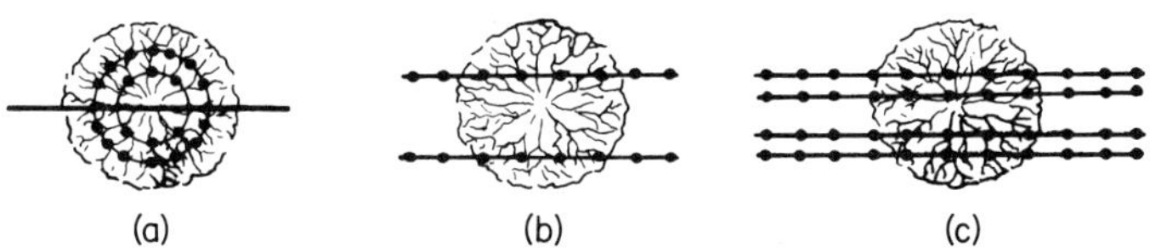

FIG. 8. Line-source emitters for tree crops: (a) two loops of perforated tubing per tree; (b) two perforated tubings per tree; (c) four perforated tubings per tree.

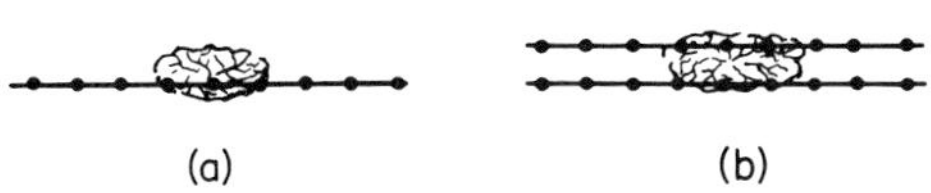

FIG. 9. Line-source emitters for vine crops: (a) single perforated tubing per vine; (b) dual perforated tubings per vine.

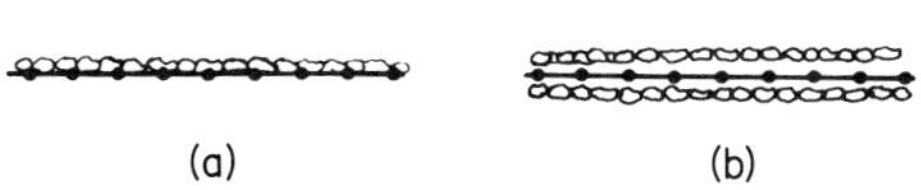

FIG. 10. Line-source emitters for row crops: (a) single perforated tubing per row; (b) single perforated tubing per two rows.

FIG. 11. Typical wetting pattern on the soil surface from a line-source, subsurface trickle irrigation system.

have indicated that lateral lines should be placed only as deeply as required for cultural operations. Subsurface depths range from a few centimeters for shallow-rooted vegetables to 30 cm or more for deep-rooted row crops or trees. Installation patterns for subsurface trickle systems are similar to those for surface systems (Figs. 5–10).

3. *Bubbler Irrigation Systems*

Like other trickle irrigation systems, bubblers can deliver frequent, small applications of water and nutrients. Bubblers resemble short-orifice emitters used for surface trickle, except that discharge rates are higher and range from 1 to 4 liters/min. Because the emitter discharge rate normally exceeds the infiltration rate of the soil, a small basin is usually formed around or between the trees as shown in Fig. 12. Larger diameter delivery lines are normally required with bubbler systems to reduce pressure head loss associated with higher flow rates and lower operating pressures (energy requirements). Rawlins (1977) reported the use of low-head bubbler system constructed from thin-walled drainage tubing for lateral lines and medium-diameter polyethylene (PE) tubes for bubblers.

FIG. 12. A small basin near the tree is often required to contain the water from a bubbler irrigation system (photograph courtesy of R. L. Roth, Yuma, Arizona).

4. *Spray Irrigation Systems*

The application of water as a small spray, fog, or mist to the soil surface is being practiced in the southeastern and western United States, as well as in many other areas of the world on tree crops (Fig. 13). Instead of water distribution being controlled by the soil (as with surface trickle, subsurface trickle, or bubbler systems), a spray system uses air travel and the soil to promote the distribution of water. Spray systems can be vulnerable to high wind and evaporation rates, particularly where a straw or leaf mulch is used around young trees with a limited crop canopy. However, the spray and bubbler systems normally have minimal filtration and other maintenance requirements.

5. *Mechanical-Move Irrigation Systems*

Mechanically moving irrigation systems that expand the bubbler concept to large-scale row crops have been developed by Phene *et al.* (1980, 1981) and Lyle and Bordovsky (1979). These systems utilize linear-move sprinkler lines that have drop structures or tubes to deliver water as a continuously moving stream to each row. Operating pressures are lower than those of most conven-

FIG. 13. Spray irrigation systems are designed to deliver water at a full- or part-circle pattern (a 180° sprayer is shown).

tional sprinkler systems, and the uniformity of water application over the field is ısually excellent. Figure 14 shows the linear-move trickle system used on the research studies at Fresno, California (Phene *et al.*, 1980). Potential advantages of these linear-move trickle systems include a possible reduction in clogging problems and a less extensive pipe network compared to solid-set trickle sys-ems.

Hose-reel and drag-type trickle systems have also been tried in Israel, Mexico, and the United States. Surface trickle irrigation lateral lines are either pulled as a set across the field, moved singly from one row to the next (Wilke, 1974), or wound onto a reel before being moved by a tractor. Possible advantages of these ystems are a decrease in initial costs and maintenance requirements for the rickle equipment. Examples of installation methods for bubbler, spray, and nechanical-move systems are shown in Figs. 15–17.

·. *Pulse Irrigation Systems*

The pulse irrigation system (Karmeli and Peri, 1974, 1977; Zur, 1976) was eveloped to obtain a low time-averaged application rate from a high emitter

FIG. 14. Linear-move trickle irrigation system being used on fully automated research studies (photograph courtesy of C. J. Phene, Fresno, California).

discharge rate. Karmeli and Peri (1974) defined pulse irrigation as a series of irrigation time cycles with an operating (water discharge) and rest phase (no discharge). Typical operating phases were 5, 10, or 15 min/hr. Zur (1976) concluded from column studies that pulses are quickly damped out, so that a constant water flow develops throughout the soil profile. The primary advantage ascribed to this system is the reduction in clogging problems, whereas a possible disadvantage is the need to develop reliable, inexpensive pulse emitters and automatic controllers (Peri and Rasmussen, 1978).

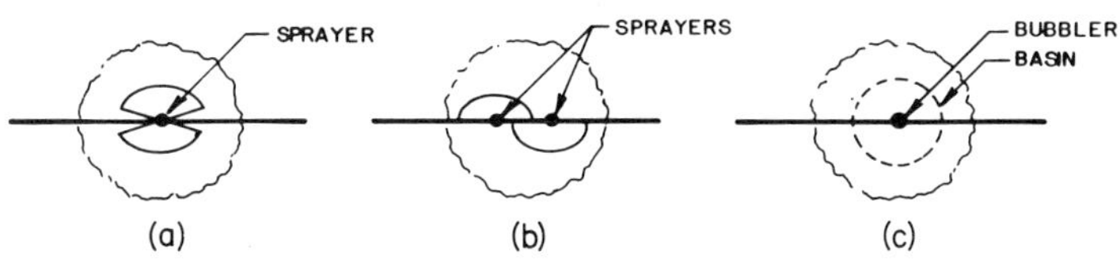

FIG. 15. Examples of installation methods for bubbler and sprayer irrigation systems on tree crops: (a) single 280° sprayer per vine; (b) two 180° sprayers per tree; (c) single bubbler per tree.

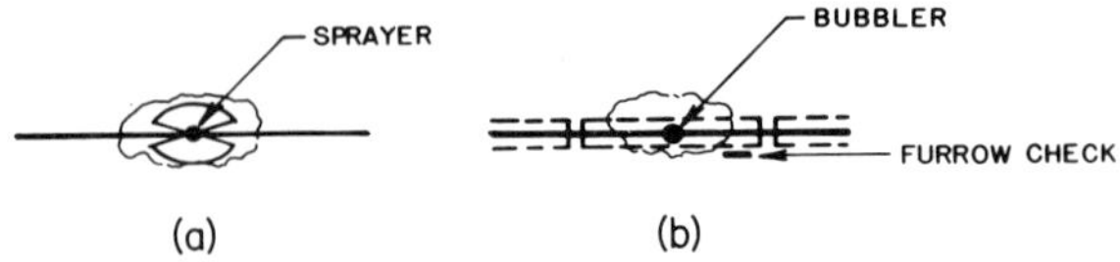

FIG. 16. Examples of installation methods for bubbler and sprayer irrigation system on vine crops: (a) single 280° sprayer per vine; (b) single sprayer per vine.

B. DISTRIBUTION LINES AND FITTINGS

Distribution lines consist of a network of graduated pipe sizes starting with a single, large main line followed by smaller submain and lateral lines. The buried main and submain lines are normally of rigid polyvinyl chloride (PVC) pipe. They can also consist of lay-flat PVC pipe (temporary surface lines for row crops), asbestos-cement pipe (main line for tree crops), or polyethylene (PE) plastic (temporary surface lines for all crops). Main and submain lines range in diameter from 4 to 15 cm and should have valved outlets for periodic pipeline flushing. The submain line may also contain pressure regulators or flow control valves, manual or automatic control valves, secondary filters for final filtration or protection against pipeline breakage, or additional safety devices.

Lateral lines are usually of PE plastic and range in diameter from 8 to 25 mm, with the 14-mm size being most common. Early versions of PE tubing developed cracking problems, caused by polyaxial stress from the insertion of fittings that were larger than the inside diameter of the lateral, and/or by the thermal or

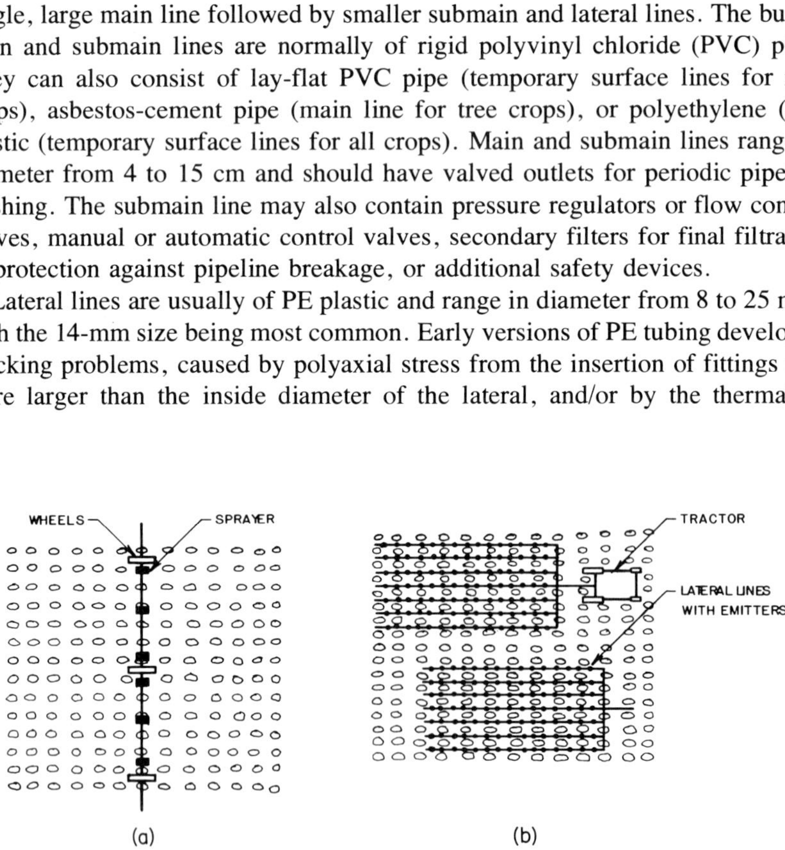

FIG. 17. Examples of installation methods for mechanical-move irrigation systems on vegetables and other row crops: (a) spray system; (b) drag or hose-reel system.

chemical environment (Shipston, 1976). Stress cracks also developed where ends of PE laterals were closed off or crimped. Improved PE pipe extrusion techniques, the use of cross-linking methods and ultraviolet decomposition inhibitors, the proper sizing of barb fittings, and the development of compression fittings have eliminated most of the problems. Also, polybutylene is sometimes used in place of PE for lateral lines to avoid stress cracking. Howell and Barinas (1980) have measured the pressure losses across emitters and fittings that are inserted in lateral lines, and have presented head loss curves for them in terms of equivalent lengths of pipe as a function of flow rate.

C. System Control, Water Measurement, and Automation

The main control station for a trickle irrigation system consists of a pump, a backflow-prevention device, a primary filter, a pressure regulator (or automatic or mechanical flow control valves), pressure gauges, a water meter, and sometimes automation and chemical injection equipment (see Fig. 2). Some of these components will be discussed in greater detail later in this chapter. The importance of installing a water measurement device on every trickle system has been too often overlooked. A water meter is needed to check initial designs, to manage or schedule irrigations, and to monitor possible maintenance problems. Frequent measurement of the flow rate to various sections of the field facilitates detection of problems before they become serious. For example, a gradual decrease in flow rate may indicate a clogged secondary filter or emitter, whereas a sudden increase in flow could suggest a break in the distribution lines. Because discharge rates for trickle emitters are normally low, smaller capacity and less costly water meters can be used for trickle than for other traditional irrigation methods.

Trickle systems are readily automated. Single and multistation timers or controllers and related solenoid valves are easily installed to eliminate the work of turning water on or off manually. Filter backwashing (Wilson, 1972) and lateral line flushing (Shearer, 1977) for system maintenance can also be automated. Soil moisture sensors (Austin and Rawlins, 1977; New and Roberts, 1973) can be used to actuate control systems. Sophisticated switching tensiometers were tested by Wendt *et al.* (1973); Phene *et al.* (1973) described the successful use of a soil matric potential (heat-dissipation) sensors to automate trickle irrigation systems. Pan evaporation measurements (Busman and Fangmeier, 1979; Phene and Campbell, 1975) have also been used with trickle irrigation systems. Mears *et al.* (1979) developed a microprocessor based controller for automatic trickle irrigation. Future trickle irrigation systems will probably use microprocessors that monitor not only soil moisture, but also hydraulic pressure, flow rate, chemical injection rate, and weather data.

VI. Advances in Soil Water Modeling

A. Principles

The governing principles of soil water modeling for trickle irrigation systems are the same as for other irrigation methods. The differences which exist are primarily in the geometry of the sources and frequency of water applied. The pertinent flow occurs only in the veritcal direction for conventional systems with negligible horizontal moisture gradients. On the other hand, only a small part of the total soil surface is wetted for trickle applications and the flow patterns vary vertically as well as laterally. Furthermore, trickle irrigation frequency is sufficiently high that the soil water holding capacity is of less importance than for flood or sprinkler irrigation systems.

In soils, water moves relative to the total soil water potential ψ_T, where

$$\psi_T = h - z + \pi \tag{3}$$

with h the pressure head (cm), z the soil depth (cm), and π the osmotic head (cm). Each term represents an energy per unit weight. Other factors can be important, but generally are minor. The pressure head h will, for the most part, be negative (i.e., the water is under a tension in the unsaturated zone), although there may be a small positive pressure (saturated) zone near the emitter or near a shallow water table. The value of h in an unsaturated soil is the soil matric potential; it is equal to the absolute pressure of the water minus the air pressure (normally the soil air pressure will be that of the atmosphere).

The flow is described by Darcy's law:

$$\mathbf{J} = -K(h)\,\nabla\psi_T \tag{4}$$

with $\mathbf{J}$ the water flux density (cm/hr), $K(h)$ hydraulic conductivity (cm/hr) and $\nabla\psi_T$ the vector gradient of the total potential. For unsaturated conditions, $K(h)$ is dependent upon the water status, a fact that is underscored by writing it as a function of pressure head h. We will assume K is a scalar quantity, which is an adaptation of Darcy's law for saturated conditions to the unsaturated case. Combining Darcy's law and assuming a continuity of mass gives Richards' equation

$$\partial\theta/\partial t = \nabla \cdot [K(h)\,\nabla\psi_T] - S \tag{5}$$

where S is the "uptake rate" of water by the plant roots ($cm^3/cm^3\ hr^{-1}$). If the osmotic potential is assumed negligible, the substitution of Eq. (3) gives

$$\partial\theta/\partial t = \nabla \cdot [K(h)\,\nabla h] - \partial K/\partial z - S \tag{6}$$

Alternatively, Eq. (6) can be expressed in terms of θ versus h. To "eliminate" θ we define a specific water capacity $C(h)$ as

$$C(h) = d\theta/dh \tag{7}$$

and it follows that

$$C(h)\,\partial h/\partial t = \nabla \cdot [K(h)\,\nabla h] - \partial K/\partial z - S \tag{8}$$

To write Eq. (6) in terms of θ, the soil water diffusivity D_s is defined as

$$D_s = K(\theta)/C(\theta) \tag{9}$$

where K and C are now taken as functions of θ to give

$$\partial\theta/\partial t = \nabla \cdot [D_s(\theta)\,\nabla\theta] - \partial K/\partial z - S \tag{10}$$

Still another choice, which will be used in the examples, is to use Kirchhoff's relation to define the matric flux potential ϕ (cm^2/hr) as

$$\phi = \int_{-\infty}^{h} K(h')\,dh' \tag{11}$$

for which Eq. (6) becomes

$$(d\theta/d\phi)(\partial\phi/\partial t) = \nabla^2\phi - \partial K/\partial z - S \tag{12}$$

Of particular interest is when K is of the special form

$$K = K_o \exp(\alpha h) \tag{13}$$

for which $\phi = K/\alpha$ and Eq. (12) becomes linear for the steady-state case. Values of K_0 and α for over 20 soils are tabulated by Bresler (1978). Additional values are given by Gilley and Allred (1974), Ababou (1981), and Warrick *et al.* (1981).

Thus, mathematical modeling of the soil moisture flow regimes for trickle systems reduces to the solution of Eq. (6) (or the equivalent), subject to appropriate input and geometric factors. The solution is difficult, in general, because of nonlinearity. Also, the two- and three-dimensional geometries of trickle systems are more complex than the one-dimensional cases typical for many other soil water regimes. In order to solve Eq. (6), we must input (1) the hydraulic properties of the soil, (2) the boundary conditions, (3) the initial conditions, and (4) the plant uptake pattern. From a practical standpoint, the input data are often sparse or difficult to gather. Scarcity and quality of input obviously affect the reliability of results.

We are generally seeking output in the form of water content or pressure heads. Alternatively, wetting fronts and streamlines can be located. Also, design criteria such as emitter spacing or discharge rate can be calculated.

B. Flow Geometries

Some typical geometries are given in Fig. 18. Water moves downward (vertically and horizontally) from the source for bare soils as in Fig. 18a. In the two-

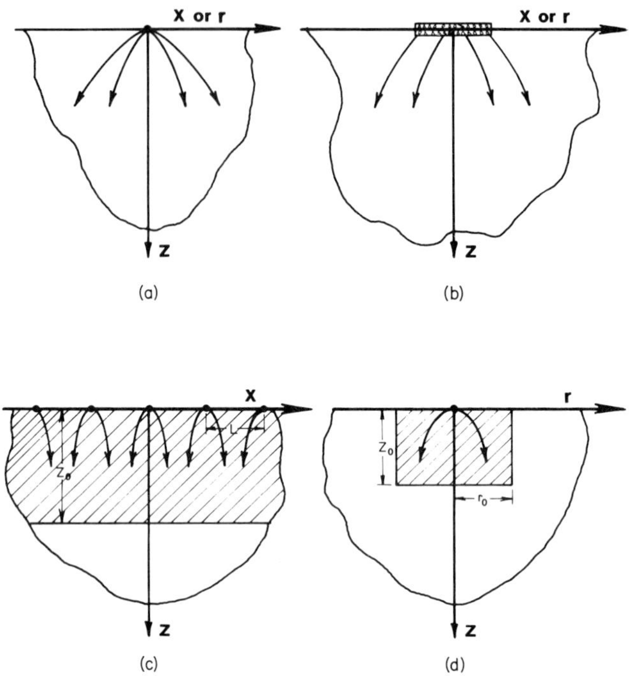

FIG. 18. Four common types of flow geometries for trickle irrigation systems: (a) point- or line-source emitters; (b) disk- or strip-source emitters; (c) two-dimensional array of line sources with uptake from a finite root zone; (d) three-dimensional point source with a cylindrical uptake.

dimensional case, the supply line is perpendicular to the x,z plane. We are assuming that the pore openings in the tubing or emitters are sufficiently close where the flow output is independent of position in the direction of the line itself. The flow pattern is axially symmetric for single sources and is the same in all r, z planes. As water is added to the soil surface, a finite area becomes wetted, as in Fig. 18b. The size of the area will be largest for soils with a low intake ability and heavy application rates; it will be smallest for high-intake soils and slow application rates. The wetted soil surface will be in a strip for the two-dimensional case and approximate a circle for the three-dimensional case.

Plant uptake is depicted in Fig. 18c and d for parallel line sources and a single point source, respectively. Water uptake may be limited to a rooting zone as shown with very little activity at deeper depths. The water at deep depths can be described as independent of x for parallel line sources. A simplifying assumption that the sink term is one dimensional throughout the zone may be justified for some cases. Perhaps the withdrawal pattern is within a cylinder depicted by Fig. 18d for a single emitter. An alternative would be an active rooting zone offset from the emitter position, such as for an emitter setting adjacent to the trunk of a tree.

C. Boundary Conditions

In order to solve for moisture distribution from Eq. (6), it is necessary to specify conditions along the surface and at infinity (i.e., large distances from the emitter). Possibilities include statements of

(1) ψ_T or h or θ
(2) J
(3) J as a specified function of ψ_T
(4) a mixture of (1), (2), and (3)

The most rigorous surface boundary condition is (3) with the soil evaporation properly related to water availability and the adjoining microclimate. Often the water flux density J is specified as zero along the x (or r) axis except at the source. The soil is assumed saturated near the source, or the velocity can be specified. Generally, h (or θ) will approach a constant for large distances from the origin. For the semi-infinite cases depicted by Fig. 18a,b,d, the appropriate constant would be $h = -\infty$ (or $\theta = 0$) if no other sources were present. The appropriate limiting value in Fig. 18c as $z \rightarrow \infty$ would be a constant h for which the hydraulic conductivity is equal to the excess of water applied over the plant uptake. This is because the hydraulic gradient of Eq. (6) approaches unity at large soil depths.

D. Solutions and Examples

Five examples will be presented illustrating the usefulness of soil water models. The first example was calculated using a finite difference solution; the second, fourth, and fifth solutions use a linearized form of Eq. (12) for steady flow; and the third solution is a comparison of field measured, finite difference and linearized results. References are made to additional similar studies for readers interested in more results.

1. *Wetting Front for Single Point-Source Emitters*

Brandt *et al.* (1971) present a finite difference solution for both line-source (two-dimensional case) and point-source emitters (three-dimensional case) corresponding to the geometry in Fig. 18b. Calculations were performed primarily for infiltration and results compared with laboratory results in Bresler *et al.* (1971) and Bresler (1977, 1978). Two soils were used, the Gilat loam and the Nahal Sinai sand. The "wetting fronts" were defined numerically for infiltration by the maximum gradient of water content.

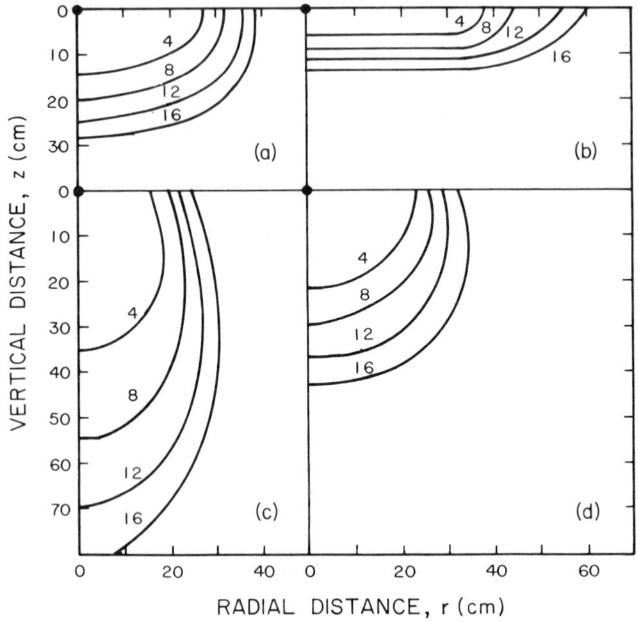

FIG. 19. Wetting fronts as a function of infiltration time and accumulated irrigation quantity (as indicated by the numbers labeled on each curve, in liters) for two soils (after Bresler, 1978): (a) Gilat loam, q = 4 liters/hr; (b) Gilat loam, q = 20 liters/hr; (c) Nahal Sinai sand, q = 4 liters/hr; (d) Nahal Sinai sand, q = 20 liters/hr; (●) single point-source emitters.

Figure 19 shows wetting front positions for both of the soils and for two infiltration rates—4 and 20 liters/hr. The numbers on each curve represent a total water applied of 4, 8, 12, or 16 liters. Obviously, the sand wetted deeper, which was due to both the ability of the sand to transmit water easily and the lower soil water holding capacity. Also, wider wetting patterns were observed with the higher application rate on both soils. This was a consequence of a larger wetted circle (Fig. 18b) for the higher rate. When the water had spread over a larger surface, the movement was not as deep for a given quantity of water. Also, it was evident from Fig. 19 that the finer soil resulted in more spreading at the surface. In addition, the higher rates resulted in a lesser wetted soil volume as the pores were more nearly filled with water.

Additional numerical solutions have been compiled by Van der Ploeg and Benecke (1974), Nassehzadeh-Tabrizi *et al.* (1978), and Ababou (1981). An analytical solution for a cylindrical soil volume with plant uptake has also been developed (Warrick *et al.*, 1980), and experimental measurements have been made by Roth (1974), Bresler and Russo (1975), and Levin *et al.* (1979b) for wetting fronts from point-source emitters.

2. *Wetting Pattern for Double Point-Source Emitters*

As previously mentioned, for steady-state conditions and K exponential with h [i.e., Eq. (13)], the form of Eq. (12) is linear. This allows a solution for multiple sources to be formed by adding together the ϕ derived for individual sources, i.e., the linearity permits superposition. For this example, consider K for a Pima clay loam of

$$K = 9.9 \exp(0.014h) \text{ (cm/day)} \tag{14}$$

The source strength q is 2 liters/hr for each emitter, one located at $r = 0$, $z = 0$ and a second at $r = 200$ cm, $z = 0$ (Fig. 20). The solution by Warrick [1974, see Eq. (18)] is

$$\phi = (\alpha q/8\pi)[\Phi(\alpha r/2, \alpha z/2) + \Phi(\alpha r'/2, \alpha z/2)] \tag{15}$$

where

$$\Phi(R, Z) = 2 \{[\exp(Z-\rho]/\rho - \exp(2Z)E_1(Z + \rho)\} \tag{16}$$

$$\rho^2 = R^2 + z^2 \tag{17}$$

and E_1 is the exponential integral as defined, for example, by Abramowitz and Stegun (1964). The r' value is the radius relative to the second emitter, i.e., $r' = r - 200$.

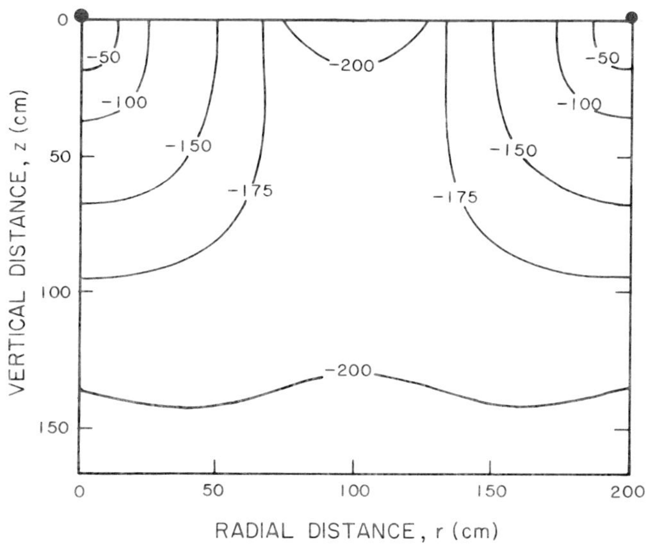

FIG. 20. Pressure head h contours for double, point-source emitters (●) spaced 200 cm apart; q = 2 liters/hr (after Warrick *et al.*, 1981).

The resulting calculation for the pressure head h contours is in Fig. 20. The value of h was wettest (in excess of $h = -50$ cm) next to the emitter source at the upper left and the upper right. The driest region was near the bottom of the figure and at the surface midregion with h less than -200 cm. Also, note the symmetry that existed about the middle of the plot, corresponding to the plane of equal distance from each source. The lines of equal pressure head h appeared somewhat like lines of advance for the unsteady case, but they are physically unrelated. These contours resulted after water had been added for a sufficiently long period such that the moisutre levels no longer change with time. Water was moving from the source and out of the system by drainage. A profile for the two-dimensional plane equally spaced from the sources (i.e., perpendicular to the center of Fig. 20) was presented earlier by Warrick (1974) for a similar two-source example.

3. *Comparison of Field Measurements and Soil Water Models*

Field measurements were made gravimetrically and reported by Ben-Asher *et al.* (1975) for moisture profiles at $r = 2$ cm during infiltration from a single

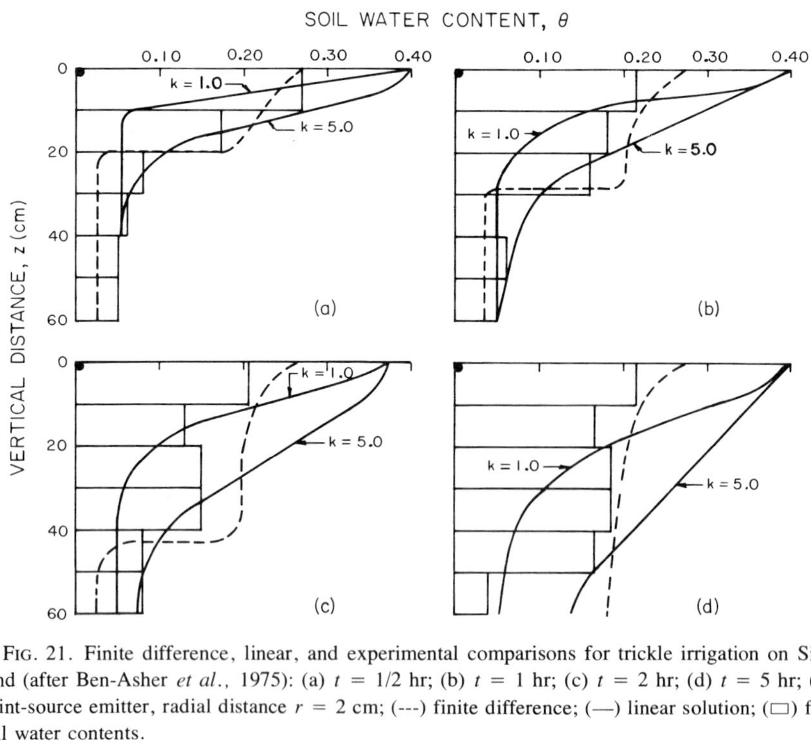

FIG. 21. Finite difference, linear, and experimental comparisons for trickle irrigation on Sinai sand (after Ben-Asher *et al.*, 1975): (a) $t = 1/2$ hr; (b) $t = 1$ hr; (c) $t = 2$ hr; (d) $t = 5$ hr; (●) point-source emitter, radial distance $r = 2$ cm; (---) finite difference; (—) linear solution; (□) field soil water contents.

emitter. The discharge rate was 0.4 liters/hr and the soil a Nahal Sinai sand. The results are presented in histogram form in Fig. 21 for $t = 0.5, 1, 2$, and 5 hr. Note the surface quickly reached a maximum value of θ in excess of 0.2 and did not change for the 0.5 to 5 hr. (The profiles differ slightly as sampling was at a slightly different position each time.) At deeper increments the water tended to increase with time. The results for the finite difference model reported (see Section VI,D,1) are shown by the "dashed lines." These are characterized by a fairly distinct wetting front which was propagated through the profile. Also shown are results from the time-dependent linearized solution of Eq. (12) by choosing $k = d\theta/d\phi$ equal to a constant. Two values were used, $k = 1$ and 5 cm/hr. Neither linearized solution gave a sharp wetting front. In general, the linearized time-dependent solution was not a good approximation of infiltration. Numerical versus linearized solutions have also been compared by Ababou (1981). Ben-Asher *et al.* (1978b) also examined linear versus nonlinear solutions for cyclic irrigation.

4. *Evaluation of Discharge Rate and Spacing for Point-Source Emitters*

Bresler (1978) has prepared a graph relating discharge to spacing for the Gilat loam and Nahal Sinai sand (see Section VI,D,1). This allows one to conveniently choose spacing and discharge relations to maintain a specified pressure head at the soil surface at a distance midway between the two sources. The solution used was that of Wooding (1968) using the steady-state linearized form of Eq. (12) for the geometry of Fig. 18b. Conductivity relationships used for the loam were $K = 0.84 \exp(0.025h)$ cm/hr and for the sand $K = 8.52 \exp(0.062h)$ cm/hr.

Figure 22 shows Bresler's results, where each curve relates distance between the emitters d_s (cm) to discharge rate q (liter/hr) for the midpoint pressure head $h_M = -10$ or -70 cm. The larger value of h_M (i.e., a wetter condition) can be attained by either increasing discharge rate or reducing the spacing. At a given discharge, the wetter condition (-10 cm) requires a closer spacing than the drier one (-70 cm). Also, the two emitters must be closer together for the sand in order to give the same midway pressure head as the loam for any given discharge rate.

A more general result is by Eq. (16), but for a point source (Fig. 18a) rather than the disk source (Fig. 18b). Results are summarized in Table I. The ϕ of Eq. (16) is related to h_M by $\phi = (K_0/\alpha) \exp(\alpha h_M)$. To compare results in Table I with Fig. 22, we consider the loam for $h_M = -70$ cm. For $\alpha d_s = 2$ and 4, d_s will be 2/0.025 and 4/0.025, or 80 and 160 cm, respectively. Use of Table I gives $(8\pi K_0/\alpha^2 q) \exp(\alpha h_M) = 2.6$ and 0.59 resulting in $q = (2.3)(10)^3$ and $(9.9)(10)^3$ cm^3/hr. These points are plotted as 2.3 and 9.9 liters/hr on Fig. 22 and they fall, for practical purposes, directly on the $h_M = -70$ cm curve from the disk source results of Bresler.

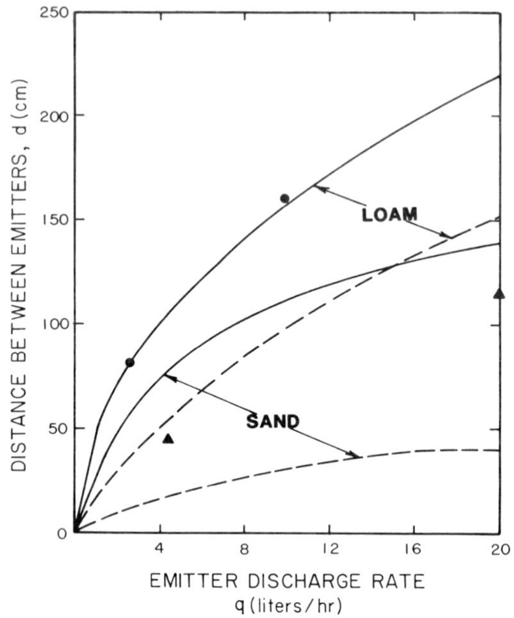

FIG. 22. Relationship between emitter spacing d_s and discharge rate q based on specified midway pressure heads h_M for two soils (after Bresler, 1978): (—) $h_M = -70$ cm (●); (---) $h_M = -10$ cm (△).

A repeat of similar steps for $h_M = -10$ cm results in $d_s = 48$ ($\alpha d_s = 1.2$) corresponding to $q = 4.2$ liters/hr and $d_s = 110$ ($\alpha d_s = 2.8$) to $q = 20$ liters/hr. These two points fall well to the right (and below) the disk source results. This is a consequence of generally higher flow rates necessary to maintain h_M at the wetter condition of -10 cm. The higher flow rate gives a larger saturated zone (see Fig. 18b). In order to achieve the higher rate, the point source would more

TABLE I

GENERALIZED SPACING, DISCHARGE RELATION FOR TWO POINT-SOURCE EMITTERS SEPARATED BY A DISTANCE d_s WITH MIDPOINT PRESSURE HEAD h_M

αd_s	$(8\pi K_0/\alpha^2 q)\exp(\alpha h_M)$	αd_s	$(8\pi K_0/\alpha^2 q)\exp(\alpha h_M)$
0.4	29.2	2.4	1.8
0.8	11.4	2.8	1.3
1.2	6.3	3.2	1.0
1.6	3.9	3.6	0.76
2.0	2.6	4.0	0.59

resemble a saturated hemisphere, than would result from a disk source with a larger flow with the same h_M.

5. *Parallel Line-Source Emitters with and without Plant Uptake*

Warrick *et al.* (1979) solved the steady linearized form of Eq. (12) for parallel line sources spaced at constant intervals (such as along equally spaced rows). For no plant uptake, they showed empirically the midpoint matric flux potential ϕ_M at the soil surface was almost exactly satisfied by

$$q_L/\phi_M = \alpha L_s \exp(0.24\alpha L_s) \tag{18}$$

without uptake for at least $\alpha L_s < 5$. In Eq. (18), L_s is the full spacing of the array and q_L is the volume discharge rate per unit time per unit length of the line. Figure 23 shows the relation of the spacing, discharge, and midpoint pressure head h_M. The curve was calculated exactly and the points were approximated by Eq. (18) with no discernable difference.

In applying Fig. 23 to find a spacing, suppose we use $K = 9.9 \exp(0.014h)$ for

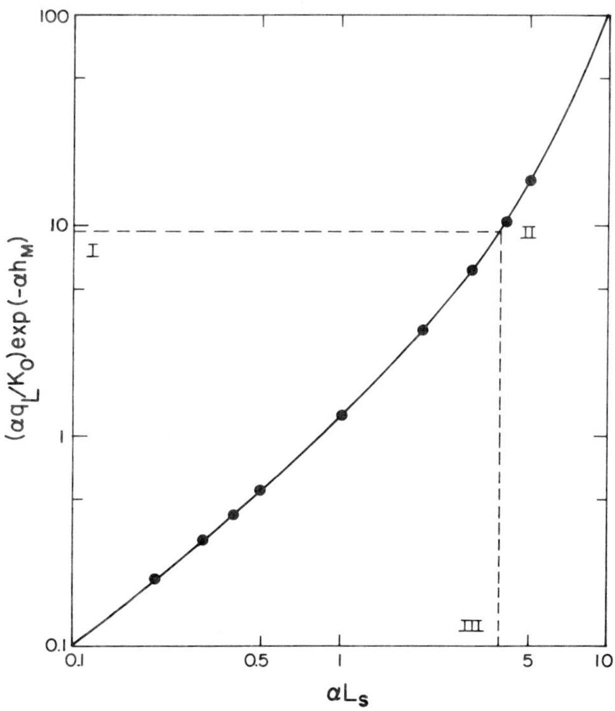

FIG. 23. Values of $(\alpha q_L/K_0)\exp(-\alpha h_M)$ midway between line-source trickle emitters as a function of the dimensionless spacing αL_s: (●) $\alpha L_s \exp(0.2403\alpha L_s)$.

the Pima clay loam. Furthermore, we assume the line-source discharge rate is 100 cm²/day and the midpoint pressure head is −300 cm (i.e., a suction of about 0.3 bar). To use Fig. 23, we first find $(\alpha q_L/K_0)\exp(\alpha h_M)$ as 9.4 (dimensionless), which is labeled "point I" of the ordinate. Proceeding right to the curve at II and down to the abscissa at III, we read $\alpha L = 3.9/\alpha = 278$ cm. To find q_L for a given spacing we would reverse the steps. For example, to evaluate h_M for a given q and L_s, we proceed from III to II to I and solve for h_M from the ordinate value.

Plant uptake increases the complexity; however, an appropriate nomograph has been developed. The plant uptake of Fig. 18c is taken to be a step function of z only:

$$S = u/z_0, \quad 0 < z < z_0; \qquad S = 0, \quad z > z_0 \tag{19}$$

Thus, z_0 is the rooting depth and u is the total uptake per unit time per unit area (i.e., an average transpiration rate). The relationship analogous to Eq. (18) is

$$h_M = (1/\alpha)\ln\{(q_L/K_0L_s)[\exp(-0.24\alpha L_s) + (\exp(-\alpha z_0) - 1)(uL_s/q\alpha z_0)]\} \tag{20}$$

Equation (20) is easily evaluated on a small calculator. Normally for a given problem we can solve for q_L, L_s, or h_M given the other two.

The relationship of Eq. (20) is presented in a nomograph form as Fig. 24. To illustrate its use, suppose we want to determine q for Pima clay loam for a spacing of 140 cm and midpoint pressure head $h_M = -300$ cm. Choose as a rooting depth $z_0 = 35$ cm and an uptake amount of $u = 0.2$ cm/day. For our input, we evaluate

$$\alpha\phi_M/u = (K_0/u)\exp(\alpha h_M) = 0.74 \tag{21}$$

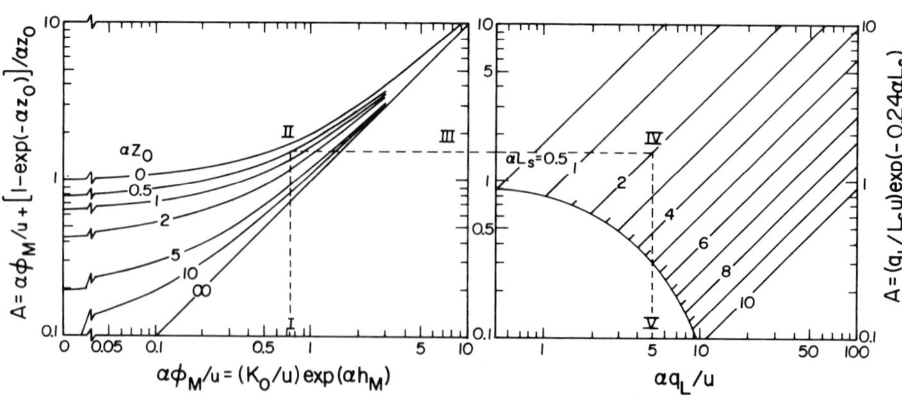

FIG. 24. Nomographs relating dimensionless αL_s, dimensionless rooting depth αZ_0, a relative uptake of $\alpha q_L/u$, and a dimensionless matrix flux potential $\alpha\phi_M/u = (K_0/u)\exp(\alpha h_M)$ between surface, line-source trickle emitters.

corresponding to the left abscissa at point I. Moving up to II we stop at $\alpha z_0 = 0.49 \simeq 0.5$. The value of the ordinate could be read, but we instead go to the right through III and pause at IV where $\alpha L_s = 1.96 \simeq 2$. Proceeding down to V gives $q_L = 4.9\ u/\alpha$ or approximately 70 cm^2/day (i.e., the discharge rate for each cm of line must be 70 cm^3/day, corresponding to an areal application rate of $q_L/L_s = 70/140 = 0.5$ cm/day). If L_s is unknown, the same initial step is used from I to II to III. Then V is located on the right abscissa, allowing IV to be found as the intersection horizontal line from point III and the vertical from V. Point IV defines αL_s, whence L_s is given by dividing by α.

Additional two-dimensional examples have been presented by Zachmann and Thomas (1973), Gilley and Allred (1974), Thomas *et al.* (1974, 1976), and Warrick and Lomen (1981). An example of plant root uptake for two-dimensional surface and subsurface sources has been completed by Oron (1981).

VII. Advances in System Design and Evaluation

Before any system is installed, the hydraulic design should be adequately evaluated, and the emission uniformity continually reevaluated for assuring maximum economical and efficient operation.

A. Hydraulics

The flow regime throughout a trickle irrigation system is hydraulically steady, spatially varied pipe flow with lateral outflows. The total discharge in the distribution network (lateral, submain, and main lines) decreases with respect to distance from the pump. The lateral and submain have similar hydraulic characteristics and are designed to maintain a small pressure variation along the lateral line. The main line is designed in terms of input pressures and minimal required pressures at any submain line. Trickle design principles are similar to those developed for sprinkler irrigation systems except that the flow rates are lower and the number of outlets (submains, laterals, and emitters) is larger.

Trickle irrigation distribution lines are normally considered to be smooth pipes, and either the Darcy–Weisbach or Hazen–Williams equation can be used to compute friction losses for the pipelines.

The Darcy–Weisbach equation is

$$H_f = 6.38 f L D^{-5} Q^2 \tag{22}$$

where H_f is the pipe friction loss (m), L is the pipe length (m), D is the inside pipe diameter (mm), Q is the pipe flow rate (liters/hr), and f is a dimensionless friction factor. An acceleration of gravity of 9.81 m/sec^2 was assumed in this equation. Watters and Keller (1978) proposed a simplified form of Eq. (22) as

$$H_f = 0.465 L D^{-4.75} Q^{1.75} \tag{23}$$

Equation (23) incorporates a friction factor estimated from the Blasius equation for smooth pipes with a water temperature of 20°C ($\nu = 1.0 \times 10^{-6}$ m^2/sec).

The empirically developed Hazen–Williams equation is

$$H_f = 0.628LD^{-4.865}[(100Q)/C]^{1.852} \tag{24}$$

where C is a dimensionless pipe roughness factor. Equation (24) is widely used because of its simplicity, although it has no correction for viscosity. Care should be taken in selecting the C values. Hughes and Jeppson (1978) showed that the selection of the proper C for Eq. (24) obviously depends on the Reynolds number Re; and Howell *et al.* (1980) suggested that the best C values for trickle irrigation systems were $C = 130$ for 14–15-mm pipe, $C = 140$ for 18–19-mm pipe, and $C = 150$ for 25–27-mm pipe. A low estimate of C will overestimate the friction loss, whereas a high estimate will result in more conservative friction loss for design purposes.

Pressure variations along a trickle irrigation line are primarily due to friction and slope. Numerous publications and books describe the basic principles, procedures, and details for proper sizing of trickle irrigation pipelines (Baars, 1976; Keller and Karmeli, 1975; Howell *et al.*, 1980). Wu *et al.* (1979) present design charts for lateral, submain, and main lines for different field slopes, shapes, and pipe sizes along with design examples. Some of the more recent developments in trickle irrigation hydraulics are as follows: Bresler (1978) introduced a design procedure using both soil water flow models and hydraulics; Keller (1980) developed a compact "polyplot" design technique for tapered submains; Braud and Soom (1981) introduced design equations including both emitter and pipeline hydraulics; and Wu and Gitlin (1981) developed double-inlet and inflow–outflow systems using network design principles.

Steady and nonsteady infiltration models (Bresler, 1978) were used to calculate the desired spacing between emitters as a function of their discharge, soil hydraulic properties, and crop sensitivity to water stress. Some of the limitations arise from the linearization procedures, and include selecting the correct value for water pressure midway between emitters, neglecting seepage below the root zone, and estimating the hydraulic parameters of the soil. Many fields have soils that are nonhomogeneous and nonisotropic, and the hydraulic conductivity usually decreases rapidly with time when ponding occurs under trickle emitters. Ideally, trickle irrigation systems should be designed with emitter spacings and discharge rates such that small depression ponds of less than 0.2 m in diameter will develop beneath or around an emitter without runoff or downslope channelization.

Varying the size of submain lines gives the designer an opportunity to reduce costs, investigate alternate designs, and develop site-specific designs (Keller, 1980). Design equations in a form that include emitter characteristics (emitter exponent, emitter variability, etc.) as well as lateral line hydraulics (length, diameter, land slope, etc.) are being developed for analysis by programmable digital calculators (Braud and Soom, 1981). Multiple-inlet systems can be designed with a combination of double-inlet (inflow from both ends of a lateral line from two submains) and inflow–outflow (inflow from a submain to a lateral line with outflow to the next submain line) in a total network design (Wu and Gitlin, 1981). This type of design can permit any length of lateral in varying field shapes and terrain. As a general recommendation, lateral lines should seldom be more than 300 m long and should be laid downslope for less than 5% slopes. If the slope exceeds 5%, laterals should be installed along the field contour, pressure-compensating emitters specified, or pressure control devices installed.

Uniformity of water application along lateral lines can also be improved by changing emitter sizes (Myers and Bucks, 1972), microtube lengths (Kenworthy, 1972) or microtube lengths and sizes (Bucks and Myers, 1973). Although sprinkler systems have used similar concepts, varying the size of trickle emitters within an irrigated field has not been widely accepted because of increased installation time and site-specific design requirements. The fixed emitter characteristics are normally adequate from a hydraulic standpoint for level terrain.

B. Emission Uniformity

Several methods have been proposed for assessing the uniformity of water application in irrigation systems. The term emission uniformity has generally been used to describe the emitter flow variation for a trickle irrigation unit or subunit. Emission uniformity can be a function of: (1) hydraulic variation caused by elevation changes and friction losses along distribution lines and (2) emitter discharge variation at a given operating pressure caused by manufacturing variability, clogging, water temperature changes, and aging. Presently, no design equation has been developed that includes all the factors which might affect emission uniformity.

With sprinkler irrigation systems, the coefficient of uniformity (Christiansen, 1942) C_u has been widely accepted and can be expressed as

$$C_u = [1 - (\Delta q/\bar{q})]100 \tag{25}$$

where $\bar{q}$ is the mean sprinkler discharge rate and Δq is the mean absolute deviation of sprinkler discharge rates. Assuming that the sprinkler discharge rates are normally distributed, Hart (1961) developed the following:

$$C_u = (1 - 0.798C_v)100 \tag{26}$$

where C_v is the coefficient of variation (statistically, the standard deviation divided by the mean) between the discharge rates.

Keller and Karmeli (1975) were the first to define an empirical design emission uniformity percentage EU for trickle irrigation systems as

$$\mathrm{EU} = [1 - 1.27(C_{\mathrm{vm}})n^{-1/2}](q_n/\bar{q})100 \tag{27}$$

Where C_{vm} is the manufacturer's coefficient of variation, n is the number of emitters per plant (minimum of one), q_n is the minimum emitter discharge rate computed from the minimum pressure, and $\bar{q}$ is the mean emitter discharge rate. The emission uniformity increases as more emitters are added to each plant.

Nakayama *et al.* (1979) developed a coefficient of design uniformity C_{ud} based on statistical analysis:

$$C_{\mathrm{ud}} = [1 - 0.798(C_{\mathrm{vm}})n^{-1/2}]100 \tag{28}$$

Admittedly, similarities exist between Eqs. (27) and (28) since both stress the importance of manufacturing variability and the number of emitters per plant. However, the original derivation of EU was based on the ratio of the discharge rate for the lowest 25% of emitters to the average discharge rate, whereas C_{ud} is based on the discharge rate deviations from the average rate. Table II presents recommended ranges of EU and C_{ud} to be used in design procedures, unless economic considerations dictate higher or lower degrees of uniformity.

Wu and Gitlin (1977) proposed another paramter called emitter flow variation q_{var}, defined as

$$q_{\mathrm{var}} = [1 - (q_n/q_m)]100 \tag{29}$$

where q_n and q_m are the minimum and maximum emitter discharge rates.

TABLE II

RECOMMENDED RANGES OF DESIGN EMISSION UNIFORMITY EU AND DESIGN UNIFORMITY COEFFICIENT C_{ud} FOR ARID AREAS

Emitter type	Crop spacing	Soil topography	EU range[e] (%)	C_{ud} range[e] (%)
Point source	Wide[a]	Uniform[c]	90–95	93–98
		Steep or undulating[d]	85–90	—
	Close[b]	Uniform	85–90	90–95
		Steep or undulating	80–90	
Line source	Close	Uniform	80–90	85–95
		Steep or undulating	75–85	—

[a] Spaced >4 m apart.
[b] Spaced <2 m apart.
[c] Slope <2%.
[d] Slope >2%.
[e] Values can be lowered up to 10% for humid areas.

Although a relationship exists between emitter flow variation and Christiansen's uniformity coefficient, the number of emitters per plant, manufacturer's variability, or other factors are not included for design comparisons.

Bralts *et al.* (1981) illustrated for a single-chamber trickle irrigation tubing that the hydraulic and manufacturing variability were independent and that the total coefficient of variation C_{vt} can be expressed as

$$C_{vt} = (C_{vh}^2 + C_{vm}^2)^{1/2} \quad (30)$$

where C_{vh} is the hydraulic coefficient of variation. They also present design nomographs for utilizing this equation.

C. Emitter Variability

As mentioned earlier, emitter discharge variations can be caused by manufacturing variability, clogging, water temperature, and aging. Solomon and Keller (1978) showed from pressure distribution histograms that the manufacturer's coefficient of variation C_{vm} is often more important than the hydraulic losses in the system design. Solomon (1979) stated that C_{vm} depends upon the emitter design, construction material, and manufacturing process. A list of C_{vm} values, along with emitter exponents e, to be used in Eq. (2) is given in Table III. These C_{vm} values were measured in the laboratory at normal operating pressures and ranged from 0.02 to 0.40 with the lower values indicating better emitter performance. Table IV gives the authors' recommendations for interpretation of C_{vm} values.

Nakayama and Bucks (1981a) found that emission uniformity was greatly reduced even when 1–5% of the emitters were completely clogged with two to eight emitters per plant. In the example of Fig. 25, the entire emitter population was divided into 20 groups, with 5% of the emitters of a specified discharge rate in the respective grouping, i.e., 5% of the lowest emitter discharge rates will be in group 1 and 5% of the highest in group 20. With four emitters per plant, an initial $C_{vm} = 0.05$, and no clogging, 5% of the highest discharging emitter combinations will overirrigate by about 5%; and because of symmetry, 5% of the lowest discharging emitters will underirrigate by 5%. With just 5% of the emitters clogged, the lowest 15% of the emitter group has an average flow rate $0.80\bar{q}$, while the highest 5% represents $1.11\bar{q}$. Bralts *et al.* (1980) indicated that complete and partial emitter plugging could be statistically included in uniformity calculations for line-source emitters. They developed design charts relating partial plugging to complete plugging and initial to adjusted emission uniformity for different degrees of plugging.

Water temperature–emitter discharge variations can occur because of temperature variations along the lateral line due to solar heating of exposed distribution lines, changes in seasonal temperatures, and fluctuations in frictional loss pat-

TABLE III

MANUFACTURER'S COEFFICIENT OF VARIATION C_{vm} AND EMITTER EXPONENT e FOR VARIOUS TRICKLE IRRIGATION EMISSION DEVICES[a]

Device No.	Type of device	Manufacturer's coefficient of variation C_{vm}	Emitter discharge exponent e
1	Single vortex	0.07	0.42
2	Multiple-flexible orifices	0.05	0.70
3	Multiple-flexible orifices	0.07	0.70
4	Ball and slotted seat—non-compensating	0.27	0.50
5	Ball and slotted seat—pressure compensating	0.35	0.15
6	Ball and slotted seat—pressure compensating	0.09	0.25
7	Small tube	0.05	0.70
8	Small tube	0.05	0.80
9	Spiral long-path—nonflushing	0.02	0.65
10	Spiral long-path—manual flushing	0.06	0.75
11	Long-path—pressure compensating	0.05	0.40
12	Long-path—pressure compensating	0.06	0.20
13	Tortuous long-path	0.02	0.65
14	Groove and flap, short-path	0.02	0.33
15	Slot and disk, short-path	0.10	0.11
16	Slot and disk, short-path	0.08	0.11
17	Porous pipe	0.40	1.00
18	Twin-wall lateral	0.17	0.61

[a] After Solomon (1979).

TABLE IV

RECOMMENDED CLASSIFICATION SYSTEM FOR MANUFACTURER'S COEFFICIENT OF VARIATION C_{vm}

Emitter type	C_{vm} ranges	Classification
Point source	<0.05	Good
	0.05–0.10	Average
	0.10–0.15	Marginal
	>0.15	Unacceptable
Line source	<0.10	Good
	0.10–0.20	Average
	>0.20	Marginal to unacceptable

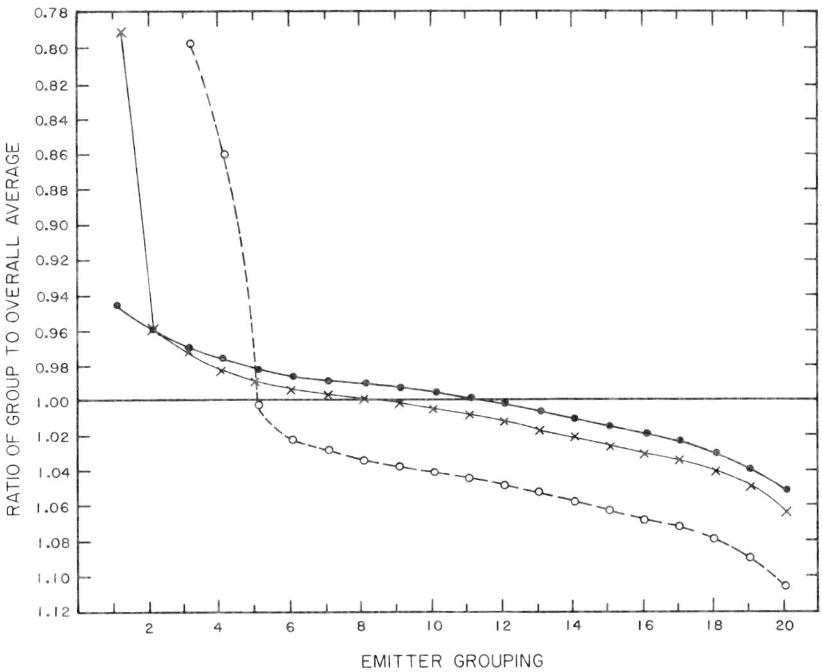

FIG. 25. Effect of complete emitter clogging on the average discharge-rate distribution with four emitters per plant: (●) no clogging; (○) 5% clogging; (×) 1% clogging.

terns within a trickle irrigation system. Gilad *et al.* (1968) and Parchomchuk (1976) have reported 12°C (54°F) and 16°C (61°F) rise in water temperature at the end of PE lateral lines. Keller and Karmeli (1975) suggested that the discharge sensitivity due to fluid temperature changes could be minimal for turbulent emitters; however, emitters having laminar and unstable flow regimes can be expected to increase in discharge rate as the temperature rises and viscosity decreases. Thus, a positive effect would result for laminar emitters from an increase in temperature that should compensate for a decrease in pressure at the ends of the lateral lines. Zur and Tal (1981) indeed found a positive temperature effect for a laminar emitter and no significant effect for one turbulent emitter; on the other hand, another turbulent emitter had a negative effect (decreasing discharge as the temperature increased) with no explanation given for this phenomenon. Parchomchuk (1976) recommended that water temperature variations be controlled or minimized by the following: (1) shading or burying of lateral lines; (2) operating the system at night; (3) adjusting the operating temperature to counteract seasonal temperature effects, but not variations in emission uniformity; and (4) using turbulent flow emitters with high discharge rates, that may not, however, always be desirable from hydraulic or design viewpoints.

Aging or deterioration can also be a factor in emitter variability, particularly since the plastic materials are subjected to expansion and contractions, drying

TABLE V

INITIAL AND FINAL COEFFICIENTS OF VARIATIONS FOR EIGHT EMITTERS AND FOUR TREATMENTS OF COLORADO RIVER WATER[a]

Emitter number and design	Chemical water treatment[b]							
	C (none)		D (chlorine and acid intermittent)		E (chlorine and acid continuous)		F (acid continuous)	
	C_{vi}[c]	C_{vf}[d]	C_{vi}	C_{vf}	C_{vi}	C_{vf}	C_{vi}	C_{vf}
1. Long-path capillary tubing	0.10	0.68	0.10	0.46	0.10	0.46	0.10	0.58
2. Long-path, spiral-grooved, manual-flush, low-discharge rate	0.06	0.40	0.06	0.31	0.06	0.27	0.06	0.15
3. Long-path spiral-grooved, manual-flush, high-discharge rate	0.06	0.54	0.06	0.51	0.06	0.29	0.06	0.16
4. Automatic, diaphragm, expandable orifice	0.07	0.60	0.07	0.50	0.07	0.27	0.07	0.19
5. Short-path, spiral-grooved, removable insert	0.06	0.59	0.06	0.87	0.06	0.42	0.06	0.62
6. Automatic, needle-flush valve	0.10	0.55	0.10	0.30	0.10	0.77	0.10	0.62
7. Automatic, ball-flush valve	0.15	0.18	0.15	0.53	0.15	0.29	0.15	0.14
8. Single vortex	0.07	0.68	0.07	0.64	0.07	0.41	0.07	0.34

[a] Unpublished data of D. A. Bucks, F. S. Nakayama, and R. G. Gilbert.

[b] In addition to chemical water treatment, water filtration included a sand filter with No. 20 silica sand followed by a 200-mesh stainless-steel screen filter. Details of water treatment are given in Table VII.

[c] Initial coefficient of variation for emitters installed in January, 1975, based on laboratory measurements.

[d] Final coefficient of variation for emitters removed in December, 1978, based on field measurements.

and wetting, presence of chemicals in the water, and prolonged exposure in the field. Prior research has not been conducted to isolate or quantify the effect of aging on emitter variability. Table V, however, does present previously unpublished data on changes in coefficients of variation for eight emitter types after four years of operation with four water treatments. The coefficient of variation C_v had increased from three- to fivefold, representing the combined effects from increased manufacturing variability, clogging, water temperature change, aging, and hydraulic variation. We believe that the reduction in performance was primarily caused by clogging and aging since operating pressures were controlled to short lengths of lateral lines. Details of this study were given in a number of publications (Nakayama *et al.*, 1978; Gilbert *et al.*, 1979, 1981a); they will be briefly discussed later in this article.

D. System Evaluation

Every trickle irrigation design needs to be checked immediately after installation and periodically evaluated at least every year. Merriam *et al.* (1980) state that the purpose of evaluating any irrigation system is fourfold: (1) to obtain information that will assist enigneers in designing other systems; (2) to determine how effectively the system can be operated and whether it can be improved; (3) to determine the efficiency of the system as it is being used; and (4) to obtain information to enable comparing various methods, systems, and operating procedures as a basis for economic decisions. We believe that the first three objectives are realistic, but doubt whether different methods, systems, or management practices can be objectively compared. Evaluations involve selecting typical or representative sites or locations and assuming that soils, topography, operating pressures, and pump flow rates are reasonably uniform. Definitions and efficiency terms certainly have different interpretations for different irrigation methods. Calculated efficiencies can seldom be more accurate than 3–5%.

Merriam and Keller (1978) described field procedures for measuring pressures and emitter discharge rates using a modified pressure gauge, stopwatch, and graduated cylinder. They recommended taking at least eight pressure readings and 32 discharge volumes at 16 different locations for point-source emitters and using a 1-m-long trough for collecting discharge readings from 16 locations for line-source emitters. A field emission uniformity EU_f was calculated as follows:

$$EU_f = (\bar{q}_{25}/\bar{q}) \tag{31}$$

where $\bar{q}_{25}$ is the average of the lowest 25% of measured discharge rates. Field worksheets as well as methods for estimating the potential and actual application efficiency based on the low-quarter concept were also discussed. Figure 26 demonstrates the systematic measurement of emitter discharge rates used in this type of system performance evaluation.

FIG. 26. Measuring the discharge rate from a point-source trickle emitter to be used in the evaluation of the irrigation system.

Increased attention is needed in the future towards development of a more complete evaluation system. Recommendations resulting from such evaluations include adjusting operating pressures, adding more pressure control, increasing or decreasing the number of emitters per plant, cleaning or replacing bad emitters, etc.

VIII. Advances in System Maintenance Practices

Plugging of emitters can be caused by physical, chemical, or biological contaminants. Hence, before the problem can be solved or reduced the exact cause or causes for the phenomenon must be determined in each case. This involves a thorough evaluation and examination of water quality and of the clogged emitters. Extensive research on emitter clogging has been conducted in the past five or so years. Additional work is still needed on specific water qualities. Many times clogging is site-specific and universal solutions are not available.

A. REASONS FOR EMITTER CLOGGING

Solomon (1977) indicated that most emitter designs represent a compromise between the opposing tendencies toward small openings (with discharge rates

TABLE VI

PRINCIPAL PHYSICAL, CHEMICAL, AND BIOLOGICAL CONTRIBUTORS TO CLOGGING OF TRICKLE SYSTEMS[a]

Physical (suspended solids)	Chemical (precipitation)	Biological (bacteria and algae)
Inorganic particles:	Calcium or magnesium carbonate	Filaments
Sand		
Silt		Slimes
Clay	Calcium sulfate	
Plastic		Microbial depositions:
	Heavy metal hydroxides, carbonates, silicates, and sulfides	
Organic particles:		Iron
Aquatic plants (phytoplankton/algae)		Sulfur
		Manganese
	Oil or other lubricants	
Aquatic animals (zooplankton)		
	Fertilizers:	
Bacteria	Phosphate	
	Aqueous ammonia	
	Iron, copper, zinc, manganese	

[a] After Bucks *et al.* (1979).

below the soil infiltration rate) and large openings (with higher discharge rates and a lower probability of clogging). Unfortunately, such a compromise will not always solve the clogging problem. Bucks *et al.* (1979) summarized the various factors contributing to emitter clogging as follows (see Table VI): physical, chemical, and biological. These three factors are closely interrelated, and controlling one may also alleviate problems caused by the other. For example, by reducing microbial slime, one also reduces the tendency of suspended particles in the water to stick onto the slime, agglomerate, and build up in the trickle lines and emitters. In addition, some small aquatic organisms such as snail eggs and larva, that are not readily observed and analyzed, can develop into large colonies in the drip lines and impose a combined physical and biological problem.

In the United States, research on emitter clogging has been conducted primarily at three separate locations with different water qualities: in the southwest with Colorado River water (Nakayama *et al.*, 1978); in the southeast with Florida waters (Ford, 1977a); and in the far west with Hawaiian waters (McElhoe and Hilton, 1974). Clogging problems in the southwest, which are related primarily to high contents of suspended solids and to water hardness, are quite different from those in the southeast, which are associated with high biological activity and with iron and sulfides, and from those in the far west, which are associated with the growth of biological slimes and filaments.

The Colorado River water study was conducted near Yuma, Arizona, for over four years on trickle-irrigated trees. Briefly, eight different types of point-source emitter designs, with initial discharge rates less than 12 liters/hr and six water treatments which varied by the degree of filtration and amount of chemical conditioning were selected (Table VII). When the experiment was terminated, emitters with discharge rates less than 50% of design were carefully dissected and examined microscopically. The predominant causes of emitter clogging and flow reduction were physical factors, followed by the combined effects of biological and chemical factors, as shown in Table VIII. Physical clogging problems usually occurred quickly where five emitter systems (Nos. 1, 5, 6, 7, and 8) failed after 278 days with screen filtration alone (treatments A and B) (Nakayama *et al.*, 1978). Even with sand and screen filtration (treatments C, D, E, and F),

TABLE VII

DESCRIPTION OF EMITTER DESIGNS AND TREATMENTS USED ON COLORADO RIVER WATER[a]

Emitter No.	Emitter design	
1	Long-path, capillary tubing	
2	Long-path, spiral-grooved, manual-flush, low-discharge rate	
3	Long-path, spiral-grooved, manual-flush, high-discharge rate	
4	Automatic, diaphragm, expandable orifice	
5	Short-path, spiral-grooved, removable insert	
6	Automatic, needle-flush valve	
7	Automatic, ball-flush valve	
8	Single vortex	
	Water treatment	
Treatment letter	Filtration	Chemical
A	Screen (50 mesh)	None
B	Screen (50 mesh)	Chlorine[b] (10 mg/liter) and acid[c] (lower pH to 7)—intermittent
C	Sand (silica No. 20) plus Screen (200 mesh)	None
D	Same as treatment C	Same as treatment B
E	Same as treatment C	Chlorine (1 mg/liter) and acid (lower pH to 7)—continuous
F	Same as treatment C	Acid (lower pH to 7)—continuous

[a] After Nakayama *et al.* (1978).
[b] Free residual chlorine.
[c] Sulfuric acid.

TABLE VIII

CAUSES OF CLOGGING OR FLOW REDUCTION AND RELATIVE PERCENT OCCURRENCE USING COLORADO RIVER WATER[a]

Causes of clogging	Percent of occurrence		Causes of clogging	Percent of occurrence	
	Individual	Total		Individual	Total
Physical:			Chemical:		
Sand grain	17	—	Carbonate precipitates	2	—
Plastic particles	26	—	Iron–magnesium precipitates	0	2
Sediment	2	—	Combined[c]:		
Body parts of insects and animals	3	—	Physical/biological	8	—
Deformed septa[b]	—	55	Physical/chemical	2	—
Biological:			Chemical/biological	6	—
Microbial slime	11	—	Physical/biological/chemical	2	18
Plant roots and algal mats	3	14	Nondetectable (probably physical)	—	11

[a] After Gilbert *et al*. (1981). Results are representative of eight emitter systems and four water treatments (C, D, E, and F), that were operated for more than four years. There were 1200 emitters installed in these water treatments and 119 with reduced flow or clogged conditions (<50% design flow) were dissected and microscopically examined for causes of flow reductions. (See Tables V and VII for descriptions of emitters and treatments.)

[b] Silicone rubber disk in emitter No. 6 were deformed by water treated with chlorine and acid (treatments D and E), which restricted flow. (See Tables V and VII for descriptions of emitters and treatments.)

[c] The observations indicated that the most likely initial cause of flow reduction was a physical factor, followed by the development of biological and chemical factors. The major physical factors involved were sand grains and plastic particles.

FIG. 27. Water samples can be taken before and after filtration. The unfiltered water is used to evaluate possible emitter clogging problems.

physical clogging due to sand grains and plastic particles occurred in 43% of the completely and partially clogged emitters (Gilbert *et al.*, 1981).

Although trickle irrigation systems are generally flushed at installation and periodically, users must be warned that extreme care is required to keep pipelines clean of all foreign materials. Figure 27 shows the difference between filtered and unfiltered Colorado River water.

At the same time that emitters were sampled for clogging causes, water samples from lateral line segments and specific emitters were collected for detailed biological analysis (Gilbert *et al.*, 1982). Of 14 primary bacterial genera identified, the most common were *Pseudomonas*, *Flavobacterium*, *Vibro*, *Brevibacterium*, *Micrococcus*, and *Bacillus*. The occurrence of *Bacillus* was enhanced by sand and screen filtration and markedly reduced by chemical conditioning of the water (treatments D, E, and F). No strictly anaerobic bacteria, such as *Clostridium* sp., were detected in the water treated with chlorine and acid (treatments B, D, and E). Indications were that pigmented bacteria, *F. lutescens* and *Cytophaga hutchinsonii*, caused the yellow color of the slime deposits in biologically clogged emitters, and their growth was supported by *Pseudomonas stutzeri*, a nonpigmented bacteria. Iron bacteria, *Sphaerotilus* spp., were not detected in any of the samples collected.

With the Florida waters, bacterial clogging has been associated with three types of slimes: (1) iron, (2) sulfur, and (3) nonspecific filaments and nonfilamentous slimes. Ochre (filamentous iron deposits) occurs when ferrous (Fe^{2+}) iron in the water is precipitated as ferric (Fe^{3+}) iron by the activity of certain filamentous bacteria such as *Gallionella, Leptothrix, Toxothrix, Crenothrix,* and *Spareotilus* (Ford and Tucker, 1975) plus certain nonfilamentous aerobic slimes such as *Pseudomonas* and *Aerobacter* (Ford, 1978). The primary clogging agents were the sticky bacterial slimes that adhere to the suspended solids and not necessarily the precipitated ferric iron itself. The iron complexing problem is particularly severe when the water pH is between 7.0 and 7.8. Aerobic sulfur slimes were formed by the transformation of hydrogen sulfide (H_2S) to elemental sulfur (S) by the filamentous bacteria *Thiothrix* and to a lesser extent *Beggiatoa* sp. (Ford and Tucker, 1975). *Thiothrix* requires only traces of oxygen for its development, and its optimum pH range is about 6.7–7.2. The presence in solution of hydrogen sulfide, however, can inhibit ochre and other nonsulfur slime clogging problems (Ford, 1976). Miscellaneous bacteria such as filamentous *Vitreoscilla* and nonfilamentous *Pseudomonas* and *Aerobacteria* have also been found to clog emitters by their sheer mass (Ford, 1977a).

Most irrigation water in Hawaii contains up to 50 mg/liter of fine suspended solids of which about 25% includes organic materials such as algae, diatoms, plant fibers, slimes, filaments, etc. (McElhoe and Hilton, 1974). Filtration alone will remove only a small portion of these particles, and an oxidizing agent is needed to prevent coagulation or flocculation of the suspended solids in the lateral lines and emitters.

B. Evaluation of Emitter Clogging

Presently, there is no foolproof method for determining whether a user will encounter clogging problems before installing a trickle irrigation system. Table IX summarizes a classification system formulated and modified by Bucks and Nakayama (1980), where the major physical, chemical, and biological factors for unfiltered irrigation water are rated in terms of little, some, or severe clogging hazards. A warning in terms of any classification system is that water quality will fluctuate with time and representative samples of the specific field situation are required. This classification system has also been used as an indication of the type of maintenance needed.

All trickle irrigation systems require some type of filtration to remove the bulk of the suspended materials (Pelleg *et al.*, 1974; Wilson, 1972, 1977; Bucks *et al.*, 1979). Even with the best water filtration, some small physical particles can pass through filters and act as a source for other clogging problems. For example, algae from a surface water source will be caught in the filter, but the algae filaments can pass through the filters and accumulate in lateral lines and emitters

TABLE IX

TENTATIVE CRITERIA FOR CLASSIFYING IRRIGATION WATERS USED IN TRICKLE IRRIGATION SYSTEMS[a]

Type of problem	Little	Some	Severe
Physical			
Suspended solids[b]	50	50–100	100
Chemical			
pH	7.0	7.0–8.0	8.0
Dissolved solids[b]	500	500–2000	2000
Manganese[b]	0.1	0.1–1.5	1.5
Iron[b]	0.1	0.1–1.5	1.5
Hydrogen sulfide[b]	0.5	0.5–2.0	2.0
Biological			
Bacteria populations[c]	10,000	10,000–50,000	50,000

[a] After Bucks and Nakayama (1980).

[b] Maximum measured concentration from a representative number of water samples using standard procedures for analysis (mg/liter).

[c] Maximum number of bacteria per milliliter can be obtained from portable field samplers and laboratory analysis. Bacterial populations do reflect increased algae and microbial nutrients.

where bacteria grow (Ford, 1978). Water analysis normally gives only the inorganic and not the organic suspended load.

When the irrigation water has a pH above 7.5 and a high degree of hardness (calcium plus magnesium), calcium or magnesium carbonate can precipitate in the filter, pipelines, or emitters. Adding acid to such water will lower the pH and help to reduce chemical precipitation. Lowering the pH will also improve the effectiveness of chlorination for bacterial control (White, 1972); however, maintaining the water pH continuously below 6 can cause corrosion problems. The amount of acid required to lower the pH to a specific value can be determined from a laboratory titration curve for a specific water source.

The water pH, temperature, and organic carbon sources are the most important factors that influence bacterial growth and slime development. The slime bacteria grow best at temperatures between 20 and 30°C, but they can still grow at lower temperatures although at slower rates. The pH is particularly important in connection with iron deposition problems. Waters should be checked for iron content at the field site with portable test kits. Iron concentrations may fluctuate with time such that concentrations should be checked periodically as part of a maintenance program when a potential problem exists (Ford, 1980). Portable testers are also available for field determination of bacterial populations. Higher total bacterial counts generally reflect increased algae nutrients and microbial growth (Bucks *et al.*, 1979).

C. Reclamation of Clogged Emitters

Reclamation of partially clogged emitters has been successful in several cases. Black (1976) removed slime with superchlorination of 1000 mg/liter, but extreme care was required to prevent injury to plants. Myers *et al.* (1976) cleaned emitters, having small amounts of slime and minimal reductions in discharge rates, by using 250 mg/liter of chlorine for at least 12 hr without causing injury to citrus trees. Ford (1980) used a 2% hydrochloric acid treatment for 15 min to remove ochre and slimes from operational emitters. Unfortunately, the soil pH was lowered from 6 to 5. Nakayama *et al.* (1977) reclaimed flushable emitters clogged with biological slime by treating the system for about 24 hr with 100 mg/liter of chlorine and adding sulfuric acid to lower the pH to 2. The discharge rates were increased from as low as 50% back to 90 and 95% of original design. After the reclamation treatment, a continuous 1 mg/liter chlorine at a pH of 7 helped to maintain the system operational for the three remaining years of the investigation.

D. Prevention of Emitter Clogging

Prevention continues to be the best solution to reducing or eliminating emitter clogging. Preventive maintenance includes water filtration, field inspection, pipeline flushing, and chemical water treatment. Water filtration and field inspection are absolutely essential. Flushing can help to minimize sediment buildup, and chemical water treatment can improve the long-term performance of a trickle irrigation system.

On the Colorado River water experiment (water treatments described in Table VII), the effectiveness of each water treatment in preventing emitter clogging can be seen in Table X in terms of the number of emitters that were operating at less than 50% of initial design at the end of the experiment. Screen filtration alone was found to be inadequate, as the overall emitter failure rate was 68%. In contrast, a combination of sand and screen filtration (treatment C) reduced the incidence of clogging to 23%, and chemically conditioning the water (treatments D, E, and F) reduced it even further. Emitter performance was best with continuous acid (treatment F), under which only an 8% clogging rate was observed. These results indicated that the most efficient treatment for prevention of emitter clogging was continuous acidification for pH control to prevent carbonate formation and precipitation coupled with sand and screen filtration. A combination of continuous or intermittent chlorine and acid treatments was nearly as effective as the acid alone in reducing emitter clogging. Figure 28 presents an example of a chemically clogged emitter where acid was effectively used to remove the precipitated calcium carbonate.

TABLE X

EFFECT OF WATER TREATMENT ON CLOGGING USING COLORADO RIVER WATER[a]

Water treatment[b]	Number of emitters clogged[c]	Percent clogged
A	205	68
B	205	68
C	68	23
D	53	18
E	41	14
F	25	8
	Total: 597	33

[a] After Gilbert *et al.* (1981). Pooled observations from 300 emitters for each of eight different designs with a total of 1800 emitters.

[b] See Tables V and VII for descriptions of treatments.

[c] Clogged and partially clogged emitters had discharge rates of less than 50% of initial design.

A complete discussion of the individual performances of the eight individual designs (Table VII) can be found in a paper by Gilbert *et al.* (1981). Generally, emitters with moving parts were more susceptible to malfunctioning and clogging. Emitters with flexible membranes either failed after a few months of use with chlorination or showed serious deterioration after four years. Five of the eight emitter systems required sand and screen (200 mesh) filtration plus chemically conditioned water to keep suspended materials from physically clogging the emitters. In this experiment, all pipelines were thoroughly flushed at least every six months and the systems were operated only minimally during periods of canal cleaning and peak suspended loads (Nakayama *et al.*, 1978).

Ford (1977b) indicated that an iron content as low as 0.1 mg/liter can contribute to iron deposition, and chlorination successfully controlled ochre when iron was less than 3.5 mg/liter and the pH was below 6.5. Sodium hypochlorite was injected near the bottom of the well, and the precipitated iron was collected in a sand filter. Bucks and Nakayama (1980) suggested for long-term operation that water with high levels of iron, manganese, or hydrogen sulfide along with bacteria (Table IX) may not be suitable for trickle irrigation.

1. *Water Filtration*

In the process of selecting the type, size, and capacity of the filter, the initial water quality (Wilson, 1972, 1977) and emitter design (Solomon, 1977) must be considered. Recommendations by emitter manufacturers on the degree of filtration required should be followed; however, where no recommendations are available, final filtration size should be one-tenth the diameter of the emitter's smallest opening. Bucks and Nakayama (1980) suggested that when physical

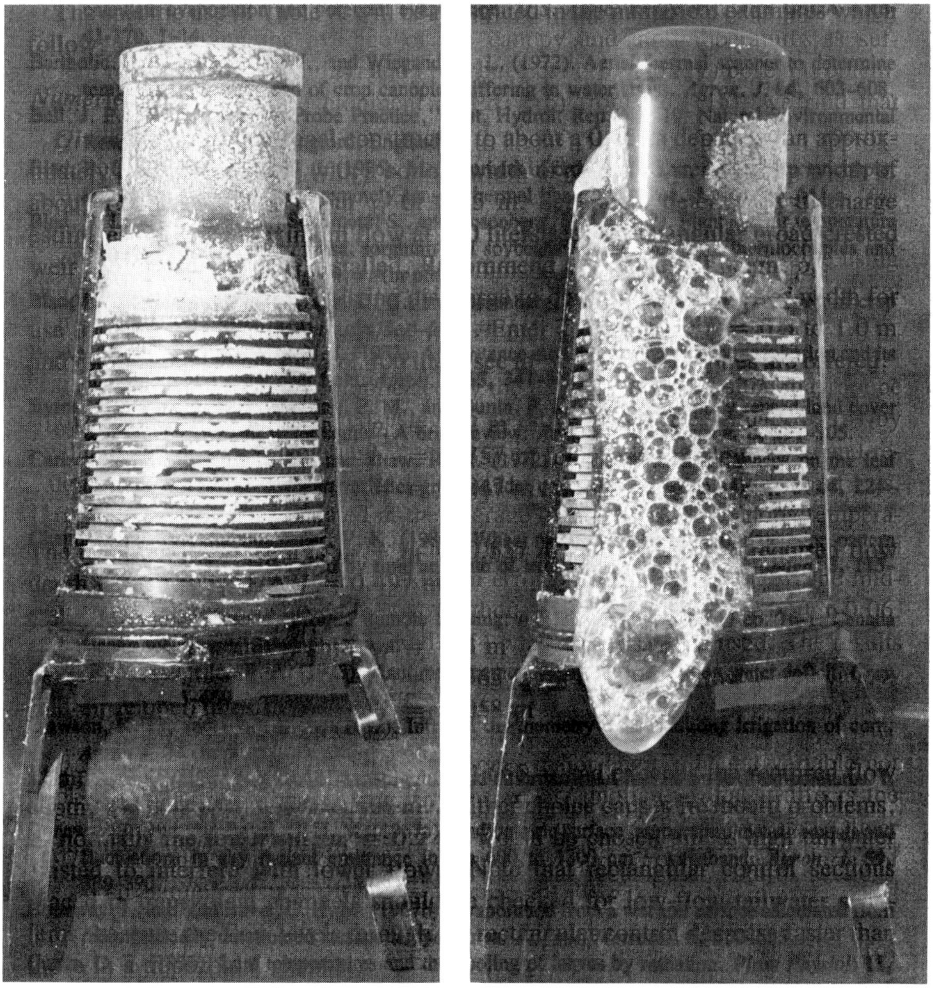

FIG. 28. (a) The precipitation of calcium carbonate from the irrigation water has chemically clogged this trickle emitter (cover removed). (b) A small amount of acid has removed the encrustation from the emitter.

factors become severe (Table IV), two or more types of filters in series may be needed. As a general rule, filtration units should be designed with at least 20% extra capacity. Pump sizes should also be increased accordingly to provide some reserve operating pressure and capacity for backwashing of filters and flushing of trickle lines (Bucks *et al.*, 1979).

Screen filters, made of stainless steel, plastic, or synthetic cloth and enclosed

FIG. 29. Screen filters must be thoroughly cleaned to maintain proper performance from a trickle irrigation system.

in a special housing, are the simplest. Aquatic algae in the water tend to cause screen blockage and can reduce the filtering capacity. Most manufacturers recommend 100 or 200 mesh (150 or 75 μm) screen filter with a few recommending a coarser screen of 30 mesh (600 μm). Screen filters, as well as other filtering systems, must be routinely cleaned and inspected to insure satisfactory operation of any trickle system (Fig. 29).

Media filters consist of fine gravel and sand of selected sizes placed in a pressurized tank. Since media filters are not easily plugged by algae, they can remove relatively large amounts of suspended solids before backwashing is needed. However, they can provide conditions favorable for increased bacterial growth. Media filters that are presently used will retain particle sizes in the range of about 25–100 μm. Media filters can be followed by a secondary filter or a rinse-away valve to prevent possible contaminants from going beyond the sand filter during the backwashing process.

Sand separators, hydrocyclones, or centrifugal filters remove suspended particles that have a specific gravity greater than water and that are larger than 75 μm, but these filters are ineffective in removing most organic solids. A sand separator can effectively remove a large amount of sand particles and can be used efficiently as a pretreatment before another type of filter.

Settling basins, ponds, or reservoirs can remove large volumes of sand and silt; but unless they have a protective covering, the water is subject to windblown contamination and algae growth that must be controlled. Thus, these structures are normally used only for pretreatment of surface water sources.

2. *Field Inspection*

Systematic inspection of a trickle system is required to spot malfunctioning emitters, pipeline leaks, and accessory equipment failures. Good maintenance practice requires that filters be cleaned, either automatically or manually, and inspected at least weekly. Operation of chemical injectors, time clocks, pressure regulators, water meters, and the main pump must also be checked and repairs made, as necessary. Field checks for malfunctioning emitters and pipeline leaks should then be carried out on a regular basis.

Water meters, whose primary use is to aid in the scheduling of irrigations, can be used to indicate inadequate pressure regulation, pipeline leaks, or clogged emitters. By dividing the accumulated flow by the operating time, operators can determine a system flow rate. When this rate changes more than 10%, the system should be thoroughly inspected (Bucks *et al.*, 1979).

3. *Pipeline Flushing*

Flushing valves should always be provided at the ends of mains and submains, as well as provisions made for flushing lateral lines. Laterals can be flushed either automatically or manually. Automatic flushing can be useful where irrigation water is extremely high in silt, clay or biological residues (McElhoe and Hilton, 1974; Shearer, 1977). Bucks *et al.* (1979) recommended that trickle lines be flushed at least once every six months for tree crops and at the beginning, middle, and end of each season for row crops. In addition, operators will need to flush the lines more frequently to remove initial sediment and plastic particles when the system is first installed. Shearer (1977) indicated that a minimum flow velocity of 0.3 m/sec is essential for flushing lateral lines.

4. *Chemical Water Treatment*

Sulfuric and hydrochloric acid are commonly used acids to reduce chemical precipitation. Phosphoric acid can also be used as a water treatment and fertilizer source (Rolston *et al.*, 1979). Chlorination is the primary means for controlling microbial activity; sodium hypochlorite (a liquid), calcium hypochlorite (a solid or powder), and chlorine gas are the basic sources. Since both sodium and calcium hypochlorite have an alkaline reaction with water, an acid may need to be added if the water is already alkaline (high pH water). Chlorine gas, on the other hand, produces an acid reaction with water so that acid additions may not be necessary. When chlorination is used, test kits designed to measure free residual chlorine, which is the excess of active chlorine over the amount required to kill bacteria, should be used (Ford, 1975). Ford (1977b) indicated that 20 mg/liter of free residual chlorine applied three times per week for three weeks to potted citrus seedings caused some drooping of leaves and stopped root growth. Plant damage would not be expected from applications of lower concentrations of chlorine in the field.

Some of the alternative chemicals to control bacteria and algae are acrolein, copper salts, hydrogen peroxide, iodine, and quaternary ammonium salts. Acrolein requires special handling and does not destroy certain iron complexes in the laboratory (Ford, 1977b). Copper salts have been widely used to control algae in settling basins, ponds, or reservoirs (Bucks *et al.*, 1979). Hydrogen peroxide can be a good bactericide for iron problems, but it is generally poor for sulfur and other slimes that do not contain iron (Fridovich, 1972). Iodine is an excellent bactericide, although it complexes with iron and is toxic to plants (Ford, 1978). Quaternary ammonium salts also kill bacteria and even snails, but they are expensive (Bucks *et al.*, 1979).

E. Injection of Water Treatment Chemicals

Chemicals for water treatments (acids, algicides, bactericides) can be safely injected through trickle irrigation systems. The most common method for injection of a liquid or water-soluble-chemical water treatment is by means of a positive-pressure metering pump. Ford (1979) has also used the vacuum-type injector for sodium hypochlorite. Final chemical concentrations for all water treatments are generally low, between 0.5 and 10 mg/liter (Bucks and Nakayama, 1980). Concentrations should be determined routinely after the solution has passed the primary filter and before it enters the main line. Occasionally, it should be checked at the end of the last lateral line to ensure that the treatment has been disseminated throughout the trickle irrigation system.

Acids and chlorine compounds should be stored separately, preferably in plastic or fiberglass storage tanks. Acid can react with hypochlorite to produce chlorine gas and heat. Hypochlorite can react with emulsifiers, fertilizers, herbicides, and insecticides, and decrease their effectiveness. Bulk chemicals, and diluted solutions as well, should be stored in a secure place. In preparing dilutions, concentrated stock of acid or other chemicals should be added to the water and not vice versa. A ready water source should be provided near the chemical tank and injector for washing off any chemicals that may contact the skin. Protective goggles, face shields, and clothing should also be worn when making the chemical dilutions. State and local regulations and codes must be followed in regards to type of backflow-prevention device used to ensure against contamination of an irrigation well or potable water supply. Where gas chlorinators are used safety devices should be installed and routinely inspected to prevent the buildup of excessive pressure in the chlorine supply bottles (Bucks and Nakayama, 1980).

IX. Advances in Systems Management

Once a trickle irrigation system is correctly designed and maintained, the real key to its success is proper management. Management practices can include the application of water amendments, scheduling of irrigations, salinity management, water placement, and adaptation of cultural practices.

A. Apllication of Fertilizer and Other Water Amendments

Fertilizers and other water amendments (herbicides, insecticides, fungicides, etc.) applied through a trickle irrigation system can improve efficiency, save labor, and increase flexibility in scheduling of applications to fit crop needs (Rolston *et al.*, 1979). However, all chemicals must meet the following criteria (Bucks and Nakayama, 1980): they must (1) avoid corrosion or clogging of any component of the system; (2) be safe for field use; (3) not decrease crop yield; (4) be soluble or emulsifiable in water; and (5) not react adversely with salts or other chemicals in the irrigation water. The total amount of a water amendment applied through a trickle system will depend for the most part on the crop requirements, growing period, soil type, time of year, and other factors; however, the final chemical concentration present in the trickle irrigation water, normally ranging from 4 to 100 mg/liter, should not be allowed to become excessive.

Nitrogen, the plant nutrient most commonly deficient for crop production, is often applied in trickle irrigation systems. Nitrate nitrogen moves readily in the soil with the irrigation water and can be applied separately or in mixtures with such compounds as ammonium sulfate, urea, and ammonium nitrate. Calcium nitrate can also be used when bicarbonates are low. Anhydrous ammonia, aqua ammonia, and ammonium phosphate in most instances cause clogging problems. Nitrogen source selection should be based on its possible reactions with the irrigation water and the soil. In general, the first four nitrogen compounds mentioned will cause few, if any, precipitation problems in trickle irrigation systems.

Nitrate will move to the periphery of the wetted soil volume with repeated trickle irrigations (Goldberg *et al.*, 1971). Phene and Beale (1976) found that nitrate concentrations remained higher in the root zone with frequent trickle irrigation of sweet corn than with the flood and sprinkler irrigation. Rolston and Broadbent (1977) found that little denitrification occurred in a clay loam soil if the soil water tension was higher (i.e., a drier soil) than 10 cbar. Phene *et al.* (1979) have shown that high-frequency nitrogen application on a shallow sand loam soil with trickle irrigation improved the efficiency of nitrogen use by potatoes more than double that of conventional (broadcasted plus banded) fertilization methods with trickle irrigation (Fig. 30). Miller *et al.* (1976) indicated that nitrogen is used more efficiently when applied through trickle irrigation than when banded and furrow irrigated or banded and trickle irrigated on tomatoes.

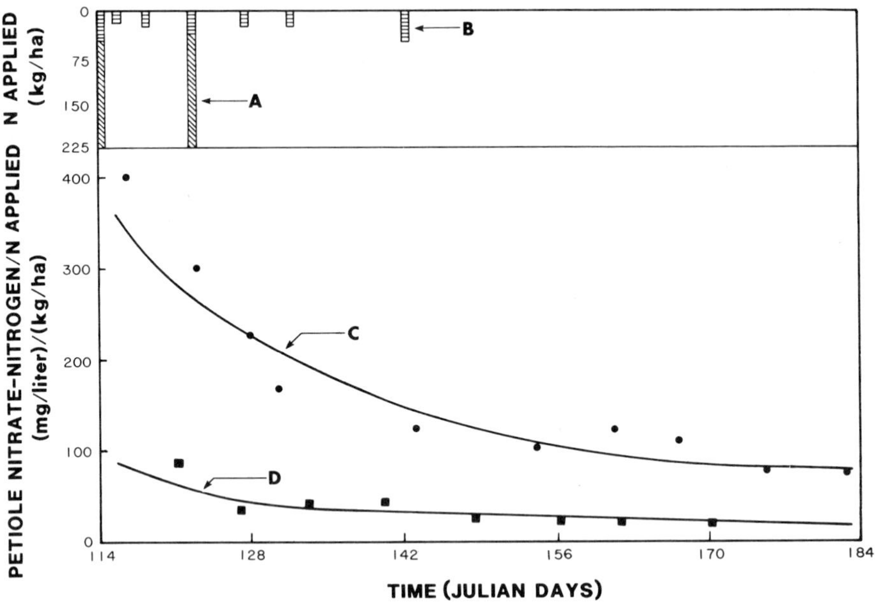

FIG. 30. Petiole NO_3-N content of potato leaf per unit of N applied for high-frequency and conventional N fertilization using trickle irrigation (after Phene *et al.*, 1979): (A) conventional N fertilization; (B) high-frequency N fertilization; (C) trickle high-frequency N fertilization; (D) trickle conventional N fertilization.

Phosphorous has not generally been recommended for application in trickle irrigation systems because of its tendency to cause clogging and its limited movement in the soil. If irrigation water is high in calcium and magnesium, precipitates of insoluble calcium and magnesium phosphates may result from the application of inorganic phosphate. Rauschkolb *et al.* (1976) applied phosphoric acid along with short pulses of sulfuric acid to keep the water pH low in a trickle system without precipitation or clogging problems. They aslo applied glycerophosphate, an organic phosphate, and orthophosphate, an inorganic phosphate, through the system. Organic phosphates will not precipitate unless (1) the compound is being hydrolyzed to the inorganic phosphate in the water or (2) the water pH is high. The glycerophosphate moved further than the orthophosphate from the trickle irrigation source in the clay loam soil, but not to an extent to significantly affect plant uptake of phosphate.

O'Neill *et al.* (1979) found that phosphorous was delivered to greater soil volume when applied as orthophosphoric acid through a trickle system than treble superphosphate applied as a soil amendment beneath each emitter. The orthophosphoric acid lowered the pH of the irrigation water enough to minimize clogging problems from phosphate precipitation over 3- and 24-day irrigation periods.

Potassium seems to cause few, if any, clogging problems in trickle irrigation systems. Common potassium sources, such as potassium sulfate, potassium chloride, and potassium nitrate, are readily soluble in water. These fertilizers move freely into the soil and some of the potassium ions are exchanged on the clay complex and are not readily leached away. However, Uriu *et al.* (1977) did demonstrate that there was some movement after the potassium ions became concentrated in the soil near the emitter.

Micronutrients such as iron, zinc, copper and manganese can be applied as chelates or sulfate salts in trickle irrigation water. Normal plant requirements for these nutrients are low, and their application through trickle systems requires careful and precise metering. Lindsey and New (1974) found that zinc EDTA applied through trickle irrigation systems for pecan trees costs less than foliar application methods, but leaf concentrations of zinc were generally lower with trickle than with foliar applications.

Generally, crop response to fertilizer applications by the trickle method has been excellent, and frequent nutrient applications have improved the fertilizer application efficiency. Reductions of 25–50% in total fertilizer applications using trickle irrigation as compared with surface broadcasting have been reported with no yield reductions (Coston *et al.*, 1978; Goode *et al.*, 1978; Miller *et al.*, 1976; Phene and Beale, 1976; Phene and Sanders, 1976; Phene *et al.*, 1979).

Other water amendments, such as fumigants or gases, can be applied through a trickle irrigation system, but success has been limited. Chesness *et al.* (1976) and Overman (1974, 1976) described several nematocides that can be effectively applied through line-source trickle irrigation systems. Zentmeyer *et al.* (1974) applied ethazol (5-ethoxy-3-trichloromethyl-1,2,4-thiadiazole) through a trickle irrigation system in an attempt to control *Phytophthora* root rot on avocado trees. Lange *et al.* (1974) found that herbicides tended to break down quickly where the soil surface remained moist so that the applications through trickle systems were essentially ineffective. Phene *et al.* (1979) indicated that application of eptam (S-ethyl dipropylthiocarbamate) herbicide at one-half rate through a trickle system was as efficient as the full rate applied conventionally. Nakayama and Bucks (1981b) applied CO_2-saturated water through subsurface trickle systems to provide additional CO_2 around the plants for improving their photosynthetic efficiency. Preliminary results showed that wheat yield increased a significant 20%, and potato and cantaloupe yield also tended to improve. Subsurface CO_2 applications greatly increased CO_2 concentrations and decreased soil pH by 1.5 units in the wetted zone around the trickle tubing, an effect which could improve the root environment where alkalinity is a problem.

Fertilizer and other chemicals have been injected into trickle irrigation systems by three principal methods: (1) a pressure differential system, (2) a venturi (vacuum), and (3) a metering pump. Descriptions of these systems were given by Howell *et al.* (1980) and Bucks and Nakayama (1980). The third method, which consists of using a rotary, gear, or piston pump to inject a measured amount of a chemical solution from a supply tank into the pressure line, is the most common. The pumps are made of corrosion-resistant materials and may be either of the proportional rate type, where the injection rate changes with the water flow in the system so that a constant dilution ratio is maintained, or of the constant rate type, where the injection rate is independent of the main line flow and the dilution ratio is variable. Typical injection rates for trickle irrigation systems range from 8 to 80 liters/hr per injection point. A metering pump should be equipped with an antisiphon valve to prevent the water supply from getting into the tank and causing concentrated chemicals to spill over onto the surrounding area. If electrical pumps are used, a pressure or flow control switch should be installed in series with the automatic controller so that chemical injection stops when the flow stops or when operating pressure decreases.

B. Irrigation Scheduling and Crop Response

Irrigation scheduling involves two decisions: (1) how much to apply (quantity), and (2) how often to apply (frequency). These decisions are critical to the management of a trickle irrigation system. The primary objective of trickle irrigation is to maintain a moist portion within the plant root zone with relatively small applications of water. Therefore, the concept implies a rather high irrigation frequency, but we believe that this does not necessarily imply daily or more frequent irrigations for the entire growing season or in every cropping situation.

There are two alternative objectives for scheduling irrigations: (1) maximum yield per unit of land area, or (2) maximum yield per unit of applied water. Most managers hope to achieve the first objective; however, as irrigation costs increase and water supplies diminish, more will tend to adopt the latter objective. In order to maximize yield, plant water potential cycles should be managed within limits conducive to maximum seasonal net photosynthesis. Hsiao (1973) suggested that stresses are mild when midafternoon plant water potentials are not less than about -12 bars, moderate in the range of -12 to -16 bars, and extreme at below -16 bars for many crops. Ritchie (1974) estimated that actual net photosynthesis probably begins to fall below potential rates when root-zone extractable water is depleted by only 20–25%. On the other hand, his data (Ritchie, 1973) showed that plants can meet most evaporative damands until the extractable water is about 70–80% depleted, or until 40–50% depleted when expressed on the more traditional "available soil water" basis. Scheduling of irrigations for cases in which water is limited can be significantly different. For maximum yield per unit amount of water applied, the actual yield and seasonal evapotranspiration (ET) are typically less than for the first objective (Downey, 1972; Hanks, 1974; Stewart *et al.*, 1975).

Irrigation quantity and frequency are related in most cases; however, most of the irrigation studies that have been conducted have failed to define the effects of both of these variables at the same time. In most studies, irrigation applications were varied while frequency of trickle irrigation was fixed; in others, the frequency was varied while quantity of trickle irrigation was fixed. Still other studies compared quantity or frequency among different irrigation methods, e.g., daily trickle with variable sprinkler irrigations.

1. *Irrigation Quantity*

The irrigation water requirement is the net amount of water that can be supplied by irrigation to satisfy ET, leaching of salts, and miscellaneous water requirements that are not provided by water stored in the soil or by precipitation that enters the soil (Jensen, 1974). Some of the miscellaneous water requirements include essential watering for germination and emergence, frost protection, wind erosion protection, plant cooling, and maintenance of harvest quality. In equation form, the irrigation water requirement WR is

$$\mathrm{WR} = \mathrm{ET} + \mathrm{LR} + \mathrm{MR} - P - \Delta S \tag{32}$$

where LR is the water required for leaching, MR is miscellaneous requirements, P is effective precipitation, and ΔS is the decrease in soil water storage that can be utilized during the period or season.

Davis and Pugh (1974) observed little difference in vegetable yields or irrigation water requirements between trickle and furrow irrigation after the runoff water was subtracted from the amount applied by the furrow method. Hanson and Patterson (1974) found water requirements to be similar for sprinkler and trickle irrigation, but higher for furrow methods. Roth *et al.* (1974) noted tremendous differences in water requirements for basin and trickle methods compared with limited-coverage sprinkler and flood methods on the first two years with trees, whereas Rodney *et al.* (1977) indicated that water requirement savings decreased as the tree grew older. Generally, one would anticipate greater savings in water requirements under trickle irrigation for widely spaced crops versus closely spaced crops, for sandy soil versus heavy-textured soil, and for an inefficient existing irrigation system versus an efficient trickle system.

Wendt *et al.* (1977) reported no significant difference in seasonal ET of sweet corn under sprinkler, furrow, and subsurface irrigation. Potato yields decreased with trickle irrigation rates of less than ET for furrow irrigation, even though the seasonal irrigation water requirement was 50% less than for furrow irrigation (Singh *et al.*, 1978). Lysimeter measurement of ET on trickle-irrigated sugarcane was about 80% of the previously measured value under sprinkler; however, Ekern (1977) explained that the sprinkler method watered over the top of a leaf litter so that direct evaporation was probably greater than with trickle irrigations, which were beneath this layer. Based on a radiation and energy balance approach, Ben-Asher *et al.* (1978a) suggested that daily ET could be smaller than previously reported for sprinkler-irrigated tomato. Ben-Asher and Sammis (1978) also suggested that additional energy can be contributed from the dry to wet areas during the ET process, thereby increasing evaporation from the smaller, more frequently wetted portion of the soil surface. Although numerous investigators have suggested that direct evaporation from the soil surface should be less with trickle irrigation because the fraction of the surface area wetted is reduced, direct measurements of soil evaporation under trickle irrigation have not been made to account for soil evaporation rates on a more continuous basis from a smaller area compared with evaporation rates from an intermittently wetted larger surface area.

For experiments where both irrigation amount and frequency were varied using trickle irrigation, the quantity of water applied had a larger effect on yield than frequency in the cases of cotton (Bucks *et al.*, 1973), cabbage (Bucks *et al.*, 1974a), grapes (Bucks *et al.*, 1974b; Bucks, 1979), and cantaloupes (Bucks *et al.*, 1981a). Figures 31 and 32 show that the highest production with trickle irrigation occurred at seasonal water applications nearly equal to or slightly greater than the measured ET under furrow irrigation. Different ratios of the estimated ET rate were applied to plots that were trickle irrigated on prior long-term measurements for surface irrigation methods. ET was simultaneously measured on small, level-basin irrigation plots so that weekly adjustments could be made for the trickle-irrigated plots. At the end of the season, an adjusted ET ratio was calculated using a water balance approach and neglecting deep percolation losses. Since water could be accurately measured and uniformly applied on the small trickle plots, it was assumed that yield would be maximum where the ET quantity was applied with precision by the trickle method. Direct measurement of ET under trickle irrigation is difficult; however, most studies to date confirm that the irrigation method does not significantly change the seasonal ET of the crop where the objective is maximizing yield per unit land area.

We believe, therefore, that any reduction in water requirement for trickle irrigation over another method should come primarily from an improvement in on-farm irrigation efficiency. On-farm irrigation efficiency refers to the ratio (expressed as a percentage) of the irrigation water requirement to the total quantity of water delivered or pumped to the field, On-farm efficiency for trickle irrigation can theoretically approach 90–95%.

In terms of limited or deficit trickle irrigation practices, Hiler and Howell (1973) reported higher water-use efficiencies for trickle and mist irrigation than for other irrigation methods, and later (Howell and Hiler, 1975) concluded that the timing of the water deficit was more critical with regard to yield effects than was the magnitude of the deficit. Sammis (1980) found that higher water-use efficiencies (yields per unit amount of water applied) were obtained for potatoes with surface and subsurface trickle than with sprinkler and furrow irrigation. Comparable water-use efficiencies were obtained with surface trickle, subsurface

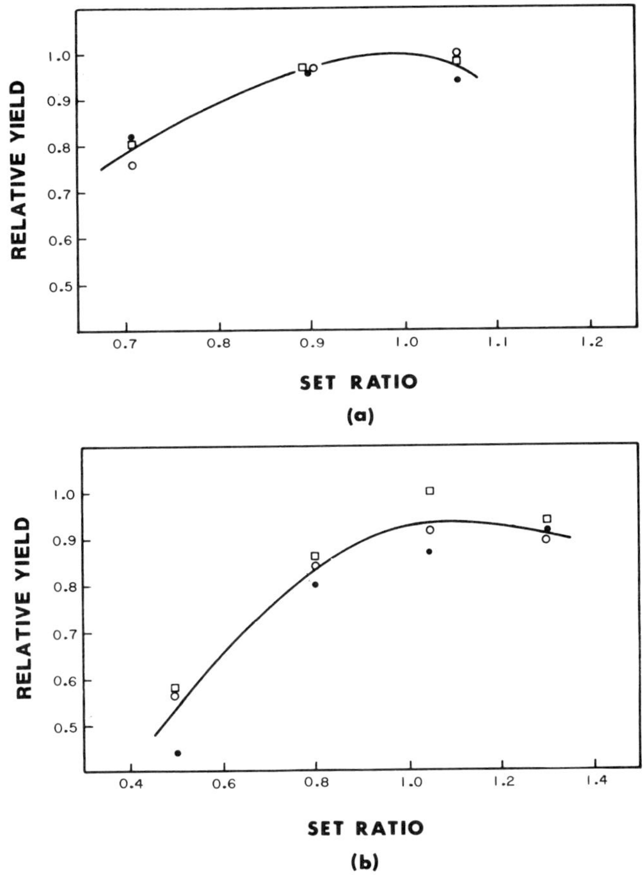

FIG. 31. Yield relationships for different trickle irrigation quantities on (a) cotton (1971) and (b) cabbage (1972) grown in central Arizona, disregarding irrigation frequency [(●) 3 days; (○) 6 days; (□) 12 days; after Bucks *et al.*, 1973, 1974].

Fig. part	Relative yield ($Y = Y_{act}/Y_{max}$)	R (means only)	Mean 1.0 SET[a] (cm)
a	$-1.56 + 5.15(ET) - 2.59(ET)^2$	0.96	102
b	$-0.38 + 2.32(ET) - 1.02(ET)^2$	0.96	38

[a] SET, seasonal evapotranspiration.

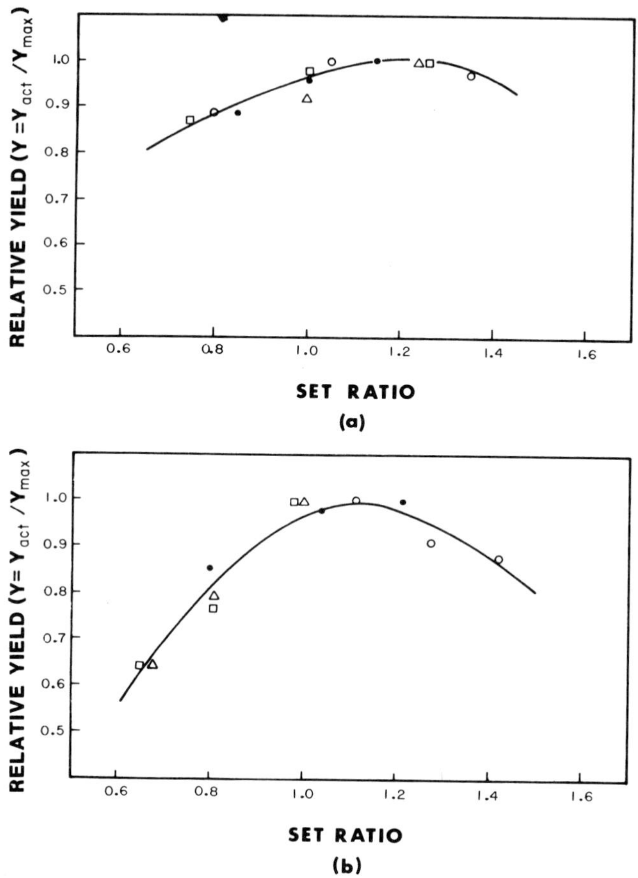

FIG. 32. Yield relationship for different trickle irrigation quantities on (a) grapes (two emitters per vine) and (b) cantaloupes grown in central Arizona, disregarding irrigation frequency (after Bucks, 1979 and Bucks *et al.*, 1981): (a) (●) 1973; (○) 1974; (□) 1975; (△) 1977; (b) (●) spring 1977; (○) fall 1977; (□) spring 1978; (△) fall 1978.

Fig. part	Relative yield ($Y = Y_{act}/Y_{max}$)	R (means only)	Mean 1.0 SET[a] (cm)
a	0.19 + 1.28(ET) − 0.51(ET)^2	0.93	38
b	−1.12 + 3.75(ET) −1.66(ET)^2	0.95	50 (spring) 40 (fall)

[a] SET, seasonal evapotranspiration.

TABLE XI

AVERAGE YIELD COMPARISON FOR THE TRICKLE IRRIGATION FREQUENCY FOR COTTON, CABBAGE, GRAPE, CANTALOUPE, CARROT, AND ONION[a]

Crop	Yield units (metric tons/ha)	Year	Irrigation frequency					Discussion of results
			Daily	3-Day	6-Day	7-Day	12-Day	
Cotton	Lint	1971	—	1.57	1.58	—	1.61	NSG[b]
Cabbage	Marketable	1972	—	57.4	61.0	—	64.0	Trend for 12-day over 6-day over 3-day
Perlette grape	Marketable	1973	14.6	15.0	14.8	—	—	NSG
		1974	9.8	10.2	10.4	—	—	Trend for 3 and 6-day over daily
		1975	11.8	12.4	11.5	—	—	NSG
		1977	9.7	—	10.9	—	—	13% incr. for 6-day over daily
Cantaloupe	Marketable (crates/ha)	Spring 1977	1264	—	—	1348	—	7% incr. for 7-day over daily
		Fall 1977	481	—	—	596	—	24% incr. for 7-day over daily
		Spring 1978	798	—	—	781	—	NSG
		Fall 1978	333	—	—	476	—	42% incr. for 7-day over daily
Carrot	Marketable	Spring 1979	61	—	—	61	—	NSG
Onion	Marketable	Spring 1979	51	—	—	44	—	15% incr. for daily over 7-day

[a] Average yields were based on three irrigation quantities and four replications on cotton, grape, and carrot; four irrigation quantities and four replications on cabbage; three irrigation quantities and four replications on cantaloupe; and three irrigation quantities and six replications on onion.

[b] NSG, no significant difference.

trickle, frequent sprinkler, and properly managed furrow irrigation methods with lettuce, but the efficiency was lower with infrequent sprinkler irrigation. For the previously mentioned cotton and cabbage studies (Bucks *et al.*, 1973, 1974a), the highest water-use efficiency was achieved when the irrigation totaled 0.9 and 0.8 times the estimated ET rates for the two crops, respectively.

2. *Irrigation Frequency*

Considerable research has been conducted on crop response to irrigation frequency. For irrigation applications considered adequate for optimum plant growth, several studies reported increased yields as frequency increased (Goldberg and Shmueli, 1970; Hiler and Howell, 1973; Freeman *et al.*, 1976), whereas other studies suggested no increase or even a decrease in yield as frequency increased (Earl and Jury, 1977; Levin *et al.*, 1974; Yagev and Choresh, 1974; Ravelo *et al.*, 1977). Table XI presents yields for different irrigation frequencies as averages of three or four irrigation quantities for six crops produced in Arizona. No significant differences in lint cotton and carrot yields resulted from irrigation, whereas less frequent irrigation tended to increase yields on cabbage, cantaloupe, and grapes. Shallow-rooted dry onions had higher yields with daily than with weekly trickle irrigations. All the crops were grown on a clay loam soil (except for the grapes, which were grown on a sandy loam soil) using a moderately saline water.

The optimum irrigation frequency will in general depend on the type of crop, soil, and water quality. Possible reasons for applying trickle irrigations on other than a daily schedule include the need for deeper watering on a deep-rooted crop, improving aeration in the effective crop root zone on fine-textured soils, and moving salts further away from the plant. On the other hand, daily or more frequent applications can provide for shallow-rooted crops, reduce deep percolation losses on coarse-textured soils, and improve the usage of saline irrigation water. The most important point is that trickle irrigation allows precise scheduling of irrigations. Although irrigation must be frequent, and the system capacity will normally require irrigating at least every 10 or 12 days for most crops, a manager has the flexibility of being able to apply daily irrigations during the hotter months and to decrease to weekly irrigations when the ET requirements of the plant are low during the cooler months.

C. Salinity Management

The primary reason given for the possible improvement in salinity management with trickle irrigation has been that frequent trickle irrigations can help maintain a high total soil water potential (high soil matrix and osmotic potential) in the effective crop root zone. Rhoades *et al.* (1974) stated that this type of controlled salinity management can (1) maximize the precipitation of carbonate

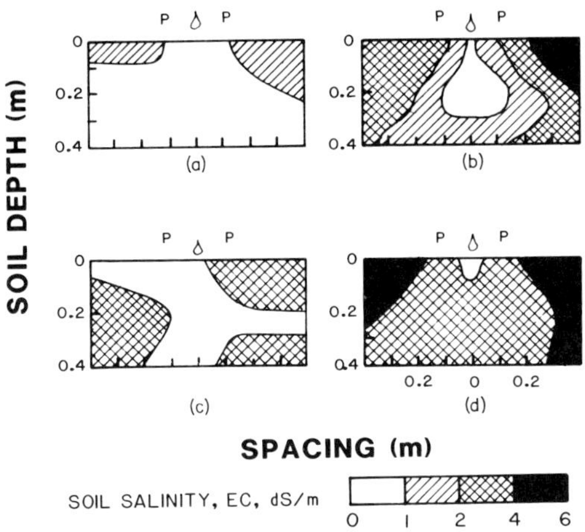

FIG. 33. Soil salinity distribution at the end of a grain sorghum experiment averaged for 2 years (after Seifert *et al.*, 1975): (◊) emitter; (P) plant.

Fig. part	Water quality (dS/m)	Water use (mm)[a]	Yield (metric tons/ha)
a	0.7	467	6.1
b	1.2	426	6.9
c	2.4	412	6.9
d	3.6	439	6

[a] Includes water applied plus use from soil moisture storage.

minerals and gypsum in the subsoil; (2) minimize soil mineral weathering and dissolution; and (3) maximize the amount of soluble salts retained in subsoil storage and not transmitted to the drainage water.

Figure 33 illustrates how increased salinity in the irrigation water can affect salt distribution with adequate water in a lysimeter study (Seifert *et al.*, 1975). Although salt concentrations increased significantly throughout the soil profile as the irrigation water salinity increased, yields did not decline until the water quality decreased to 3.6 dS/m (an electrical conductivity of 1 decisiemens (dS)/m equals 1 mmho/cm). Bernstein and François (1973) found with trickle irrigation that brackish water (having an EC of 3.8 dS/m) reduced yields by only 14% compared to low-salinity water (0.7 dS/m); but with furrow and sprinkler irrigation methods, the corresponding yield reductions were 54 and 94%, respectively. When irrigation frequency was increased for furrow and sprinkler irrigation, the brackish water reduced yields by 18 and 59%, respectively. Similarly, Goldberg and Shumeli (1970) reported small reductions in tomato yields between waters

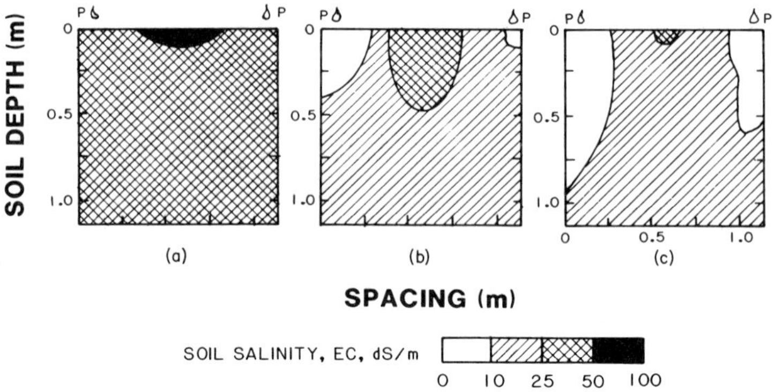

FIG. 34. Soil salinity distribution at the end of a tomato experiment applying different amounts of water (after Tscheschke *et al.*, 1974): (◊) emitter; (P) plant.

Fig. part	Water quality (dS/m)	Water applied (mm)	Yield (kg)
a	5.5	95 < ET	4.6
b	5.5	121 ≈ ET	4.8
c	5.5	145 > ET	5.1

with an EC of 0.4 and 3 dS/m with daily trickle irrigations. Singh *et al.* (1978) showed that saline water of 3 dS/m did not limit yields, whereas yields did decline by 91% for potato and 35% for tomato crops with 10 dS/m.

Another reason given for the fact that more saline water can apparently be used with trickle than with sprinkler irrigation is that the former does not cause foliar damage. When tomatoes and cucumbers growing on a sandy soil were trickle- and sprinkler-irrigated with water having an EC of 3.6 dS/m, Gornat *et al.* (1973) found that tomato yields under trickle irrigation were more than double those obtained for sprinkler irrigation. The sprinkler-irrigated cucumbers gave no yield at all. Bernstein and François (1975) also attributed yield reductions with sprinkler irrigation to salt accumulation on the bell pepper leaves rather than to osmotic shock caused by the flushing of salts that accumulated at the soil surface between irrigations into the root zone.

A third reason given for the potential improvement in salinity management with trickle irrigation is that leaching can be closely controlled. Tscheschke *et al.* (1974) showed the importance of adequate leaching (Fig. 34) with irrigations every other day on tomatoes grown in lysimeters using water having an EC of 5.5 dS/m. Salt concentrations were the highest with irrigation amounts below the crop ET, whereas irrigation amounts above ET increased soil matric potentials and yields. Patterson and Wierenga (1974) gave an example of how a preirrigation using a line-source trickle system can be used for adequate leaching before planting. With closely controlled leaching, Hoffman *et al.* (1978) indicated that citrus yields were not reduced with continuous 5, 10, or 15% leaching fractions on a sandy soil after two years. Obbink and Alexander (1977) observed that additional emitters per apple tree could improve salt leaching and root growth.

D. Water Placement and Plant Rooting

What are the optimal wetted soil volume, plant rooting volume, and number of emitters per plant for a particular soil, crop, or water quality? This is a difficult question to answer, considering the dynamics of soil water movement, plant water uptake, nutrient availability, soil aeration, etc. However, there would appear to be a minimum wetted soil volume and root distribution for optimum plant growth. Research has provided some insight into the expected wetting and plant rooting patterns under trickle irrigation, but the question previously asked has not yet been answered in its entirety.

Earl and Jury (1977) and Jury and Earl (1977) found greater movement of water and roots, both laterally and vertically, with weekly trickle than with daily trickle irrigations. These results were attributed to appreciable surface ponding which occurred with weekly irrigation amounts that totaled those of the daily applications. Levin *et al.* (1979b) found good agreement in the field with a computer simulation model which indicated that 1 liter/hr emitter discharge rate and a 2 liter/hr pulsed rate retained more water in a 0–60-cm soil depth 12 hr after the start of an irrigation than did discharge rates of 2, 4, and 8 liters/hr applied continuously on a sandy soil. Roth (1974) observed that approximately the same volume of soil was wetted in three bare, sandy soil sites in Arizona using three emitter discharge rates and time periods.

Levin *et al.* (1979a) showed that soil water and root system distribution in an apple orchard extended over a larger volume when trickle irrigations were applied twice weekly with 8 liters/hr emitter discharge rate rather than every day or once a week with 4 liters/hr rate. The daily—compared with weekly—irrigations caused an appreciably narrower soil water distribution, but a similar root distribution pattern. Earlier, Black and West (1974) suggested that wetting substantially less than the entire root system on young trees with daily trickle irrigations could produce just as good a regime for plant growth as with a biweekly irrigation interval. Greenhouse experiments (Frith and Nichols, 1974) indicated that water applications to less than the total soil volume did not reduce the ability of the trees to take up water and nutrients. Citrus roots in Arizona were found to adapt and proliferate rapidly where the water was applied (Rodney *et al.*, 1977). Higher concentrations of roots under a smaller total area were found for trickle and basin irrigation compared with lower concentrations under a larger area for sprinkler and flood irrigation methods. Bielorai (1977), working in the Western Negev of Israel, obtained somewhat higher grapefruit yields with sprinkler irrigation wetting 70% of the soil root volume and with trickle irrigation every two or three days wetting 30–40% of the root zone than with weekly trickle irrigations that moistened 30–40% of the root zone.

Silberbush *et al.* (1979), using a two-dimensional model of oxygen diffusion rates and root development, indicated that roots tend to concentrate at the periphery of the wetted soil volume with fewer roots growing near the trickle emitter. They believed this to be the general behavior in the field as well. Bernstein and François (1973) found that most of the vegetable roots occur near the soil surface with trickle systems, except when salt accumulation inhibits root development. With deeper rooted row crops, Bucks *et al.* (1981a) observed more extensive rooting and possibly improved soil aeration in a fine-textured soil with weekly compared to daily trickle irrigations. On a coarse-textured soil, yields increased in all three years with two over one emitter per grapevine (Bucks, 1979), a trend which was attributed to the greater percentage wetting of the soil root volume as well as to the better uniformity of water application.

Unanimous agreement on the optimal percentage of soil wetting is not possible for all field situations. Nevertheless, we can define a soil wetting percentage term as the ratio of the soil volume which is at our near field capacity after a trickle irrigation to the total potential root zone of the crop, expressed as a percentage. We would suggest from field experience in arid regions, where crop production depends principally on irrigation water, that a minimum of 33% is needed, and up to 66% may be preferred for better margins of safety and plant nutrient availability. In fact, many trickle managers in our opinion have not maintained adequate soil water storage in the root zone prior to the crop's peak ET rate. In humid areas, where considerable rainfall occurs and irrigation supplies only a portion of the crops' peak water needs, a lower soil wetting percentage of between 25 and 50% may be adequate for most crops.

E. Adaptation of Cultural Practices

Methods for improving cultural practices to match trickle irrigation concepts have not been fully explored. More hedge plantings for tree crops, higher plant populations for row crops, more use of herbicides through trickle systems instead of conventional cultivation practices, and increased use of plastic, natural, or synthetic mulches over the trickle emitters are examples of possible practices to be tested. Kays *et al.* (1976) found that pepper and squash could be replanted through the original holes in the plastic mulch while the surface trickle irrigation was undisturbed. Strawberries are produced in southern California on steep slopes using line-source trickle irrigation systems, where cultural practices are being

FIG. 35. Strawberries are being produced on steep terrain in southern California, where the trickle irrigation systems have improved cultural practices.

modified to include greater bed heights and plastic mulches for labor conservation and weed control (Fig. 35).

Minimum tillage practices can be readily applied under trickle irrigation, since dry areas are available for controlled traffic. Bucks *et al.* (1981a) found that minimum tillage coupled with subsurface trickle systems offered an excellent potential for consecutive or multiple cropping of small fruit and vegetable crops. Water penetration problems caused by soil compaction may not be as great a concern with trickle as with other irrigation methods because water can be applied slowly and frequently. Deep plowing and chiseling in hardpan soils may be required only occasionally to maximize crop production.

X. Future Developments and Research in Trickle Irrigation

Significant advancements in trickle irrigation principles have been made within the past decade. Many of the earlier emitter manufacturing problems have been corrected; and many of the design, maintenance, and management practices have been improved so that trickle irrigation systems continue to offer unique potentialities for irrigated agriculture in the future. Howell *et al.* (1980) and Bucks *et al.* (1981b) made several predictions for the next decade, which are summarized as follows:

(1) Surface trickle, bubbler, and spray irrigation systems will continue to be widely used on tree crops, whereas subsurface trickle systems will be used on vegetable crops. Mechanical-move, subsurface trickle with multiple cropping, and conventional trickle systems will also be tried on commercial row crops.

(2) Marginal soils, which include hardpans, sands, and steep terrains, will continue to provide new opportunities for applying trickle irrigation techniques.

(3) Future emitter designs will stress pressure-compensating, self-flushing, and other reduced-clogging designs with minimal moving parts made of durable materials.

(4) Additional automatic controllers for trickle irrigation systems will be developed to enhance both water and salinity management and to monitor maintenance requirements, using microprocessor feedback systems.

(5) Application of fertilizers and other chemicals, such as bactericides, systemic pesticides, nematocides, and herbicides, will probably expand. In some cases, chemically treated water may be delivered by water districts or farmer groups to trickle-irrigated farms.

The intensive research effort in trickle irrigation will possibly diminish in the next few years, since the need to keep pace and provide solutions to field problems and developments has slowed. Future trickle irrigation will probably be directed along the following lines:

(1) Design research will be concerned with optional lateral line lengths, network designs, and more integrated design procedures that include soil properties and all factors affecting emission uniformity as well as hydraulic characteristics.

(2) Soil water modeling research will continue to improve models with an emphasis on three dimensional water and salt flow and hydraulics of trickle irrigation systems, and to verify models with data obtained under actual field conditions (including water extraction by plants) with emphasis on improved understanding of water, plant root, and oxygen distribution under trickle irrigation.

(3) Maintenance research should be directed towards long-term solutions to iron, manganese, and other bacterial or slime problems.

(4) Management research should become less concerned with comparing trickle to other irrigation methods and more concerned with optimizing irrigation scheduling, cultural practices, and crop production methods under trickle irrigation. Additional chemicals and other water amendments are to be screened for

taking full advantage of the unique aspects of trickle irrigation for applying agricultural chemicals. Long-term studies on the merits of using high-salinity water and on specific-ion toxicity to plants are needed. The soil water evaporation process under trickle irrigation needs to be studied both qualitatively and quantitatively, and this could lead to methods for evaporation reduction by chemical or other treatments.

List of Symbols

c	Emitter discharge coefficient (dimensionless)
C	Hazen–Williams roughness factor (dimensionless)
$C(h)$, $C(\theta)$	Specific water capacity (L^{-1})
C_u	Coefficient of uniformity for sprinkler and trickle systems (%)
C_{ud}	Coefficient of design uniformity for trickle systems (%)
C_v	Coefficient of variation (dimensionless)
C_{vf}	Final coefficient of variation for emitters (dimensionless)
C_{vh}	Hydraulic coefficient of variation for emitters (dimensionless)
C_{vi}	Initial coefficient of variation for emitters (dimensionless)
C_{vm}	Manufacturer's coefficient of variation for emitters (dimensionless)
C_{vt}	Total coefficient of variation for emitters (dimensionless)
d	Inside emitter diameter (L)
d_s	Spacing distance between two adjacent point-source emitters (L)
D	Inside pipe diameter (L)
D_s	Soil water diffusivity ($L^2\ T^{-1}$)
e	Emitter discharge exponent (dimensionless)
E_1	Exponential integral function (dimensionless)
ET	Evapotranspiration rate or seasonal amount ($L\ T^{-1}$ or L)
EU	Design emission uniformity for trickle systems (%)
EU_f	Field emission uniformity for trickle systems (%)
f	Darcy–Weisbach friction factor (dimensionless)
h	Soil pressure head (for unsaturated condition the soil matric potential) (L)
h_M	Pressure head at the midpoint distance between emitters (L)
H	Operating pressure head (L)
H_f	Pipe friction loss (L)
J	Water flux density (Darcian velocity ($L\ T^{-1}$)
k	Constant equal to $dK/d\theta$ ($L\ T^{-1}$)
K, $K(h)$, $K(h')$, $K(\theta)$	Hydraulic conductivity ($L\ T^{-1}$)
K_0	Empirical constant defining unsaturated hydraulic conductivity ($L\ T^{-1}$)
L	Pipe length (L)
L_s	Spacing distance between parallel line-source emitters (L)
LR	Leaching requirement (L)
MR	Miscellaneous water requirements (L)
P	Effective precipitation (L)
q	Discharge rate for a point-source emitter ($L^3\ T^{-1}$)
$\bar{q}$	Mean emitter discharge rate ($L^3\ T^{-1}$)
$\bar{q}_{25}$	Mean lowest 25% emitter discharge rates ($L^3\ T^{-1}$)
q_m	Maximum emitter discharge rate ($L^3\ T^{-1}$)
q_L	Discharge rate per unit lateral length for a line-source emitter ($L^2\ T^{-1}$)
q_n	Minimum emitter discharge rate ($L^3\ T^{-1}$)
Q	Pipe flow rate ($L^3\ T^{-1}$)
r, r'	Radial distance (L)
Re	Reynolds number (dimensionless)
S	Sink function, volume of uptake

per unit soil volume per unit time (T^{-1})
ΔS Soil water storage change (L)
t Time (T)
u Plant uptake rate ($L\ T^{-1}$)
v Emitter flow velocity ($L\ T^{-1}$)
WR Irrigation water requirement (L)
x Horizontal distance (L)
z Vertical distance (L)
z_0 Rooting depth for a one-dimensional sink function (L)
α Empirical constant defining unsaturated hydraulic conductivity (L^{-1})
θ Volumetric soil water content (dimensionless)
ψ_T Total soil water potential (L)
ϕ Soil matric flux potential ($L^2\ T^{-1}$)
ϕ_M Soil matric flux potential at the midpoint distance between emitters, ($L^2\ T^{-1}$)
Φ Dimensionless form of the matric flux potential (dimensionless)
π Soil osmotic head (L)
∇ Vector gradient operator (L^{-1})
∇^2 Laplacian operator (L^{-2})
ν Kinematic viscosity ($L^2\ T^{-1}$)

References

Ababou, R. (1981). Modélisation des transferts hydriques dans le sol en irrigation localisée. Ph.D. Dissertation, University of Science and Medicine and National Polytechnic Institute, Grenoble, France.

Abramowitz, M., and Stegun, I. A. (1964). "Handbook of Mathematical Functions," Natl. Bur. Stand. Appl. Math. Ser., Vol. 55. US Govt. Printing Office, Washington, D.C.

Aljibury, F. K. (1974). Water use in drip irrigation. *Proc. Int. Drip. Irrig. Congr., 2nd, 1974,* pp. 341-350.

Austin, R. S., and Rawlins, S. L. (1977). Optoelectronic level detector for mercury manometer. *Agric. Eng.* **29,** 29-30.

Baars, C. (1976). "Design of Trickle Irrigation Systems." Agric. Univ., Wageningen, The Netherlands.

Batty, J. C., Hamad, S. N., and Keller, J. (1975). Energy inputs to irrigation. *J. Irrig. Drain. Div., Am. Soc. Civ. Eng.* **101**(IR4), 293-307.

Ben-Asher, J., and Sammis, T. W. (1978). Radiation and energy balance of trickle-irrigated lemon grove. *Agron. J.* **70**(4), 568-572.

Ben-Asher, J., Diner, N., Brandt, A., and Goldberg, D. (1975). Measurement of hydraulic conductivity and diffusivity for predicting the process of soil and water infiltration from a trickle source. *Water Resour. Bull.* **11**(6), 1187-1197.

Ben-Asher, J., Fuchs, M., and Goldberg, D. (1978a). Radiation and energy balance of sprinkler and trickle irrigated fields. *Agron. J.* **70**(3), 415-419.

Ben-Asher, J., Lomen, D. O., and Warrick, A. W. (1978b). Linear and nonlinear models of infiltration from a point source. *Soil Sci. Soc. Am. J.* **42**(1), 3-6.

Bernstein, L., and François, L. E. (1973). Comparison of drip, furrow, and sprinkler irrigation. *Soil Sci.* **115**(1), 73-76.

Bernstein, L., and François, L. E. (1975). Effect of frequency of sprinkling with saline waters compared with daily drip irrigations. *Agron. J.* **67**(2), 185-190.

Bessesen, A., ed. (1972). 1972 irrigation survey. *Irrig. J.* **22**(6), 9-18.

Bester, D. H., Lotter, D. C., and Veldman, G. H. (1974). Drip irrigation on citrus. *Proc. Int. Drip. Irrig. Congr., 2nd, 1974,* pp. 58-64.

Bielorai, H. (1977). The effect of drip and sprinkler irrigation on grapefruit yield, water use and salinity. *Proc. Int. Soc. Citric., 1977* Vol. 1, pp. 99-103.

Black, J. D. F. (1976). Trickle irrigation—A review. Part Two. *Hortic. Abstr.* **46,** 59-73.

Black, J. D. F., and West, D. W. (1974). Water uptake by an apple tree with various portions of the root system supplied with water. *Proc. Int. Drip Irrig. Congr., 2nd, 1974,* pp. 432-433.

Blass, S. (1964). Sub-surface irrigation. *Hassadeh* **45,** 1.
Bralts, V. F., Wu, I. P., and Gitlin, H. M. (1980). Drip irrigation uniformity considering emitter plugging. *Pap.—Am. Soc. Agric. Eng.* **80-2077,** 1-28.
Bralts, V. F., Wu, I. P., and Gitlin, H. M. (1981). Manufacturing variation and drip irrigation uniformity. *Trans. ASAE* [*N.S.*] **24**(1), 113-119.
Brandt, A., Bresler, E., Diner, N., Ben-Asher, I., Heller, J., and Goldberg, D. (1971). Infiltration from a trickle source. I. Mathematical models. *Soil Sci. Soc. Am. Proc.* **35**(5), 675-682.
Braud, H. J., and Soom, A. A. (1981). Trickle irrigation lateral design on sloping fields. *Trans. ASAE* [*N.S.*] **24**(4), 941-944, 950.
Bresler, E. (1977). Trickle-drip irrigation: Principles and applications to soil-water management. *Adv. Agron.* **29,** 343-393.
Bresler, E. (1978). Analysis of trickle irrigation with application to design problems. *Irrig. Sci.* **1,** 3-17.
Bresler, E., and Russo, D. (1975). Two-dimensional solutes transfer during nonsteady infiltration: Laboratory test of mathematical model. *Soil Sci. Soc. Am. Proc.* **39**(3), 585-586.
Bresler, E., Heller, J., Diner, N., Ben-Asher, I., Brandt, A., and Goldberg, D. (1971). Infiltration from a trickle source. II. Experimental data and theoretical predictions. *Soil Sci. Soc. Am. Proc.* **35**(5), 683-689.
Bucks, D. A. (1979). Trickle irrigation management for grape production. Ph.D. Dissertation, University of Arizona, Tucson.
Bucks, D. A., and Myers, L. E. (1973). Trickle irrigation: Application uniformity from simple emitters. *Trans. ASAE* [*N.S.*] **16**(6), 1108-1111.
Bucks, D. A., and Nakayama, F. S. (1980). Injection of fertilizer and other chemicals for drip irrigation. *Proc. Agri-Turf Irrig. Conf., 1980,* pp. 166-180.
Bucks, D. A., Erie, L. J., and French, O. F. (1973). Limiting quantities and varying frequencies of trickle irrigation on cotton. *Prog. Agric. Ariz.* **25**(4), 13-16.
Bucks, D. A., Erie, L. J., and French, O. F. (1974a). Quantity and frequency of trickle and furrow irrigation for efficient cabbage production. *Agron. J.* **66**(1), 63-57.
Bucks, D. A., Erie, L. J., Nakayama, F. S., and French, O. F. (1974b). Trickle irrigation management on grapes. *Proc. Int. Drip. Irrig. Congr., 2nd, 1974,* pp. 503-507.
Bucks, D. A., Nakayama, F. S., and Gilbert, R. G. (1977). Clogging research on drip irrigation. *Proc. 4th Annu. Int. Drip. Irrig. Assoc. Meet.* pp. 25-31.
Bucks, D. A., Nakayama, F. S., and Gilbert, R. G. (1979). Trickle irrigation water quality and preventive maintenance. *Agric. Water Manage.* **2**(2), 149-162.
Bucks, D. A., Erie, L. J., French, O. F., Nakayama, F. S., and Pew, W. D. (1981a). Subsurface trickle irrigation management with multiple cropping. *Trans. ASAE* [*N.S.*] **24**(6), 1482-1492.
Bucks, D. A., Nakayama, F. S., and French, O. F. (1981b). Keys to successful trickle: Management and maintenance. *Proc. Natl. Agric. Plast. Congr., 15th, 1980,* pp. 3-8.
Busman, J. D., and Fangmeier, D. D. (1979). A digital controller for trickle irrigation. *Agric. Eng.* **60**(7), 30-31.
Chesness, J. L., Dryden, J. R., and Brady, U. E., Jr. (1976). Nematicide application through porous subsurface irrigation tubing. *Trans. ASAE* [*N.S.*] **19**(1), 105-107.
Childs, S. W., and Hanks, R. J. (1975). Model of soil salinity effects on crop growth. *Soil Sci. Soc. Am. Proc.* **39**(4), 617-622.
Christiansen, J. E. (1942). Irrigation by sprinkling. *Bull.—Calif. Agric. Exp. Stn.* **670,** 1-124.
Cole, T. E. (1971). Subsurface and trickle irrigation, a survey of potentials and problems. *Oak Ridge Natl. Lab.* [*Bull.*] **ORNL-NDIC-9,** 1-65.
Coston, D. C., Ponder, H. G., and Kenworthy, A. L. (1978). Fertilizing peach trees through a trickle irrigation system. *Commun. Soil Sci. Plant Anal.* **9,** 187-191.
Dan, C. (1974). The irrigation of olives by drip and other irrigation methods. *Proc. Int. Drip Irrig. Congr., 2nd, 1974,* pp. 491-496.

Dasberg, S., and Steinhardt, R. (1974). Water distribution in an orchard irrigated by sprinkler or trickle irrigation as measured by the neutron method. *In* "Isotopes and Radiation Techniques in Soil Physics and Irrigation Studies," pp. 467–474. IAEA, Vienna.

Davis, S. (1974). History of drip irrigation. *Agribusiness News* **10**(7), 1.

Davis, S. (1975). Drip irrigation. *In* "Sprinkler Irrigation" (C. H. Pair, W. H. Hinz, C. Reid, and K. R. Frost, eds.), pp. 508–528. Irrig. Assoc., Silver Spring, Maryland.

Davis, S., and Nelson, S. D. (1970). Subsurface irrigation easily automated. *J. Irrig. Drain. Div., Am. Soc. Civ. Eng.* **96**(IR1), 47–51.

Davis, S., and Pugh, W. J. (1974). Drip irrigation: Surface and subsurface compared with sprinkler and furrow. *Proc. Int. Drip. Irrig. Congr., 2nd, 1974,* pp. 109–114.

Downey, L. A. (1972). Water-yield relations for nonforage crops. *J. Irrig. Drain. Div., Am. Soc. Civ. Eng.* **98**(IR1), 107–115.

Earl, K. D., and Jury, W. A. (1977). Water movement in bare and cropped soil under isolated trickle emitters. II. Analysis of cropped soil experiments. *Soil Sci. Soc. Am. J.* **41**(5), 856–861.

Ekern, P. C. (1977). Drip irrigation of sugarcane measured by hydraulic lysimeters. *Univ. Hawaii Tech. Rep.* **109,** 1–52.

Ford, H. W. (1975). The use of chlorine in drip irrigation systems. *Proc. Fla. State Hortic. Soc.* **88,** 1–4.

Ford, H. W. (1976). Controlling slimes of sulfur bacteria in drip irrigation systems. *Hort Science* **11**(2), 133–135.

Ford, H. W. (1977a). The importance of water quality in drip/trickle irrigation systems. *Proc. Int. Soc. Citric. 1977* Vol. 1, pp. 84–87.

Ford, H. W. (1977b). Controlling certain types of slime clogging in drip/trickle systems. *Proc. Int. Agric. Plast. Congr., 7th, 1977,* pp. 118–123.

Ford, H. W. (1978). Bacterial clogging in low pressure irrigation systems. *Proc. Tech. Irrig. Conf., 1978,* pp. 239–245.

Ford, H. W. (1979). The use of surface water for low pressure irrigation systems. *Fla. Univ., AREC Res. Rep.* **GS75-2-HWF-100,** 1–6.

Ford, H. W. (1980). The problem of dripper clogging and methods for control. *Proc. Symp. Drip Irrig. Hortic. Foreign Experts Participating, 1980,* pp. 199–206.

Ford, H. W., and Tucker, D. P. H. (1974). Clogging of drip systems from metabolic products of iron and sulfur bacteria. *Proc. Int. Drip Irrig. Congr., 2nd, 1974,* pp. 212–214.

Ford, H. W., and Tucker, D. P. H. (1975). Blockage of drip irrigation filters and emitters by iron-sulfur-bacterial products. *Hort Science* **10**(1), 26–64.

Frasier, G. O., Jr., ed. (1977). Drip/trickle survey and projections. *In* "Water and Irrigation," pp. 10–11. Int. Drip Irrig. Assoc., Bloomington, California.

Freeman, B., Blackwell, J., and Garzoli, K. V. (1976). Irrigation frequency and total water application with trickle and furrow systems. *Agric. Water Manage.* **1**(1), 21–31.

Fridovich, I. (1972). Superoxide radical and superoxide dismutase. *Acc. Chem. Res.* **5,** 321–326.

Frith, G. J. T., and Nichols, D. G. (1974). Effects of nitrogen fertilizer applications to part of a root system. *Proc. Int. Drip Irrig. Congr., 2nd, 1974,* pp. 434–436.

Gilad, Y., Peleg, D., and Tirosh, G. (1968). "Irrigation Equipment Tests," No. 81268. ICWA, Israel.

Gilbert, R. G., Nakayama, F. S., and Bucks, D. A. (1979). Trickle irrigation: Prevention of clogging. *Trans. ASAE* [*N.S.*] **22**(3), 514–519.

Gilbert, R. G., Nakayama, F. S., Bucks, D. A., French, O. F., and Adamson, K. C. (1981). Trickle irrigation: Emitter clogging and flow problems. *Agric. Water Manage.* **3**(3), 159–178.

Gilbert, R. G., Nakayama, F. S., Bucks, D. A., French, O. F., Adamson, K. C., and Johnson, R. M. (1982). Trickle irrigation: Predominant bacteria in treated Colorado River water and biologically clogged emitters. *Irrig. Sci.* (in press).

Gilley, J. R., and Allred, E. R. (1974). Infiltration and root extraction from subsurface irrigation laterals. *Trans. ASAE [N.S.]* **17**(5), 927–933.

Goldberg, D., and Shmueli, M. (1970). Drip irrigation—A method used under arid and desert conditions of high water and soil salinity. *Trans. ASAE [N.S.]* **33**(1), 38–41.

Goldberg, D., Gornat, B., and Bar, Y. (1971). The distribution of roots, water, and minerals as a result of trickle irrigation. *J. Am. Soc. Hortic. Sci.* **96**(6), 645–648.

Goldberg, D., Gornat, B., and Rimon, D. (1976). "Drip Irrigation: Principles, Design, and Agricultural Practices," Drip Irrig. Sci. Publ. Kfar Shumaryahu, Israel.

Goode, J. E., Higgs, K. H., and Hyrycz, K. J. (1978). Trickle irrigation of apple trees and the effects of liquid feeding with NO_3^- and K^+ compared with normal manuring. *J. Hortic. Sci.* **53,** 307–316.

Gornat, B., Goldberg, D., Rimon, D., and Ben-Asher, J. (1973). The physiological effects of water quality and methods of application on tomato, cucumber, and pepper. *J. Am. Soc. Hortic. Sci.* **98**(2), 202–205.

Gustafson, D. C. (1976). Drip irrigation: Where it was in '75. *Irrig. J.* **26**(3), 24–26.

Hanks, R. J. (1974). Model for predicting plant yield as influenced by water use. *Agron. J.* **66**(5), 660–664.

Hanson, E. G., and Patterson, T. C. (1974). Vegetable production and water-use efficiency as influenced by drip, sprinkler, subsurface, and furrow irrigation methods. *Proc. Int. Drip Irrig. Congr., 2nd, 1974,* pp. 97–102.

Hart, W. E. (1961). Overhead irrigation pattern parameters. *Agric. Eng.* **42**(7), 354–355.

Hiler, E. A., and Howell, T. A. (1973). Grain sorghum response to trickle and subsurface irrigation. *Trans. ASAE* [N.S.] **16**(4), 792–893.

Hillel, D. (1972). The field water balance and water use efficiency. *In* "Optimizing the Soil Physical Environment Toward Greater Crop Yields" (D. Hillel, ed.), pp. 79–100. Academic Press, New York.

Hillel, D. (1980). "Applications of Soil Physics," p. 385. Academic Press, New York.

Hoffman, G. J., Dirksen, C., Ingvalson, R. D., Mass, E. V., Rhoades, J. D., and van Schilfgaarde, J. (1978). Minimizing salt in drain water by irrigation management. Design and initial result of Arizona field studies. *Agric. Water Manage.* **1**(3), 233–252.

Howell, T. A., and Barinas, F. (1980). Pressure losses across trickle irrigation fittings and emitters. *Trans. ASAE* [N.S.] **23**(4), 928–933.

Howell, T. A., and Hiler, E. A. (1974). Designing trickle irrigation laterals for uniformity. *J. Irrig. Drain. Div., Am. Soc. Civ. Eng.* **100**(IR4), 443–454.

Howell, T. A., and Hiler, E. A. (1975). Optimization of water use efficiency under high frequency irrigation I. Evapotranspiration and yield relationships. *Trans. ASAE* [N.S.] **18**(5), 873–878.

Howell, T. A., Stevenson, D. S., Aljibury, F. K., Gitlin, H. M., Wu, I. P., Warrick, A. W., and Raats, P. A. C. (1980). Design and operation of trickle (drip) systems. *In* "Design and Operation of Farm Irrigation Systems" (M. E. Jensen, ed.), Monogr. 3, pp. 663–717. Am. Soc. Agric. Eng., St. Joseph, Michigan.

Howell, T. A., Bucks, D. A., and Chesness, J. L. (1981). Advances in trickle irrigation. *Proc. Natl. Irrig. Symp., 2nd, 1980,* pp. 69–94.

Hsiao, T. C. (1973). Plant response to water stress. *Annu. Rev. Plant Physiol.* **24,** 519–570.

Hughes, T. C., and Jeppson, R. W. (1978). Hydraulic friction loss in small diameter plastic pipelines. *Water Resour. Bull.* **14**(5), 1159–1166.

Isobe, M. (1974). Investigations in sugarcane fertilization by drip irrigation in Hawaii. *Proc. Int. Drip Irrig. Congr., 2nd, 1974,* pp. 405–410.

Jensen, M. E., ed. (1974). "Consumptive Use of Water and Irrigation Water Requirements," Tech. Comm. Irrig. Water Req. Irrig. Drain. Div., Am. Soc. Civ. Eng., New York.

Jury, W. A., and Earl, K. D. (1977). Water movement in bare and cropped soil under isolated trickle emitters. I. Analysis of bare soil experiments. *Soil Sci. Soc. Am. J.* **41**(5), 852–856.

Karmeli, D., and Peri, G. (1974). Basic principles of pulse irrigation. *J. Irrig. Drain. Div., Am. Soc. Civ. Eng.* **100**(3), 309–319.

Karmeli, D., and Peri, G. (1977). Pulse irrigation—Design and operating sets. *J. Irrig. Drain. Div., Am. Soc. Civ. Eng.* **103**(4), 34–41.

Kays, S. J., Johnson, A. W., and Jaworski, C. A. (1976). Multiple cropping with trickle irrigation. *Hort Science* **11**(2), 135–136.

Keller, J. (1980). Trickle irrigation manifold design. *Pap.—Am. Soc. Agric. Eng.* **80-2058,** 1–14.

Keller, J., and Karmeli, D. (1974). Trickle irrigation design parameters. *Trans. ASAE* [N.S.] **17**(4), 678–784.

Keller, J., and Karmeli, D. (1975). "Trickle Irrigation Design." Rain Bird Sprinkler Mfg. Corp., Glendora, California.

Kemper, W. D., and Noonan, L. (1970). Runoff as affected by salt treatments and soil texture. *Soil Sci. Soc. Am. Proc.* **34**(1), 126–130.

Kenworthy, A. L. (1972). Trickle irrigation . . . the concept and guidelines for use. *Mich. Agric. Exp. Stn. Res. Rep.* **165,** 1–19.

Lange, A., Aljibury, F., Fischer, B., Humphrey, W., and Otto, H. (1974). Weed control under drip irrigation and vineyard crops. *Proc. Int. Drip Irrig. Congr., 2nd, 1974,* pp. 422–424.

Lemon, E. R. (1956). The potentialities for decreasing soil moisture evaporation loss. *Soil Sci. Soc. Am. Proc.* **20**(1), 120–125.

Lemos, P., and Lutz, J. F. (1957). Soil crusting and some factors affecting it. *Soil Sci. Soc. Am. Proc.* **21**(5), 485–491.

Levin, I., Assaf, R., and Bravdo, B. (1974). Soil moisture distribution and depletion in an apple orchard irrigated by tricklers. *Proc. Int. Drip Irrig. Congr., 2nd, 1974,* pp. 252–257.

Levin, I., Assaf, R., and Bravdo, B. (1979a). Soil moisture and root distribution in an apple orchard irrigated by tricklers. *Plant Soil* **52,** 31–40.

Levin, I., van Rooyen, P. C., and van Rooyen, F. C. (1979b). The effect of discharge rate and intermittent water application by point-source irrigation on soil moisture distribution pattern. *Soil Sci. Soc. Am. J.* **43**(1), 8–16.

Lindsey, K. E., and New, L. L. (1974). Application of fertilizer materials through drip irrigation systems in west Texas. *Proc. Int. Drip. Irrig. Congr., 2nd, 1974,* pp. 400–404.

Lyle, W. M., and Bordovsky, J. P. (1979). Traveling low energy precision irrigator. *Proc. Irrig. Drain. Spec. Conf., Am. Soc. Civ. Eng.* pp. 121–131.

McElhoe, B. A., and Hilton, H. W. (1974). Chemical treatment of drip irrigation water. *Proc. Int. Drip. Irrig. Congr., 2nd, 1974,* pp. 215–220.

McNeill, E., ed. (1980). 1980 irrigation survey. *Irrig. J.* **30**(6), 72A–72H.

Marsh, A. W., and Gustafson, C. D. (1971). Irrigating a drop at a time. *Crops Soils* **23**(6), 9–11.

Marsh, A. W., Branson, R. L., Davis, S., Gustafson, C. D., and Aljibury, F. K. (1975). "Drip Irrigation." *Univ. Calif., Berkley, Leafl.* **2740,** 1–4.

Mears, D. B., Phene, C. J., and Soenke, S. D. (1979). A microprocessor-based irrigation controller. *Pap.—Am. Soc. Agric. Eng.* **79-5525,** 1–15.

Merriam, J. L., and Keller, J. (1978). "Farm Irrigation System Evaluation: A guide for management," 3rd ed. Utah State University, Logan.

Merriam, J. L., Shearer, M. N., and Burt, C. M. (1980). Evaluating irrigation systems and practices. *In* "Design and Operation of Farm Irrigation Systems" (M. E. Jensen, ed.), Monogr. 3, pp. 721–760. Am. Soc. Agric. Eng., St. Joseph, Michigan.

Miller, R. J., Rolston, D. E., Rauschkolb, R. S., and Wolfe, D. W. (1976). Drip application of nitrogen is efficient. *Calif. Agric.* **30**(11), 16–18.

Myers, J. M., Harrison, D. S., and Phillips, J. W. (1976). Soil moisture distribution in a sprinkler irrigated orange grove. *Proc. Fla. State Hortic. Soc.* **89,** 23–28.

Myers, L. E., and Bucks, D. A. (1972). Uniform irrigation with low-pressure trickle systems. *J. Irrig. Drain. Div., Am. Soc. Civ. Eng.* **98**(IR3), 341–346.

Nakayama, F. S., and Bucks, D. A. (1981a). Emitter clogging effects on trickle irrigation uniform ity. *Trans. ASEA* [N.S.] **24**(1), 77–80.

Nakayama, F. S., and Bucks, D. A. (1981b). Use of subsurface trickle system for carbon dioxide enrichment. *Proc. Natl. Agric. Plast. Congr., 15th, 1980,* pp. 13–18.

Nakayama, F. S., Bucks, D. A., and French, O. F. (1977). Reclaiming partially clogged trickle emitters. *Trans. ASAE* [N.S.] **20**(2), 270–280.

Nakayama, F. S., Gilbert, R. G., and Bucks, D. A. (1978). Water treatments in trickle irrigation system. *J. Irrig. Drain. Div., Am. Soc. Civ. Eng.* **104**(IR1), 23–34.

Nakayama, F. S., Bucks, D. A., and Clemmens, A. J. (1979). Assessing trickle emitter application uniformity. *Trans. ASAE* [N.S.] **22**(4), 816–821.

Nassehzadeh-Tabrizi, A., Jeppson, R. W., and Willardson, L. S. (1978). Unsaturated flow through heterogeneous soils. *Bull.—Utah Water Res. Lab.* **PRWG 72-4,** 1–238.

Nelson, S. D., and Davis, S. (1974). Soil salinity distribution in sprinkler- and subsurface-irrigated citrus. *Trans. ASAE* [N.S.] **17**(1), 140–143.

New, L., and Roberts, R. E. (1973). Automatic drip irrigation for greenhouse tomato production. *Bull. Tex. Agric. Ext. Serv. Misc. Publ. 1082,* pp. 1–16.

Obbink, J. G., and Alexander, D. McE. (1977). Observations of soil water and salt movement under drip and flood irrigation in an apple orchard. *Agric. Water Manage.* **1**(2), 179–190.

O'Neill, M. K., Gardner, B. R., and Roth, R. L. (1979). Orthophosphoric acid as a phosphorous fertilizer in trickle irrigation. *Soil Sci. Soc. Am. J.* **43**(2), 283–296.

Oron, G. (1981). Simulation of water flow in the soil under sub-surface trickle irrigation with water uptake by roots. *Agric. Water Manage.* **3**(3), 179–193.

Overman, A. J. (1974). Nematicides in linear drip irrigation for full-bed mulch of tomato. *Proc. Soil Crop Sci. Soc. Fla.* **34,** 197–200.

Overman, A. J. (1976). Efficacy of soil fumigants applied via a drip irrigation system. *Proc. Fla. State Hortic. Soc.* **89,** 143–145.

Paldi, H. (1974). Drip irrigation and automation tools in efficient use of water policy. *Proc. Int. Drip Irrig. Congr., 2nd, 1974,* pp. 29–32.

Parchomchuk, P. (1976). Temperature effect on emitter discharge rates. *Trans. ASAE* [N.S.] **19**(4), 690–692.

Patterson, T. C., and Wierenga, P. J. (1974). Irrigation return flow as influenced by drip irrigation. *Proc. Int. Drip Irrig. Congr., 2nd, 1974,* pp. 376–381.

Pelleg, D., Lahav, N., and Goldberg, D. (1974). Formation of blockages in drip irrigation systems. *Proc. Int. Drip Irrig. Congr., 2nd, 1974,* pp. 203–208.

Peri, G., and Rasmussen, R. O. (1978). A pulse irrigation system for improved irrigation water application efficiency. *Proc. Natl. Agric. Plast. Congr., 14th, 1978,* pp. 114–121.

Phene, C. J., and Beale, O. W. (1976). High-frequency irrigation for water nutrient management in humid regions. *Soil Sci. Soc. Am. J.* **40**(3), 430–436.

Phene, C. J., and Campbell, R. B. (1975). Automating pan evaporation measurements for irrigation control. *Agric. Meteorol.* **15,** 181–191.

Phene, C. J., and Sanders, D. C. (1976). High-frequency trickle irrigation and row spacing effects on yield and quality of potatoes. *Agron. J.* **68**(4), 602–607.

Phene, C. J., Hoffman, G. J., and Austin, R. S. (1973). Controlling automated irrigation with soil matric potential sensor. *Trans. ASAE* [N.S.] **16**(4), 773–776.

Phene, C. J., Fouss, J. L., and Sanders, D. C. (1979). Water-nutrient-herbicide management of potatoes with trickle irrigation. *Am. Potato J.* **56,** 51–59.

Phene, C. J., Beck, R. D., Soenke, S. D., and Howell, T. A. (1980). A laser guidance/speed controller for traveling irrigation system. *Proc. Electr. Laser 80 Congr.*, pp. 23–31.

Phene, C. J., Howell, T. A., Beck, R. D., and Sanders, D. C. (1981). A traveling trickle irrigation system for row crops. *Proc. Tech. Irrig. Conf.* (in press).

Rauschkolb, R. S., Rolston, D. E., Miller, R. J., Carlton, A. B., and Burau, R. G. (1976). Phosphorous fertilization with drip irrigation. *Soil Sci. Soc. Am. J.* **40**(1), 68–72.

Ravelo, C. J., Hiler, E. A., and Howell, T. A. (1977). Trickle and sprinkler irrigation of grain sorghum. *Trans. ASAE* [N.S.] **20**(1), 96–99.

Rawlins, S. L. (1973). Principles of managing high frequency irrigation. *Soil Sci. Soc. Am. Proc.* **37**(4), 626–629.

Rawlins, S. L. (1977). Uniform irrigation with a low-head bubble system. *Agric. Water Manage.* **1**(2), 167–178.

Rawlins, S. L., and Raats, P. A. C. (1975). Prospects for high frequency irrigation. *Science* **188,** 604–610.

Rhoades, J. D., Oster, J. D., Ingvalson, R. D., Tucker, J. M., and Clark, M. (1974). Minimizing the salt burdens in irrigation drainage waters. *J. Environ. Qual.* **3**(4), 311–316.

Ritchie, J. T. (1973). Influence of soil water status and meterological conditions on evaporation from a corn canopy. *Agron. J.* **65**(6), 893–897.

Ritchie, J. T. (1973). Influence of soil water status and meteorological conditions on evaporation from **14**(2), 183–198.

Rodney, D. R., Roth, R. L., and Gardner, B. R. (1977). Citrus response to irrigation methods. *Proc. Int. Soc. Citric., 1977* Vol. 1, pp. 106–110.

Rolston, D. E., and Broadbent, F. E. (1977). *Environ. Prot. Agency Bull.* **EOA-600/2-77-233,** 1–75.

Rolston, D. E., Rauschkolb, R. S., Phene, C. J., Miller, R. J., Uriu, K., Carlson, R. M., and Henderson, D. W. (1979). Applying nutrients and other chemicals to trickle-irrigated crops. *Univ. Calif., Berkeley, Bull.* **1893,** 1–14.

Rose, C. W. (1961). Rainfall and soil structure. *Soil Sci.* **91**(1), 49–54.

Roth, R. L. (1974). Soil moisture distribution and wetting pattern from a point source. *Proc. Int. Drip Irrig. Congr., 2nd, 1974,* pp. 246–251.

Roth, R. L., Rodney, D. R., and Gardner, B. R. (1974). Comparison of irrigation methods, rootstocks, and fertilizer elements on Valencia orange trees. *Proc. Int. Drip Irrig. Congr., 2nd 1974,* pp. 103–108.

Sammis, T. W. (1980). Comparison of sprinkler, trickle, subsurface, and furrow irrigation methods for row crops. *Agron. J.* **72**(5), 701–704.

Schaper, J. C. (1977). Keeping trickle irrigation systems operational. *Proc. Irrig. Drain. Spec. Conf., Am. Soc. Civ. Eng.* pp. 394–404.

Seginer, I. (1969). Wind variation and sprinkler water distribution. *J. Irrig. Drain. Div., Am. Soc. Civ. Eng.* **95**(IR2), 261–274.

Seifert, W. J., Hiler, E. A., and Howell, T. A. (1975). Trickle irrigation with water of different salinity levels. *Trans. ASAE* [N.S.] **18**(1), 89–94.

Shani, M. (1974). Trickle irrigation. *Proc. Int. Drip Irrig. Congr., 2nd, 1974,* pp. 91–95.

Shearer, M. N. (1977). Minimum screening and automatic flushing. *Proc. 4th Annu. Int. Drip. Irrig. Assoc. Meet.* pp. 32–36.

Shipston, C. (1976). Familiarization with polyethylene. *Drip/Trickle Irrig.* **1**(1), 6–10.

Shoji, K. (1977). Drip irrigation. *Sci. Am.* **237**(5), 62–68.

Silberbush, M., Gornat, B., and Goldberg, D. (1979). Effect of irrigation from a point source (trickling) on oxygen flux and on root extension in the soil. *Plant Soil* **52,** 507–514.

Singh, S. D., Gupta, J. P., and Singh, P. (1978). Water economy and saline water use by drip irrigation. *Agron. J.* **70**(6), 948–951.

Solomon, K. (1977). Evaluation criteria for trickle irrigation emission devices. *Proc. 4th Annu. Int. Drip Irrig. Assoc. Meet.* pp. 65–76.

Solomon, K. (1979). Manufacturing variation of trickle emitters. *Trans. ASAE* [N.S.] **22**(5), 1034–1038, 1043.

Solomon, K., and Keller, J. (1978). Trickle irrigation uniformity and efficiency. *J. Irrig. Drain. Div., Am. Soc. Civ. Eng.* **104**(IR3), 293–306.

Stewart, J. T., Misra, R. D., Pruitt, O. W., and Hagan, R. M. (1975). Irrigating corn and grain sorghum with a deficient water supply. *Trans. ASAE* [N.S.] **8**(2), 270–280.

Thomas, A. W., Kruse, E. G., and Duke, H. R. (1974). Steady infiltration from line sources buried in soil. *Trans. ASAE* **[N.S.]** **17**(1), 125–128, 133.

Thomas, A. W., Duke, H. R., Zachmann, D. W., and Kruse, E. G. (1976). Comparisons of calculated and measured potentials from line sources. *Soil Sci. Soc. Am. J.* **40**(1), 10–14.

Tscheschke, P., Alfaro, J. F., Keller, J., and Hanks, R. J. (1974). Trickle irrigated soil water potential as influenced by management of highly saline water. *Soil Sci.* **117**(4), 226–231.

Uriu, K., Carlson, R. M., and Henderson, D. W. (1977). Application of potassium fertilizer to prunes through a drip irrigation system. *Proc. Int. Agric. Plast. Congr., 7th, 1977,* pp. 211–214.

Van Bavel, C. H. M., Ahmed, J., Bhuiyan, S. I., Hiler, E. A., and Smajstrla, A. G. (1973). Dynamic simulation of automated subsurface irrigation systems. *Trans. ASAE* [N.S.] **16**(6), 1095–1099.

Van der Ploeg, R. R., and Benecke, P. (1974). Unsteady, n-dimensional moisture flow in soil: A computer simulation program. *Soil Sci. Soc. Am. Proc.* **38**(6), 881–885.

Warrick, A. W. (1974). Time dependent linearized infiltration: I. Point sources. *Soil Sci. Soc. Am. Proc.* **38**(3), 383–386.

Warrick, A. W., and Lomen, D. O. (1981). Two-dimensional linearized moisture flow with water extraction. *J. Hydrol.* **49,** 235–245.

Warrick, A. W., Amoozegar-Fard, A., and Lomen, D. O. (1979). Linearized moisture flow from line sources with water extraction. *Trans. ASAE* [N.S.] **22**(3), 549–553, 559.

Warrick, A. W., Lomen, D. O., and Amoozegar-Fard, A. (1980). Linearized moisutre flow with root extraction for three dimensional, steady conditions. *Soil Sci. Soc. Am. J.* **44**(5), 911–914.

Warrick, A. W., Lomen, D. O., and Tonellato, P. J. (1981). Soil moisture flow for point and line sources—computer programs for linearized solutions. *Ariz., Agric. Eng. Soil Sci., Bull.* **81–2,** 1–78.

Watters, G. F., and Keller, J. (1978). Trickle irrigation tubing hydraulics. *Pap.—Am. Soc. Agric. Eng.* **78-2015,** 1–18.

Wendt, C. W., Harbert, H. P., Bausch, W., and Wilke, O. C. (1973). Automation of drip irrigation systems. *Pap.—Am. Soc. Agric. Eng.* **73-2502,** 1–15.

Wendt, C. W., Onken, A. B., Wilke, O. C., Hargrove, R., Bausch, W., and Barnes, L. (1977). Effect of irrigation systems on the water requirements of sweet corn. *Soil Sci. Soc. Am. J.* **41**(4), 785–788.

White, G. C. (1972). "Handbook of Chlorination." Van Nostrand-Reinhold, Princeton, New Jersey.

Wilke, O. C. (1974). Mobile drip irrigation systems. *Proc. Int. Drip. Irrig. Congr., 2nd, 1974,* pp. 188–192.

Willoughby, P., and Cockroft, B. (1974). Changes in root patterns of peach trees under trickle irrigation. *Proc. Int. Drip. Irrig. Congr., 2nd, 1974,* pp. 439–442.

Wilson, D. L. (1972). Filtration, filters and water treatment. *Proc. Drip Irrig. Semin., 3rd, 1972,* pp. 17–23.

Wilson, D. L. (1977). Filtration in drip irrigation. *Proc. Int. Agric. Plast. Congr., 7th, 1977,* pp. 160–165.

Wooding, R. A. (1968). Steady infiltration from a circular pond. *Water Resour. Res.* **4**(6), 1259–1273.

Wu, I. P., and Gitlin, H. M. (1977). Drip irrigation design in metric units. *Misc. Publ.* **144,** Univ. of Hawaii, Honolulu, 19 p.

Wu, I. P., and Gitlin, H. M. (1981). Preliminary concept of a drip irrigation network design. *Trans. ASAE* [N.S.] **24**(2), 330–334.

Wu, I. P., Howell, T. A., and Hiler, E. A. (1979). Hydraulic design of drip irrigation systems. *Hawaii, Agric. Exp. Stn., Bull.* **105,** 1–80.

Yagev, E., and Choresh, Y. (1974). Drip irrigation in citrus orchards. *Proc. Int. Drip Irrig Congr., 2nd, 1974,* pp. 456–461.

Yaron, B., Shalhevet, J., and Shimshi, D. (1973). Patterns of salt distribution under trickle irrigation. *In* "Physical Aspects of Soil Water and Salts in Ecosystems" (A. Hadas, D. Swartzendruber, P. E. Rijtema, M. Fuchs, and B. Yaron, eds.), pp. 389–394. Springer-Verlag, Berlin and New York.

Zachmann, D. W., and Thomas, A. W. (1973). A mathematical investigation of steady infiltration from line sources. *Soil Sci. Soc. Am. Proc.* **37**(4), 495–500.

Zaslavsky, D. (1972). Analysis of growth parameters and their fluctuations in searching for increased yields. *In* "Optimizing the Soil Physical Environment Toward Greater Crop Yields" (D. Hillel, ed.), pp. 223–232. Academic Press, New York.

Zaslavsky, D., and Mokady, R. S. (1967). Non-uniform distribution of phosphorous fertilizer: An analytical approach. *Soil Sci.* **104**(1), 1–6.

Zentmeyer, G. A., Guillement, F. B., and Johnson, E. L. V. (1974). The relation between drip irrigation, phytophthora root rot of avocado and fungicides. *Proc. Int. Drip Irrig. Congr., 2nd, 1974,* pp. 512–514.

Zur, B. (1976). The pulsed irrigation principles for controlled soil wetting. *Soil Sci.* **122**(5), 282–291.

Zur, B., and Tal, S. (1981). Emitter discharge sensitivity to pressure and temperature. *J. Irrig. Drain. Div., Am. Soc. Civ. Eng.* **107**(IR1), 1–9.

INDEX

E

F

G

H

I

L